QD 83 B84813

Busev, A I

Handbook of the analytical chemistry of rare elements [by] A. I. Busev, V. G. Tiptsova, and V. M. Ivanov. Translated by J. Schmorak. Ann Arbor [Mich.] Ann Arbor-Humphrey Science Publishers, 1970.

xi, 402 p. 25 cm.

On label mounted on t. p.: Distributed by Keter Inc., New York.
Translation of Prakticheskoe rukovodstvo po analiticheskoĭ khimii redkikh ėlementov.
Includes bibliographies.

1. Chemistry, Analytic — Laboratory manuals. 2. Nonferrous metals—Analysis. I. Tipt͡sova, Valentina Gavrilovna. II. Ivanov, Vadim Mikhaĭlovich. III. Title.

QD83.B84813 543'.0028 75-104379
ISBN 0-250-39960-1 MARC

Library of Congress 71 [7]

HANDBOOK OF THE

ANALYTICAL CHEMISTRY OF RARE ELEMENTS

Handbook of the Analytical Chemistry of

RARE ELEMENTS

A.I. BUSEV, V.G. TIPTSOVA, and V.M. IVANOV

Translated by J. SCHMORAK

ISRAEL PROGRAM FOR SCIENTIFIC TRANSLATIONS

Fairleigh Dickinson
University Library

Teaneck, New Jersey

168098

DISTRIBUTED BY:
KETER INC.
104 EAST 40TH ST.
NEW YORK, N.Y. 10016

ANN ARBOR–HUMPHREY SCIENCE PUBLISHERS

ANN ARBOR · LONDON · 1970

ANN ARBOR-HUMPHREY SCIENCE PUBLISHERS, INC.
Drawer No. 1425, 600 S. Wagner Road, Ann Arbor, Michigan 48106

ANN ARBOR-HUMPHREY SCIENCE PUBLISHERS, LTD.
5 Great Russell Street London W.C. 1

© 1970 Ann Arbor-Humphrey Science Publishers, Inc.

Library of Congress Catalog Card Number 75-104379
SBN 250 39960 1

This book is a translation of
PRAKTICHESKOE RUKOVODSTVO PO ANALITICHESKOI KHIMII REDKIKH ELEMENTOV
Izdatel'stvo "Khimiya"
Moskva 1966

QD
83
B84813

Foreword

This book is intended for use by laboratory workers, industrial chemists, and graduate students specializing in the chemistry of rare elements.

Knowledge of the theoretical principles of photometric, polarographic, potentiometric and other analytical techniques is a prerequisite for successful application of the material. A brief analytical description of the element precedes the description of laboratory methods.

The book describes the methods of determination of lithium, rubidium, cesium, beryllium, scandium, REE, yttrium, vanadium, niobium, tantalum, molybdenum, titanium, zirconium, hafnium, uranium, thorium, tungsten, rhenium, technetium, gallium, indium, thallium, germanium, bismuth, selenium and tellurium. It includes the most important organic reagents for these elements, masking compounds, and solubility products of certain sparingly soluble compounds. Methods for the isolation of rare elements by solvent extraction are also given.

Only the simplest, fastest, most selective and most reliable of the numerous analytical methods for determining rare elements have been included in this book. The methods do not require the use of unconventional reagents and can be applied in scientific research as well as in industrial laboratories.

The preparation of the analytical reagents is not described in detail, since in most cases it presents no difficulties.

Each chapter is provided with a list of references where additional information on the analytical procedures described can be found. Another source of useful information are the books on analytical chemistry listed on pages 16–21.

Platinum metals and inert gases, although formally regarded as "rare" elements, are not considered in this book. Determination of rare elements by spectroscopy, neutron activation methods and other techniques which have been excluded will be found in special textbooks. The spectroscopic methods for determining both rare and commonly occuring elements are, in principle, the same. Discussion of flame-photometric methods can also be found in special textbooks on spectroscopic analysis.

Numerous constants quoted in this book were taken from L. Meites, ed., *Handbook of Analytical Chemistry*, McGraw-Hill Book Co., New York–Toronto–London, 1963.

The theoretical part of this book and the analytical treatment of rare elements were written by A.I. Busev. V.G. Tiptsova compiled the methods for the determination of vanadium, niobium, tantalum, tungsten, rhenium, gallium, indium, thallium, germanium, selenium and tellurium. The methods for the determination of lithium, rubidium, cesium, beryllium, scandium, lanthanum, cerium and the lanthanide elements, thorium, uranium, titanium, zirconium, molybdenum and bismuth were compiled by V.M. Ivanov. The entire book was edited by A.I. Busev.

We wish to acknowledge the help of Yu.A. Chernikhov, V.G. Goryushina and T.V. Cherkashina who reviewed the book and offered valuable advice, and the assistance of I.P. Alimarin and V.M. Peshkova in reading the manuscript. Special thanks are due also to M.A. Semenova and A.N. Buseva for their valuable help in preparing the manuscript for print. Any critical comments from readers will be most welcome.

Table of Contents

EXPLANATORY LIST OF ABBREVIATIONS OF U.S.S.R. INSTITUTIONS AND JOURNALS APPEARING IN THIS TEXT

Abbreviation	*Full name (transliteration)*	*Translation*
GIREDMET	Gosudarstvennyi Naucho-Issledovatel'skii Institut Redkikh Metallov	State Rare Metals Scientific Research Institute
KhGU	Khar'kovskii Gosudarstvennyi Universitet	Kharkov State University
MGU	Moskovskii Gosudarstvennyi Universitet	Moscow State University
VKhO	Vsesoyuznoe Khimicheskoe Obshchestvo	All-Union Chemical Society
Zav. Lab.	Zavodskaya Laboratoriya	Industrial Laboratory
ZhAKh	Zhurnal Analiticheskoi Khimii	Journal of Analutical Chemistry
ZhNKh	Zhurnal Neorganicheskoi Khimii	Journal of Inorganic Chemistry
ZhAKh	Zhurnal Analiticheskoi Khimii	Journal of Analytical Chemistry
ZhFKh	Zhurnal Fizicheskoi Khimii	Journal of Physical Chemistry

Rare Elements in Industry and in Geochemistry

About sixty elements in various groups of the Periodic Table are classified as rare elements. They include elements with a fairly high natural abundance as well as elements which are truly "rare". According to the accepted convention, a rare element is any element which has been produced on a commercial scale for a relatively short period of time, and whose practical uses are therefore of fairly recent origin.*

Thus the term *rare element* is to be interpreted as a *technologically new* or *relatively new element.*

The following elements are currently considered rare elements:

I. *Metals*:

1) Light elements: Li, Rb, Cs and Be;
2) rare-earth elements: La, Ce, Pr, Nd, Pm, Sm, Eu, Gd, Tb, Dy, Ho, Er, Tu, Yb, Lu, Sc, and Y;
3) dispersed elements: Ga, In, Tl, Ge, Hf, and Re;
4) high-melting elements: Ti, Zr, Hf, V, Nb, Ta, Mo, W, and Re;
5) radioactive elements: Ra, Po, Ac, Th, Pa, U, Np, Pu, Am, Cm and other transuranium elements;
6) minor elements: Bi;
7) noble elements: Pt, Ru, Os, Rh, Ir, Pd, Au, and Ag.

* See SAZHIN, N. P. and G. A. MEERSON. Rare Elements in Modern Technology.—*Khimicheskaya Nauka i Promyshlennost'*, I, No. 5, 482. 1956.

II. *Non-metals*:

1) B, Se, Te;

2) inert gases: He, Ne, Ar, Kr, Xe.

Table 1 shows the abundance in the Earth's crust (lithosphere) of *technologically* rare elements and of common elements (for comparison). It will be seen from the table that some technologically rare elements occur in the Earth's crust in large

Table 1

AVERAGE CHEMICAL COMPOSITION OF THE LITHOSPHERE* (AFTER VINOGRADOV)

(Thickness 16 km, excluding oceans and atmosphere)

Element	Composition of lithosphere		Element	Composition of lithosphere	
	atomic %	wt. %		atomic %	wt. %
Oxygen	58.0	47.2	Cerium	$6 \cdot 10^{-4}$	$4.5 \cdot 10^{-3}$
Silicon	20.0	27.6	Gallium	$4 \cdot 10^{-4}$	$1.5 \cdot 10^{-3}$
Aluminum	6.6	8.80	Neodymium	$3.5 \cdot 10^{-4}$	$2.5 \cdot 10^{-3}$
Hydrogen	3.0	0.15	Scandium	$3 \cdot 10^{-4}$	$6 \cdot 10^{-4}$
Sodium	2.4	2.64	Lanthanum	$2.5 \cdot 10^{-4}$	$1.8 \cdot 10^{-3}$
Iron	2.0	5.10	Germanium	$2 \cdot 10^{-4}$	$7 \cdot 10^{-4}$
Calcium	2.0	3.6	Niobium	$2 \cdot 10^{-4}$	$1 \cdot 10^{-3}$
Magnesium	2.0	2.10	Lead	$1.6 \cdot 10^{-4}$	$1.6 \cdot 10^{-3}$
Potassium	1.4	2.6	Arsenic	$1.5 \cdot 10^{-4}$	$5 \cdot 10^{-4}$
Titanium	$2.5 \cdot 10^{-1}$	$6 \cdot 10^{-1}$	Gadolinium	$1 \cdot 10^{-4}$	$1 \cdot 10^{-3}$
Carbon	$1.5 \cdot 10^{-1}$	$1 \cdot 10^{-1}$	Cesium	$9.5 \cdot 10^{-5}$	$7 \cdot 10^{-4}$
Barium	$5.7 \cdot 10^{-2}$	$5 \cdot 10^{-2}$	Praseodymium	$9 \cdot 10^{-5}$	$7 \cdot 10^{-4}$
Phosphorus	$5 \cdot 10^{-2}$	$8 \cdot 10^{-2}$	Samarium	$9 \cdot 10^{-5}$	$7 \cdot 10^{-4}$
Manganese	$3.2 \cdot 10^{-2}$	$9 \cdot 10^{-2}$	Thorium	$7 \cdot 10^{-5}$	$8 \cdot 10^{-4}$
Sulfur	$3.0 \cdot 10^{-2}$	$5 \cdot 10^{-2}$	Molybdenum	$6 \cdot 10^{-5}$	$3 \cdot 10^{-4}$
Fluorine	$2.8 \cdot 10^{-2}$	$2.7 \cdot 10^{-2}$	Dysprosium	$5 \cdot 10^{-5}$	$4.5 \cdot 10^{-4}$
Chlorine	$2.6 \cdot 10^{-2}$	$4.5 \cdot 10^{-2}$	Erbium	$5 \cdot 10^{-5}$	$4 \cdot 10^{-4}$
Nitrogen	$2.5 \cdot 10^{-2}$	$1 \cdot 10^{-2}$	Hafnium	$5 \cdot 10^{-5}$	$3.2 \cdot 10^{-4}$
Lithium	$1.9 \cdot 10^{-2}$	$6.5 \cdot 10^{-3}$	Bromine	$4 \cdot 10^{-5}$	$1.6 \cdot 10^{-4}$
Strontium	$1 \cdot 10^{-2}$	$4 \cdot 10^{-2}$	Ytterbium	$3 \cdot 10^{-5}$	$3 \cdot 10^{-4}$
Chromium	$8 \cdot 10^{-3}$	$2 \cdot 10^{-2}$	Thallium	$3 \cdot 10^{-5}$	$3 \cdot 10^{-4}$
Rubidium	$7 \cdot 10^{-3}$	$3.1 \cdot 10^{-2}$	Uranium	$2 \cdot 10^{-5}$	$3 \cdot 10^{-4}$
Vanadium	$6 \cdot 10^{-3}$	$1.5 \cdot 10^{-2}$	Tantalum	$1.8 \cdot 10^{-5}$	$2 \cdot 10^{-4}$
Zirconium	$4 \cdot 10^{-3}$	$2 \cdot 10^{-2}$	Europium	$1.8 \cdot 10^{-5}$	$1.2 \cdot 10^{-4}$
Copper	$3.6 \cdot 10^{-3}$	$1 \cdot 10^{-2}$	Selenium	$1.5 \cdot 10^{-5}$	$6 \cdot 10^{-5}$
Nickel	$3.2 \cdot 10^{-3}$	$8 \cdot 10^{-3}$	Holmium	$1.5 \cdot 10^{-5}$	$1.3 \cdot 10^{-4}$
Zinc	$1.5 \cdot 10^{-3}$	$5 \cdot 10^{-3}$	Tungsten	$1 \cdot 10^{-5}$	$1 \cdot 10^{-4}$
Cobalt	$1.5 \cdot 10^{-3}$	$3 \cdot 10^{-3}$	Lutetium	$1 \cdot 10^{-5}$	$1 \cdot 10^{-4}$
Beryllium	$1.2 \cdot 10^{-3}$	$6 \cdot 10^{-4}$	Terbium	$1 \cdot 10^{-5}$	$1.5 \cdot 10^{-4}$
Tin	$7 \cdot 10^{-4}$	$4 \cdot 10^{-3}$	Thulium	$8 \cdot 10^{-6}$	$8 \cdot 10^{-5}$
Yttrium	$6 \cdot 10^{-4}$	$2.8 \cdot 10^{-3}$	Cadmium	$7.6 \cdot 10^{-6}$	$5 \cdot 10^{-5}$
Boron	$6 \cdot 10^{-4}$	$3 \cdot 10^{-4}$	Antimony	$5 \cdot 10^{-6}$	$4 \cdot 10^{-5}$

Table 1 (continued)

Element	Composition of lithosphere		Element	Composition of lithosphere	
	atomic %	wt. %		atomic %	wt. %
Iodine	$4 \cdot 10^{-6}$	$3 \cdot 10^{-5}$	Gold	$5 \cdot 10^{-8}$	$5 \cdot 10^{-7}$
Bismuth	$1.7 \cdot 10^{-6}$	$2 \cdot 10^{-5}$	Rhodium	$1.7 \cdot 10^{-8}$	$1 \cdot 10^{-7}$
Silver	$1.6 \cdot 10^{-6}$	$1 \cdot 10^{-5}$	Iridium	$8.5 \cdot 10^{-9}$	$1 \cdot 10^{-7}$
Indium	$1.5 \cdot 10^{-6}$	$1 \cdot 10^{-5}$	Rhenium	$8.5 \cdot 10^{-9}$	$1 \cdot 10^{-7}$
Mercury	$7 \cdot 10^{-7}$	$7 \cdot 10^{-6}$	Radium	$9 \cdot 10^{-12}$	$1 \cdot 10^{-10}$
Osmium	$5 \cdot 10^{-7}$	$5 \cdot 10^{-6}$	Protactinium	$8 \cdot 10^{-12}$	$1 \cdot 10^{-10}$
Palladium	$1.6 \cdot 10^{-7}$	$1 \cdot 10^{-6}$	Actinium	$5 \cdot 10^{-15}$	$6 \cdot 10^{-10}$
Tellurium	$1.3 \cdot 10^{-7}$	$1 \cdot 10^{-6}$	Polonium	$2 \cdot 10^{-15}$	$2 \cdot 10^{-14}$
Ruthenium	$1 \cdot 10^{-7}$	$5 \cdot 10^{-7}$	Plutonium	$7 \cdot 10^{-17}$	$1 \cdot 10^{-15}$
Platinum	$5 \cdot 10^{-8}$	$5 \cdot 10^{-7}$	Radon	$5 \cdot 10^{-17}$	$7 \cdot 10^{-16}$

* The quantitative determinations of helium, argon, neon, krypton and xenon are not reliable.

amounts. For example, titanium, which is usually regarded as a technologically rare element, is the tenth most abundant element in the Earth's crust, and lithium is the eighteenth most abundant element.†

Some very familiar elements, on the other hand, (mercury, antimony, cadmium, lead, tin), have a low abundance in the Earth's crust.

A distinction should be made between the concept of *technologically rare element* and *low-abundance element*. In geochemistry, various chemical elements are regarded as rare because of their low abundance in nature. The geochemically rare elements include a group of dispersed elements (rhenium, radium, polonium, etc.) which do not form independent minerals and are encountered as impurities in minerals and ores of other elements.

As a rule, the abundance of chemical elements in celestial bodies and on Earth depends on the stability of the atomic nuclei in stellar interiors. The stability of atomic nuclei steeply falls off as the atomic number increases to 28, and then it continues decreasing more slowly. The relatively low abundance of the light elements—lithium, beryllium, boron, and others—is due to the large cross-section of the reaction between these nuclei and protons, neutrons, and other particles. The low abundance of the heavy elements—thorium, uranium, and the transuranium elements—is due to α-decay and spontaneous fission.

The number of technologically rare elements is steadily decreasing owing to the increase in the production volume and improved production technology. According to some authorities, titanium no longer can be regarded as a rare metal, since its production is now quite considerable; moreover, it belongs to the most abundant

† Many of the so-called rare earths are in fact fairly abundant in nature.

elements in the Earth's crust. This example shows that the concept of *technologically rare element* will in future become obsolete.

In the last twenty years there was a considerable increase in the production of certain rare metals and their compounds (titanium, zirconium, niobium, germanium, indium, gallium, cerium, lithium, etc., and their hydrides, borides, iodides, carbides, and a wide range of alloys). Some rare metals and rare-metal compounds are now being produced in highest grades of purity for use in nuclear, semiconductor, and metallurgical industries (uranium, thorium, zirconium, etc.).

The most important uses of rare elements are the following:

1. *Nucleonics*:

Uranium, thorium (nuclear fuel), zirconium, beryllium (construction materials for nuclear reactors), bismuth.

2. *Electronics*:

a) germanium, as a semiconductor in solid-state rectifiers and amplifiers, radio instruments, radar equipment, remote-control systems, automatic control of machines, computers, etc.;

b) other elements (more than 15) and their salts, oxides, carbides, borides, etc.

3. *Electric and radio industry*:

a) tungsten, molybdenum, tantalum, and niobium, as filaments and electron-emitting components in lamps and radio tubes and in the manufacture of high-temperature electric furnaces and thermocouples;

b) zirconium, titanium, and tantalum, for the absorption of gases in vacuum instruments;

c) selenium, for the production of photoelectric devices;

d) other elements (more than 15) and their compounds.

4. *Chemical industry*:

a) vanadium and some rare earths, as catalysts;

b) selenium, tellurium, lithium, etc., in organic synthesis, and in the plastics and glass industries;

c) lithium, in the production of lubricants which are stable in a wide range of temperatures;

d) tantalum, for use in the construction of corrosion-resistant equipment.

5. *Production of special steels and alloys*:

a) tungsten, molybdenum, vanadium, niobium, titanium, zirconium, beryllium, indium, rare earths, cobalt, selenium, tellurium, etc. (some of them alloyed with iron), for alloying, deoxidation and modification of numerous ferrous and non-ferrous alloys;

b) titanium and its alloys, as construction materials;

c) molybdenum, niobium, titanium, zirconium and their carbides, borides, and silicides, in the production of heat-resistant alloys;

d) tungsten, titanium and tantalum carbides, in the production of hard alloys.

Many rare elements display valuable and virtually unique properties.

The Importance of the Analytical Chemistry of Rare Elements

Analytical chemistry is of major importance in the development of the modern rare-element industry.

The sources of rare elements (ores, minerals, alloys, etc.) are very numerous and of varied composition. Analytical methods are extensively employed in quality control of starting materials, industrial processes, and finished products, and also in prospecting for new sources of rare elements.

Analytical chemistry of rare elements also plays a very important part in geochemistry. Thus, determination of the relative content in rocks of rare elements which have similar properties, such as niobium–tantalum, zirconium–hafnium, tungsten–molybdenum, sulfur–selenium, rubidium–thallium, aluminum–gallium, nickel–cobalt, radium–cadmium, etc., leads to significant results.

Traces of some rare elements (Mo, V, etc.) are important in biochemical processes.

Chemical analysis using spectroscopic, photometric, polarographic and other rapid, accurate, and sensitive techniques is essential for the production of rare elements with minimum impurity contents for atomic, semiconductor, chemical and other modern industries.

Atomic, semiconductor, and metallurgical industries require nuclear fuel, and also construction and semiconductor materials of a very high purity. Thus, pure zirconium is one of the best construction materials for nuclear reactors, and even a trace of hafnium present as an impurity renders it unsuitable for the purpose. Semiconductor properties, especially those of germanium and silicon, are displayed only in impurity-free specimens. The impurity tolerance is not more than 1 impurity atom in 10^{10} atoms of germanium; it is even smaller in the case of silicon.

High-purity materials are also needed in the production of electronic instruments (oscillographs, electronic computers, radar equipment). Of the numerous construction materials used in electron-beam and other vacuum instruments, the most important is pure titanium, which displays plastic properties. The production of pure titanium is impossible without a properly organized analytical control. Thermoelectric and photoelectric cathodes, and cathodophosphors in electron-beam instruments require high-purity titanium, as well as many other high-purity materials (e.g. reagents, metallic nickel, iron, tungsten, molybdenum, tantalum and niobium).

These examples illustrate the great importance of the determination of trace impurities in rare elements. High-purity materials cannot be manufactured without routine control of chemical composition at all the key stages of the production process.

Further development of the production of high-purity rare metals and their analytical control will undoubtedly yield new interesting results.

Quantitative Methods in the Analytical Chemistry of Rare Elements

Analytical reagents for rare elements include group reagents, selective reagents, and specific reagents.

Group reagents give similar reactions with a large number of ions and are used to separate individual groups of elements from a multicomponent mixture. The different classifications of cations in analytical work are based on the action of group reagents.

Selective reagents give similar reactions only with few ions. For example, hydrogen peroxide in sulfuric acid solution forms soluble colored compounds with Ti^{4+}, vanadate, and molybdate ions only.

No specific reagents for the ions of the individual elements are known. However, under certain conditions the selectivity of group reagents can be improved by binding the interfering ions as stable complexes. For example, ferric ions, which impart a yellow color to the solution, interfere with the detection of titanium (IV) by hydrogen peroxide (orange color). When a moderate amount of phosphoric acid is added to the sample solution, the ferric ions are converted to a colorless complex which no longer interferes with the detection of titanium. Other methods for improving selectivity of the reagents are also known.

Because of the poor selectivity of the various reagents employed in the analytical chemistry of rare elements, separation methods based on differences in the properties of their compounds acquire special importance.

As a result of systematic search for new organic reagents, the number of sensitive specific reactions for rare elements steadily increases.

The content of a rare element, which is present in the sample in the free form or as some compound, can be determined by various methods which involve determination of mass, volume, density, absorption and emission of infrared, visible and ultraviolet radiation, X-ray methods, light-scattering measurements, etc. The principal quantitative methods are listed in Table 2.

There is a functional dependence between the content (concentration) of the given

Table 2

CLASSIFICATION OF THE MAIN QUANTITATIVE ANALYTICAL METHODS

Physical parameter being measured	Method	Determinable amounts*
Mass	Gravimetric, including electro-gravimetric	Macro and micro
	Mass-spectroscopy	Micro and ultramicro
Volume	Titrimetric	Macro and micro
	Gas-volumetric	Macro and micro
	Determination of volume of precipitates	Macro, micro and ultramicro
Density	Densimetry	Macro
IR absorption	IR spectroscopy	Macro
Absorption and emission of visible and UV light; X-ray emission and absorption; light scattering	Spectroscopy	Semimicro and micro
	Atomic absorption spectroscopy; flame photometry	Semimicro and micro
	X-ray spectroscopy	Macro and semimicro
	Photometry (colorimetry, spectrophotometry, turbidimetry, nephelometry)	Semimicro and micro
	Luminescence and fluorescence	Micro
Refractive index	Refractometry	Macro
Rotation of the plane of polarization	Polarimetry, spectropolarimetry	Macro
Diffusion current in oxidation or reduction at the electrode	Polarography	Semimicro and micro
Quantity of electricity consumed in the electrode reaction	Coulometry	Micro and ultramicro
Electrode potential	Potentiometry	Macro and micro
Electric conductance	Conductometry	Macro and micro
Dielectric constant	—	Macro
Radioactivity	Activation analysis	Micro and ultramicro

* The accepted lower limits of the quantititative analytical methods are as follows:

Method	*Minimum determinable amount*
Macro method	10^{-2} g
Micro method	10^{-2} mg
Ultramicro method	10^{-2} μg

element and the parameter being measured; this dependence may be expressed as a mathematical formula or by a curve.

Of major practical importance are methods in which the concentration of the element is a linear function of the physical parameter being measured. This group includes gravimetric, titrimetric, spectroscopic, some spectrophotometric, and other methods. In many cases the linear dependence is valid only in a certain limited range. In spectroscopic methods, the linear dependence between the concentration of the element and the intensity of the analytical spectral lines is observed only between certain limits. In many photometric methods, the proportionality between the concentration and the light absorption of the solution (Beer's law) is maintained within a definite limited range. In activation analysis, the amount of the radioactive isotopes formed by neutron irradiation is proportional to the number of counts per second.

In semimicrochemical methods the quantities involved are one order of magnitude smaller than in macro analysis and one order of magnitude greater than in micro analysis. There is, however, no generally accepted classification of analytical methods according to the amount of the element to be determined or the size of the sample.

The most important methods in the analytical chemistry of rare elements are gravimetric, mass-spectroscopic, titrimetric, spectroscopic, X-ray, photometric, polarographic and activation analysis. Other analytical methods are used only occasionally.

Gravimetric methods

In gravimetric determination the amount of the material is measured by weighing. The methods are based on the law of conservation of the total mass in a chemical reaction.

Using appropriate chemical reactions, the element to be determined is quantitatively reduced to its elementary form or, alternatively, it is converted into a compound of known composition with suitable properties. The product contains the entire amount of the element taken for determination. If the composition of the compound is known, the amount of the element in the sample is readily calculated.

Compounds which are ultimately weighed in gravimetric analysis must satisfy a number of requirements; their composition must conform as closely as possible to the chemical formula, and the compound should be stable, i.e., its composition should not change in the course of the various operations involved, such as washing, drying or weighing.

Gravimetric methods are used in the analytical chemistry of rare elements whenever high accuracy is required; they are, however, very time-consuming. In practice, gravimetric methods are employed to determine beryllium in ores and ore-processing products; zirconium in concentrates and alloys; thorium in metallic tungsten and other objects; niobium and tantalum in various materials; molybdenum in alloys and concentrates; tungsten in ferrotungsten; rhenium in tungsten alloys, etc.

Volumetric (titrimetric) methods

These methods are based on chemical reactions which follow a definite chemical equation.

The most reliable results are obtained by methods using stoichiometric reactions which proceed to completion at a sufficiently fast rate. Titrimetric methods for rare elements involve various redox reactions, precipitation reactions, and complexation reactions. Complexometric methods are particularly important in analytical practice.

Special chemical compounds—indicators—are employed to determine the equivalence point. If chemical indicators are not available, the equivalence point is found by various instrumental methods.

Near the equivalence point the concentration of the substance to be determined rapidly decreases, and the concentration of the reagent rapidly increases. At the same time, a sudden change is observed in the electrode potential, the conductance of the solution, the diffusion current, or the light absorption, as the case may be. The abrupt change in these properties pinpoints the exact position of the equivalence point during the titration. Changes in radioactivity, freezing point, density, refractive index, surface tension, and viscosity are also sometimes used. Potentiometry and amperometry are the most important instrumental methods for the determination of the equivalence point.

The amperometric method is used to an ever increasing extent for establishing the equivalence points of precipitation reactions, complexation reactions, and redox reactions. It is based on the variation of the diffusion current during the titration, associated with the reduction or oxidation processes on a dropping mercury electrode or on a rotating platinum electrode. The method is used for the determination of low concentrations in dilute solutions (0.01–0.0001 moles/liter).

Conductometric and high-frequency titration methods are of minor importance in the analytical chemistry of rare elements.

Titrimetric methods are more selective than most gravimetric methods and are substantially faster. They are, however, somewhat less accurate. A higher accuracy can be attained by weighing the titrant in a special gravimetric buret, instead of measuring its volume.

Examples of titrimetric analyses include the ferric periodate determination of lithium in silicate ores; arsenate determination of beryllium in minerals, concentrates, and alloys; determination of titanium in ferrotitanium and titanium–nickel alloys; determination of germanium in industrial concentrates; determination of vanadium in alloys; determination of molybdenum in alloys and concentrates; determination of selenium and tellurium in ores and in metallurgical processing products; potentiometric determination of rhenium in alloys, etc.

Complexometric titration is of major practical importance for the determinations of gallium, indium, thallium, scandium, thorium, zirconium, and several other elements.

The amount of the substance is sometimes inferred from the volume of the precipitate. This method is very important in ultramicro analysis, e.g., for transuranium elements, when the amount of the precipitate is small and its isolation from the solution for the actual weighing is very difficult. The volume of the precipitate can be determined under the microscope, and the mass of the precipitate is then calculated from the known volume and density.

Electrochemical methods

Polarographic methods involve electrolysis on a dropping mercury electrode or a platinum wire microelectrode.

Polarography is used both in research work and in fast industrial analysis. This method is now increasingly resorted to in the analysis of rare elements.

Polarographic methods are applied to the determination of indium in polymetallic ores and their processing products, of titanium and germanium in different materials, and of niobium in metallic tantalum and tantalum-containing materials. Polarography is also frequently used in the determination of molybdenum, selenium and tellurium.

Oscillographic polarography is employed in the determination of niobium in its alloys with tantalum and of ytterbium in the presence of erbium.

The coulometric method is both sensitive and accurate. It is used in the determination of uranium and some other elements.

Photometric methods

Photometric methods are subdivided into several categories: *colorimetry* and *spectrophotometry* measure the absorption of visible or ultraviolet radiation by a solution; *turbidimetry* is concerned with the absorption of light by a suspension; *nephelometry* measures the scattering of light by a suspension.

The measurement of the optical density with a spectrophotometer is generally the final stage of the determination. Prior to this, the component to be determined is usually separated from the accompanying elements by precipitation, distillation, extraction, or electrolysis. The separation method should ensure maximum selectivity and a minimum number of separations, in order to reduce the probability of contamination by reagents; this is particularly important when determining very small amounts.

Spectrophotometric methods are applied to determine traces of elements in various industrial products, such as semiconductors, metals, and alloys.

Photometric methods are employed for determining small amounts of a wide range of rare elements: beryllium in tungsten and in alloys; gallium, indium, thallium, rare earths, and germanium in various objects; titanium in rocks, ores, and alloys, and in metallic tungsten and zirconium; thorium in rocks, zircon, and other materials; zirconium in various materials; vanadium in ores, minerals, alloys, steels, metallic zirconium; niobium in rocks and minerals; tantalum in metallic zirconium, hafnium, and niobium; bismuth in metallic molybdenum; molybdenum in titanium-based alloys, steels, and minerals; selenium and tellurium in ores and minerals; rhenium in molybdenum-containing products and in alloys with tantalum or tungsten.

Large amounts of rare elements can be accurately determined by differential spectrophotometry. The method is successfully employed in the determination of titanium, tantalum and rhenium.

Spectroscopic methods

The X-ray spectrum of any chemical element comprises very few lines. Although highly selective, the X-ray method has low sensitivity (the element to be determined must be present in a concentration of at least 0.1%). It is used for separate determination of elements with similar chemical properties in mixtures, e. g., niobium and tantalum, or zirconium and hafnium or rare earths.

Under fixed excitation conditions, the atoms of each chemical element emit light in spectral lines of definite wavelengths. In conventional spectroscopic analysis, different lines observed in the spectrum reveal the presence of the corresponding elements in the sample, and the intensities of the lines are indicatory of the quantitative content of each element.

The spectrum is obtained with a prism spectrograph or a diffraction grating spectrograph (a diffraction grating is a plate with fine parallel lines ruled on its surface—up to 250,000 lines in 25 cm).

Spectroscopic analysis is widely employed in research and in industrial laboratories.

The method is used for the determination of lithium, beryllium, scandium, rare earths, gallium, indium, thallium, germanium, zirconium, hafnium, niobium, and tantalum in rocks, ores, ore-dressing products, and alloys.

Flame photometry is used for the determination of lithium, rubidium, cesium and strontium in rocks, ores, and minerals; lanthanum, europium, ytterbium and yttrium in a mixture of total rare-earth oxides; indium, gallium and thallium in concentrates and industrial semifinished products.

Methods using radioactivity

Elements with natural radioactivity (radium, radon, uranium, thorium, potassium, rubidium, samarium, and others) can be determined by measuring the intensity of the radiation emitted by their atoms or by the radioactive decay products in equilibrium with their atoms (after separation). Thorium is successfully determined in ores, minerals, rocks and soils in this way.

Artificial radioactive isotopes (labeled atoms or tracers) are used to monitor the degree of separation of elements by precipitation, extraction, adsorption, electrolysis, and other techniques; they also comprise a means for measuring the solubilities of salts and the completeness of precipitation. Tracers are used in studies of adsorption, coprecipitation and extraction and in the determination of the stability of complex compounds.

The method of isotopic dilution (making use of radioactive isotopes) is applied in certain other cases. Concentrations of 0.001–0.3% of lithium in rocks and 0.002–0.02% of rubidium in granite, diabase, and sea water have been determined by this method.

Impurities in different materials are detected by means of the radioactive isotopes which are formed when a sample is bombarded with elementary particles, mainly neutrons (radioactivation or neutron activation analysis).

Neutron activation does not involve contamination of the sample with extraneous elements. Elements producing artificial radioisotopes can be determined with an accuracy of less than 10%.

Neutron activation is employed in the determination of impurities in pure chemical compounds, metals, and alloys, in purity control of the reactor materials, in the determination of lithium and rubidium in rocks, of beryllium in minerals and hydrometallurgical products, and in the determination of rare earths.

The sensitivity of neutron activation analysis increases with the increase in the intensity of the activating neutron beam. Irradiation with $5 \cdot 10^{11}$ neutrons/cm^2/sec will determine 0.0001 μg of manganese, rhenium, iridium, indium, samarium, europium, holmium, and lutecium.

Neutron activation analysis is usually more sensitive than spectrographic, spectrophotometric, and amperometric methods, or chemical methods involving color reactions.

The sensitivities in the determination of indium by the different methods are listed below, for comparison.

Method	*Sensitivity for indium, μg/ml*
Activation analysis	
neutron flux 10^{13} cm$^{-2} \cdot$ sec^{-1}	0.000005
neutron flux 10^{11} cm$^{-2} \cdot$ sec^{-1}	0.0001
Spectroscopic (spark excitation, copper electrode)	1
Flame photometry	1
Chemical reactions, maximum sensitivity (8-hydroxyquinoline)	0.2
Amperometric titration	100

Separation of elements

The element to be determined is generally first converted to a form which is particularly convenient for its identification or measurement. Group separation methods and methods for the isolation of individual elements are highly important in this respect.

Separation methods take advantage of the differences in the properties of the different compounds and transfer the element into another phase by precipitation, distillation, or extraction.

The isolation of an element, or a group of similar elements, by precipitation in the form of sparingly soluble compounds is one of the oldest and most frequently used methods. The precipitation of numerous accompanying ions can be prevented by converting them to stable complexes.

Elements which are normally not precipitated by the reagent if present alone in solution are often captured and firmly retained by the precipitate forming in the course of the separation. This effect, known as *coprecipitation*, interferes with the quantitative separation of the elements. Solubilization and reprecipitation of the washed precipitate does not always yield a pure precipitate. Coprecipitation has an advantageous use in isolating trace amounts of elements. An efficient method has been developed in which trace amounts of several elements are isolated by coprecipitation with organic compounds.

The element to be determined can be selectively isolated from solution by electrolysis, in the elementary form or as the oxide. Internal electrolysis is employed to isolate small amounts of certain metals.

In the separation of elements by extraction, the aqueous solution of the substance to be analyzed is shaken with an appropriate organic solvent, which is immiscible with water. If one of the substances is readily soluble in the organic solvent, it will pass into the organic phase. However, small amounts of the substance will remain in aqueous solution. A repeated extraction will result in a more complete separation. In this way, ferric ions can be separated from the ions of aluminum, bismuth, calcium, cadmium, chromium, cobalt, lead, manganese, nickel, osmium, palladium, titanium and uranium (extraction with ether from hydrochloric acid solution).

The extraction method also provides an effective technique for the concentration of small amounts of elements which have to be determined.

Elements are often separated by distillation. Thus, fluorine is distilled off as hydrogen fluoride or as fluosilicic acid, and germanium may be distilled off as the tetrachloride.

One of the most effective methods of separation is chromatography. We distinguish between adsorption, partition, ion exchange, and precipitation chromatography.

In adsorption chromatography the separation of mixtures is based on differences in adsorption capacities of different substances. The adsorption of the solvent should be much lower than that of the components to be analyzed, so as to make full use of the separating power of the sorbent.

Partition chromatography is based on differences in the distribution coefficients of the substances to be separated between two immiscible phases, one of which (the stationary phase) is locked in the pores of the sorbent material and the second (the mobile phase) is provided by the eluent solution. The sorbent retaining the liquid phase is a hydrophilic substance of the type of silica gel (with large pores which reduce the adsorption effect) or starch. The stationary phase is usually water, methanol, or nitromethane. Partition chromatography is used, for example, to separate niobium and tantalum from titanium.

In a modification of partition chromatography, macerated paper or paper sheets are used as the sorbent (paper chromatography). A successful separation depends on the difference between the migration rates of the individual components and the rate of migration of the solvent. Paper chromatography is used in the separation of rare earths.

In ion-exchange chromatography, the separation of mixtures is based on the specific tendencies of different ions to be exchanged with the labile ions of the sorbent. Synthetic organic ion exchangers (cation and anion exchangers) are usually used; alumina, silica gel, and other natural materials are employed less frequently.

Ion-exchange chromatography is used to separate zirconium from hafnium and other elements, rhenium from molybdenum, titanium from molybdenum, thorium from rare earths, etc.

Certain elements may be separated by strongly basic anion exchangers. Indium has successfully been separated from aluminum, iron, and arsenic by passing a hydrochloric acid solution of the sample through a column with a strongly basic anion exchanger.

* * *

The future development of the analytical chemistry of rare elements depends to a large extent on complete understanding of the theoretical principles of analytical methods.

The analytical potential of the chemical reactions of individual rare elements must be studied by modern techniques.

Proximate methods of inorganic analysis are also very important. This analytical technique is relatively neglected, despite its importance in mineralogy, metallurgy, metal science, silicate industry, and chemical technology, all of which require quantitative data on the actual mode of occurrence of the compounds in natural and industrial products.

Principal Literature on the Analytical Chemistry of the Rare Elements

The analytical chemistry of the rare elements is now in a stage of rapid development. Papers on the subject can be found in every issue of any analytical chemistry journal. There are also many compilations, proceedings and reports which contain much valuable information.

The most important journals on the analytical chemistry of rare elements are *Zhurnal Analiticheskoi Khimii, Zavodskaya Laboratoriya, Analytical Chemistry, The Analyst, Zeitschrift für analytische Chemie, Analytica chimica acta, Talanta,* and others.

Valuable information on rare elements is found in the standard handbook of inorganic chemistry *Gmelin's Handbuch der anorganischen Chemie,* 8th ed., appearing since 1924; in monographs on the analytical chemistry of individual rare elements, published since 1953 by the USSR Academy of Sciences; in Fresenius and Jander's *Handbuch der analytischen Chemie,* Pt. II—*Qualitative Nachweisverfahren,* and Pt. III—*Quantitative Bestimmungs- und Trennungsmethoden* (Springer-Verlag); and also in the *Treatise on Analytical Chemistry,* Pt. II—*Analytical Chemistry of the Elements,* edited by KOLTHOFF, ELVING and SANDELL (Interscience, New York).

Literature which appeared between 1941 and 1952 in the Russian language on the analytical theory and practice, including the analytical chemistry of rare elements, has been listed in A.I. BUSEV, *Analiticheskaya khimiya. Literatura na russkom yazyke* (1941–1952), Izd. AN SSSSR, 1956 (8495 references). The book includes a subject index for rapid location of all the references to a given subject which were published during the relevant period. Similar lists of references for the periods 1930–1940 and 1953–1965 are in preparation.

Bibliographical listings are found in the surveys published every year in *Analytical Chemistry* and in various monographs.

Comprehensive bibliographies are available on polarography, microchemical analysis, fluorimetric analysis, etc., which also contain material on the analytical chemistry of rare elements.

A brief review of the most important reference books, textbooks and other publications on the analytical chemistry of rare elements is given below.

MORACHEVSKII, YU.V. and I.A. TSERKOVNITSKAYA. *Osnovy analiticheskoi khimii redkikh elementov (Fundamental Principles of the Analytical Chemistry of Rare Elements).*—Izd. LGU. 1964.
This is a university textbook, reviewing the more important chemical properties and reactions of rare elements on which the analytical methods are based and discussing compilation of analytical schemes; natural compounds and applications of rare elements are briefly described.

SONGINA, O.A. *Redkie metally (Rare Metals).*—3rd ed., Izd. "Metallurgiya". 1964.
The book comprises a brief description of the analytical chemistry of rare elements.

PATROVSKY, V. *Analycka chemie vzacynch prvku.*—Prague, Statni Nakl. Tech. Lit. 1956.
The subjects treated include the preparation of samples for analysis, methods for the detection of rare metals, their concentration and quantitative determination in mineral raw materials and certain products of the metallurgical industry by titrimetric, gravimetric, photometric and other methods.

Vernadskii Institute of Geochemistry and Analytical Chemistry of the USSR Academy of Sciences has put out *Metody opredeleniya i analiza redkikh elementov (Determination and Analysis of Rare Elements).*—Izd. AN SSSR. 1961.
This is a practical textbook for the determination and analysis of rare elements. The book describes methods which have proved to be reliable in industrial and scientific analytical laboatories. Extensive treatment is given to the results obtained by analytical chemists in the Soviet Union, including papers which appeared in departmental publications. Modern advances in the analytical chemistry of most rare elements are surveyed in special articles, including results obtained outside the Soviet Union. An extensive list of references is given. Detailed descriptions of the accepted physical, physicochemical and chemical methods for the analysis of raw materials, semifinished products and pure metals are given for each element.

BUSEV, A.I. *Analiticheskaya khimiya vismuta (Analytical Chemistry of Bismuth).*—Izd. AN SSSR. 1953.

BUSEV, A.I. *Analiticheskaya khimiya indiya (Analytical Chemistry of Indium).*—Izd. AN SSSR. 1958.

RYABCHIKOV, D.I. and E.K. GOL'DBRAIKH. *Analiticheskaya khimiya toriya (Analytical Chemistry of Thorium).*—Izd. AN SSSR. 1960.

BUSEV, A.I. *Analiticheskaya khimiya molibdena (Analytical Chemistry of Molybdenum).*—Izd. AN SSSR. 1962.

KORENMAN, I.M. *Analiticheskaya khimiya talliya.—(Analytical Chemistry of Thallium).*—Izd. AN SSSR. 1960.

AVTOKRATOVA, T.D. *Analiticheskaya khimiya ruteniya (Analytical Chemistry of Ruthenium).*—Izd. AN SSSR. 1962.

These monographs contain descriptions of gravimetric, titrimetric, spectroscopic and other methods of detection and determination of the respective elements and techniques for their separation. They include extensive and exhaustive lists of references for some of the elements. [Most of these monographs are available in English translation as part of the *Analytical Chemistry of the Elements* series, published by Ann Arbor–Humphrey Science Publishers, 1969–1970.]

SCHOELLER, W.R. and A.R. POWELL. *Analysis of Minerals and Ores of Rare Elements.*—London, Griffin. 1955.
The material presented in this book includes chemical and certain other methods for analyzing ores and minerals of 22 rare elements as well as lanthanides and platinum metals; the preparation and pro-

perties of analytically important compounds of the rare elements, their identification, separation and determination, are also described. Schemes of complete analyses of ores and minerals are given. Each chapter begins with a brief account of the more important minerals of rare elements.

HILLEBRAND, W.F., G.E. LUNDELL, H.A. BRIGHT, and J.I. HOFFMAN. *Applied Inorganic Analysis.*—New York, Wiley. 1930.

A description of the more important methods of isolation and quantitative determination of the rare and common elements. The book deals with methods for solubilizing the sample material, separating a given element from other elements and quantitative (gravimetric, titrimetric, photometric etc.) methods of determination. Determination conditions for any given element in the presence of other elements are discussed, and relevant references given.

KNIPOVICH, YU.N. and YU.V. MORACHEVSKII, Editors. *Analiz mineral'nogo syr'ya (Analysis of Mineral Raw Materials).*—3rd ed., Goskhimizdat. 1959.

Methods used in the sampling of rocks and minerals are described. A brief account is given of the naturally occurring compounds of each element. Methods for solubilizing the ores of rare elements, separating the elements, determining individual componenets and for total analysis are described. References relevant to the text are given.

FAINBERG, S.YU. and N.A. FILIPPOVA.—*Analiz rud tsvetnykh metallov (Analysis of Ores of Non-ferrous Metals).*—3rd ed., Metallurgizdat. 1963.

This book describes methods for separating and determining the following rare elements: bismuth, molybdenum, tunsgten, selenium, tellurium, thallium, indium, gallium, germanium and rhenium in different technological materials; polymetallic ores, concentrates and certain other materials are analyzed.

NAZARENKO, V.A. and N.S. POLUEKTOV.—*Polumikrokhimicheskii analiz mineralov i rud (Semimicrochemical Analysis of Minerals and Ores).*—Goskhimizdat. 1950.

Procedures and techniques of quantitative semimicro-determinations of rare and common elements in minerals and ores are reviewed. The book also contains instructions on how to erect laboratories under field conditions.

ZHELEZNOVA, E.I., V.G. SOCHEVANOV, and V.I. TITOV. *Metody opredeleniya radioaktivnykh elementov v mineral'nom syr'e (Methods of Determination of Radioactive Elements in Mineral Raw Materials).*—2nd ed., Gosgeoltekhizdat. 1961.

A description of quantitative methods for the determination of uranium, thorium, ionium and radium in minerals, rocks, ores, and waters by chemical methods, radioactivity measurements and spectroscopy.

Analytical Chemistry of Uranium (Vernadskii Institute of Geochemistry and Analytical Chemistry. *Analytical Chemistry of the Elements series*).—Izd. AN SSSR. 1962.

The chemical and analytical properties of uranium, of gravimetric, titrimettic, photometric, electrochemical and radiochemical methods for determining uranium in natural and industrial products are discussed and methods for the detection and separation of uranium from other elements are indicated. A description is also given of a number of methods for the determination of impurities in pure preparations of uranium. [English translation available as part of the *Analytical Chemistry of the Elements* series, Ann Arbor–Humphrey Science Publishers. 1969.]

MARKOV, V.K., E.A. VERNYI, A.V. VINOGRADOV, S.V. ELINSON, A.G. KLYGIN, and I.V. MOISEEV. *Uran, metody ego opredeleniya (Methods for the Determination of Uranium).*—2nd ed., Atomizdat. 1964.

A description of gravimetric, titrimetric, photometric, fluorimetric, electrochemical and radiometric methods for the determination of uranium.

MUKHINA, Z.S., E.I. NIKITINA, L.M. BUDANOVA, R.S. VOLODARSKAYA, L.YA. POLYAK, and A.A. TIKHONOVA. *Metody analiza metallov i splavov (Method of Analysis of Metals and Alloys).*—Oborongiz. 1959.

The book describes methods for analyzing steels, cast irons, refractory alloys, ferroalloys, slags, aluminum-based alloys, magnesium and copper. Methods for determining a large number of the alloying elements in these materials are also given.

CHARLOT, G. *Les méthodes de la chimie analytique.*—Paris, Masson. 1961.

Methods for the determination of almost all elements as well as references are given.

DYMOV, A.M. *Tekhnicheskii analiz rud i mineralov (Technical Analysis of Ores and Minerals).*—5th ed., Metallurgizdat. 1949.

A description of rapid and empirical analyses of different materials used in the metallurgical industry, viz., iron, titanium, tungsten ores, ferroalloys, conventional and special alloys and others.

Metody khimicheskogo analiza mineral'nogo syr'ya (Vesoyuznyi nauchno-issledovatel'skii institut mineral'nogo syr'ya Ministerstva geologii i okhrany nedr) (Methods of Chemical Analysis of Minerals. All-Union Research Institute for Mineral Raw Materials of the Ministry of Geology and Preservation of Natural Resources).—Gosgeoltekhizdat.

No. 1:1955. *Analytical methods for tungsten, vanadium, molybdenum, titanium, zirconium.*

No. 2:1956. *Polarographic methods for molybdenum, antimony, indium and thallium.*

No. 3:1957. *Analytical Methods for beryllium, lithium, niobium, tantallum and titanium.*

No. 4:1958. *Analytical Methods for beryllium, gallium, hafnium, germanium, indium, lithium, REE, selenium, tellurium, thallium, zirconium.*

No. 5:1959. *Analytical methods for beryllium, gallium, indium, molybdenum, niobium, strontium, REE, rhenium, tantalum, zirconium.*

No. 6:1960. *Analytical methods for REE.*

No. 7:1963. *Analytical methods for beryllium, germanium, rhenium, scandium, tantalum.*

Publications of the Gosudarstvennyi nauchno-issledovatel'skii institut redkikh i malykh metallov (GIREDMET) (State Research Institute of Rare and Minor Metals): 1) *Compendium of GIREDMET papers,* 1931–1956, Pt. II, *Analytical Methods.*—Metallurgizdat. 1959. 2) *Nauchnye trudy* (1959) *Gosudarstvennogo nauchno-issledovatel'skii i proektnogo instituta redkometallicheskoi promyshlennosti (Compendium of Papers of the State Research and Planning Institute of the Rare Metal Industry,* 1959, Pt. III, *Analysis of Materials in the Rare Metal Industry.*—Metallurgizdat. 1961. 3) Pt. X, *Analytical Methods.*—Metallurgizdat. 1963. 4) Pt. XIII, *Analytical Methods.*—Metallurgizdat. 1964.

SAMSONOV, G.V., A.T. PILIPENKO, T.N. NAZARCHUK, et al. *Analiz tugoplavkikh soedinenii (Analysis of Refractory Compounds).*—Metallurgizdat. 1962.

A detailed description of chemical analytical techniques for all known refractory compounds (carbides, nitrides, borides, silicides, phosphides) of the transition elements.

YAKOVLEV, P.YA., A.A. FEDOROV, and N.V. BUYANOV. *Analiz materialov metallurgicheskogo proizvodsta (Opredelenie mikroprimesei) (Analysis of Metallurgical Industry Materials. Determination of Mirco-Impurities).*—Metallurgizdat. 1961.

Analytical methods for the determination of bismuth, cerium, niobium, vanadium and titanium are presented.

MUKHINA, Z.S. and E.N. NIKITINA. *Uskorennye metody analiza titaniya i ego splavov (Rapid Methods for Analyzing Titanium and its Alloys).*—Oborongiz. 1961.

Methods for the determination of molybdenum, titanium, niobium, zirconium, beryllium, cerium, vanadium, tungsten, rhenium and bismuth in titanium and titanium alloys are described.

BABKO, A.K. and A.T. PILIPENKO. *Kolorimetricheskii analiz (Colorimetric Analysis).*—Goskhimizdat. 1951.

The book discusses certain photometric methods for determining rare elements.

SANDELL, E.B. *Colorimetric Determination of Traces of Metals.*—New York, Interscience. 1959.

Methods of photometric determination for a number of rare elements are presented.

Colorimetric (Photometric) Methods for Determining Nonmetals. [Russian translations, edited by A.I. Busev.].—Izdatinlit. 1963.

A description of the more important photometric methods for determining selenium and tellurium, boron, and other elements.

KRYUKOVA, T.A., S.I. SINYAKOVA, and T.V. AREF'EVA. *Polyarograficheskii analiz (Polarographic Analysis).*—Goskhimizdat. 1959.

The book surveys polarographic methods for determining numerous rare elements.

SONGINA, O.A. *Amperometricheskoe (polyarometricheskoe) titrovanie v analize mineral'nogo syr'ya. Metodicheskoe rukovodstvo (Amperometric (Polarometric) Titration in the Analysis of Mineral Raw Materials. A Laboratory Textbook).*—Gosgeoltekhizdat. 1957.

The book describes amperometric determinations of a large number of common and rare elements and includes theory, apparatus and experimental technique for amperometric titration.

KONSTANTINOVA-SHLEZINGER, M.A., Editor. *Lyuminestsentnyi analiz (Fluorimetric Analysis).*—Fizmatgiz. 1962.

A description of methods for fluorimetric determination of numerous rare elements, including apparatus and working techniques. [Translated into English by the Israel Program for Scientific Translations, Jerusalem. IPST catalog No. 2148.]

SHCHERBOV, D.P. *Flyuorimetriya v khimicheskom analize mineral'nogo syr'ya (Fluorimetry in Chemical Analysis of Mineral Raw Materials).*—Izd. "Nedra". 1965.

A brief text and reference book for laboratory use, comprising methods for the quantitative fluorimetric determination of beryllium, gallium, indium, thallium, rhenium, zirconium and selenium in mineral raw materials. Fluorimetric methods of detection and determinations for V, W, Ge, Y, Li, Mo, Nb, Sc, Ta, Te, Ti, Th, U, Zr and REE are also presented.

SAVVIN, S.B. *Arsenazo III. Metody fotometricheskogo opredeleniya redkikh i aktinidnykh elementov (Arsenazo III. Photometric Determination of Rare and Actinide Elements).*—Atomizdat. 1966.

The book discusses the theoretical and practical aspects of the utilization of Arsenazo III and some of its analogs for determining thorium, uranium, zirconium, hafnium, scandium, REE, plutonium, neptunium, protactinium, niobium and strontium. Experimental procedures are given.

MORRISON, G.H. and H. FREISER. *Solvent Extraction in Analytical Chemistry.*—New York, Wiley. 1957.

A description of the extraction methods for separating rare and common elements; fundamentals of the extraction theory, apparatus and general experimental techniques are included.

SAMUELSON, O. *Ion Exchangers in Analytical Chemistry.*—New York, Wiley. 1953.

The book presents examples of separations of some rare elements; numerous references are included.

POLUEKTOV, N.S. *Metody analiza po fotometrii plameni (Flame-Photometric Analytical Methods).*— Goskhimizdat. 1959.

A description of apparatus and methods for the quantitative determinations of alkali, alkaline earth and other metals by way of their flame radiation intensities.

RUSANOV, A.K. *Spektral'nyi analiz rud i mineralov (Spectroscopic Analysis of Ores and Minerals).*— Gosgeolizdat. 1948.

The book discusses apparatus and methods of spectroscopic analysis. Included is also practical information on the determination of more than 50 elements in ores; tables of spectral lines and an atlas of arc spectra of elements are appended.

N.H. FURMAN, Editor. *Standard Methods of Chemical Analysis,* Vol. I, *The Elements.*—Princeton, Toronto, London, New York, Van Nostrand. 1962.

The work includes a description of analytical methods recommended for rare and other elements.

ALIMARIN, I.P., Editor. *Metody analiza veshchestv vysokoi chistoty (Analysis of High Purity Materials).*— Nauka. 1965.

A description of methods for determining traces of rare and other elements in silicon, germanium, gallium, indium, thallium, arsenic, antimony, phosphorus, aluminum, lead, bismuth, zinc, cadmium, sulfur, selenium, tellurium, iodine, boron, graphite, reagents and other substances. [Translated into English by the Israel Program for Scientific Translations, Jerusalem. IPST catalog No. 2189.]

Lithium Li

Lithium has a valency of +1. The standard electrode potential in aqueous medium at 25°C for the reaction $Li = Li^+ + e$ is -3.045 V (relative to standard hydrogen electrode). This low value is probably due to the strong hydration of lithium ions in aqueous solutions. Lithium cannot be isolated by electrolysis of aqueous solutions.

Lithium in some of its properties resembles magnesium and alkaline-earth metals and differs considerably from sodium, potassium, and other alkali metals. Thus, lithium hydroxide LiOH is a much weaker base ($pK = 13.7$) than the hydroxides of other alkali metals.

The lithium phosphate (Li_3PO_4), carbonate (Li_2CO_3), fluoride (LiF), and oxalate ($Li_2C_2O_4$) are relatively sparingly soluble in water.* Lithium carbonate and lithium phosphate are soluble in solutions of ammonium salts.

Lithium chloride, unlike the chlorides of other alkali metals, is soluble in anhydrous organic solvents (ethanol, n-propanol, isoamyl alcohol, acetone, pyridine, ethanol–ether mixture of 1:1). $LiNO_3$, like $Ca(NO_3)_2$, is soluble in ethanol and ether.

Organic solvents are used to separate lithium ions from sodium and potassium ions (p. 26). Li^+, Na^+, and K^+ ions can also be quantitatively separated by chromatography on ion exchangers [1, 2] (p. 24).

* Solubilities of lithium salts in g/100 g water are:

Temperature, °C	0	18	20	25	35	50	75	100
Li_3PO_4, g	0.022	—	0.030	—	—	—	—	—
Li_2CO_3, g	1.53	—	—	1.27	—	1.01	0.85	0.72
LiF, g	0.120	0.27	—	0.133	0.135	—	—	—

Lithium is precipitated as the phosphate Li_3PO_4 by adding disubstituted sodium phosphate:

$$3Li + HPO_4^{2-} \rightleftharpoons Li_3PO_4 + H^+.$$

The white crystalline precipitate of Li_3PO_4 is soluble in acids; for this reason the hydrogen ions liberated in the course of the reaction are eliminated by the addition of ammonia.

The formation of the phosphate Li_3PO_4 is the basis of analytical detection of lithium and of its gravimetric (p. 27) and photometric determinations. In the gravimetric determination lithium is weighed as Li_2SO_4. Lithium sulfate is obtained by evaporating LiCl solution or LiF precipitate with sulfuric acid (p. 27).

Ferric periodate and potassium periodate $K_2Fe(IO)_6$ precipitate lithium ions as $LiKFe(IO)_6$ almost quantitatively. Detection of lithium and its gravimetric, titrimetric, and photometric determinations are based on the formation of this compound (p. 28).

The ferric periodate titrimetric method is used to determine lithium in silicate minerals and ores where its concentration is between a few hundredths of one per cent and a few per cent [3]. Lithium is isolated as the ferric periodate and the determination is completed by the iodometric method. When carried out by an experienced worker, this method gives satisfactory results.

Zinc uranyl acetate precipitates the greenish yellow $LiZn(UO_2)_3(CH_3COO)_9 \cdot 6H_2O$. This compound is used for the detection and gravimetric determination of lithium.

Thoron forms an orange-colored compound with lithium ions in 2% NaOH solution (pH 13.5) [4]. It reacts [5] with Li^+ ions in the molar ratio 1 : 1. The instability constant of the complex is $(2.4 \pm 0.3) \cdot 10^{-3}$. The molar extinction coefficient is $10{,}680 \pm 80$ at 470 mμ.

Solutions of thoron and its lithium compound [5] have approximately the same absorption maxima (470 or 458 mμ [6]). The greatest difference between the optical densities of thoron and its lithium compound in aqueous solutions is noted [5] at pH 13.5.

In 70% acetone the absorption maximum of the lithium compound with thoron shifts to 480–490 mμ and the optical density of the solution increases [6]. The formation of the colored compound is then complete within 30 minutes [6].

Thoron is used in the detection and photometric determination of lithium [6,7].

Dipivaloylmethane $(CH_3)_3CCOCH_2COC(CH_3)_3$ forms a chelate compound with lithium ions in alkaline medium in quantitative yield; the compound is soluble in organic solvents, but sparingly soluble in water [8]. If an ethereal solution of the reagent is employed, lithium can be separated from the ions of other alkali metals by extraction. If sodium ions are present in very large amounts, a small correction must be introduced; potassium ions do not interfere.

Lithium 8-hydroxyquinolate emits a green fluorescence in weakly alkaline solutions in 95% ethanol [9]. The sensitivity of detection is 5 μg of Li in 25 ml. Na^+ and K^+ ions do not interfere; Mg^{2+}, Ca^{2+}, and Zn^{2+} ions interfere.

Lithium salts color the flame a characteristic raspberry color (spectrum line at 670.0 mμ). Volatile strontium salts give a similar coloration.

Flame-photometric [10] and spectrographic methods are the most reliable in the determination of lithium. Chemical methods are of secondary importance.

Lithium has been determined polarographically [11,12].

For a review of the detection and determination of lithium, see [13].

Chromatographic separation of lithium from potassium and sodium [1, 14]

Lithium, potassium, and sodium cations are sorbed on SDV-3 cation exchanger in the H^+-form. Lithium ions are eluted with 0.12 *N* hydrochloric acid in 80% methanol; sodium is then eluted with 0.24 *N* hydrochloric acid in 80% methanol, and lastly potassium ions are eluted with 0.6 *N* aqueous hydrochloric acid. If aqueous solutions of hydrochloric acid are employed, the zones of the different ions overlap and the ions cannot be quantitatively separated.

Lithium ions are quantitatively eluted with 660 ml of 0.12 *N* hydrochloric acid in 80% methanol; another 120ml of this solvent should be run to fill the gap between the emergence of lithium and sodium ions. For the elution of sodium ions 1450 ml of 0.24 *N* hydrochloric acid in 80% methanol are required. Under these conditions lithium, potassium, and sodium ions are separated quantitatively.

Note. Since methanol is toxic, the aliquot of the sample solution must be withdrawn with a syringe.

REAGENTS

Cation exchanger SDV-3 in the H^+-form, grain size 0.10–0.25 mm.
Hydrochloric acid, 0.12 and 0.24 *N* solutions in 80% methanol.
Hydrochloric acid, 2 and 0.6 *N* aqueous solutions.

PROCEDURE

A chromatographic column 170 mm high and 15 mm in diameter is filled with 10 g of SDV-3 cation exchanger in the H^+-form. The column is washed with 50 ml of 0.12 *N* hydrochloric acid in 80% methanol, while keeping the resin covered at all times with at least 1 cm of liquid. The rate of emergence of the solution is adjusted at 5–6 ml/min.

The sample solution containing not more than 30 mg of lithium, 100 mg of sodium and 200 mg of potassium is diluted to the mark in a 50-ml volumetric flask with

0.12 *N* hydrochloric acid in 80% methanol. Ten ml of the resulting solution are introduced into the chromatographic column. The column is washed with 0.12 *N* hydrochloric acid in 80% methanol, collecting the eluate in a 500-ml volumetric flask. After 500 ml have been collected, the solution is mixed and its lithium content determined photometrically (p. 29).

The column is washed with 50 ml of 2 *N* hydrochloric acid and then with water until neutral to methyl orange. The eluate, which contains sodium and potassium ions, is discarded.

Note. The extent of the separation of lithium from sodium and potassium can be controlled by spectroscopy or by flame coloration.

Gravimetric determination of lithium in silicate rocks [15, 16]

The silicate rock is treated with hydrofluoric acid after which calcium oxide is added; the ions of alkali metals remain in solution while the ions of other metals precipitate out.

The precipitate is filtered and washed, the filtrate evaporated with hydrochloric acid, the residue ignited, and the sum of alkali metal chlorides is weighed.

Lithium chloride is separated from the chlorides of other alkali metals by dissolving it in anhydrous acetone (100 g of acetone will dissolve 3.94 g of LiCl at 25°). Sodium and potassium chlorides are practically insoluble in acetone. The solution of LiCl in acetone is evaporated and the residue is ignited and weighed as Li_2SO_4.

REAGENTS

Hydrofluoric acid, concentrated (38–40% H_2F_2).
Calcium oxide, freshly ignited. Prepared by igniting $CaCO_3$, free from alkali metals, in a platinum crucible over a blowtorch flame.
Phenolphthalein, 1% ethanolic solution.
Calcium hydroxide, freshly prepared 0.05% solution (0.5 g of calcium oxide is dissolved in one liter of water).
Ammonium carbonate, solution saturated at room temperature.
Ammonium oxalate, saturated solution.
Hydrochloric acid, $d = 1.19$ g/cm^3.
Anhydrous acetone (held over calcium chloride and distilled at 56–57°C).
Sulfuric acid, $d = 1.84$ g/cm^3.

PROCEDURE

Separation of total alkali chlorides

A 0.5 g sample of carefully ground silicate in a platinum dish is moistened with water and 5–10 ml of hydrofluoric acid are added. The mixture is stirred with a platinum rod and is heated on a water bath until the silicate is fully decomposed.

The resulting solution is evaporated to dryness on a sand bath. To the dry residue are added 30–40 ml of hot water and 5–8 drops of phenolphthalein solution. Calcium oxide is then added in small portions, with constant stirring, until a permanent pink color develops (each fresh portion of calcium oxide is added only after the previous portion has dissolved). Another 0.4–0.5 g of calcium oxide are added to the solution, the dish is covered with a watch glass, placed on a water bath, and held for about two hours with frequent stirring, while water is added to make up the evaporated volume.

Macerated filter paper is added to the dish, the mixture is stirred thoroughly and the precipitate is filtered through White Ribbon filter paper, while ensuring that the filtrate is clear. The precipitate on the filter paper is washed 7–8 times with the hot solution of calcium hydroxide. The filtrate is combined with the wash waters, heated, and 3–4 ml of an ammonium carbonate solution is added to precipitate calcium ions.

The mixture is boiled for 5 minutes. The solution is filtered through a Blue Ribbon paper into a platinum dish, the filtrate evaporated to dryness on a sand bath, and the residue ignited in order to expel ammonium salts. The residue is dissolved in 1–3 ml of hot water; 0.5 ml of ammonium oxalate solution is added and the mixture is left for one hour on a water bath at 50–60°C. The precipitate is separated on Blue Ribbon filter paper and washed with hot water. The filtrate is combined with the wash waters and evaporated to dryness in a platinum dish. The dry residue is treated with a few drops of hydrochloric acid, evaporated to dryness and carefully fused. The treatment with hydrochloric acid and the subsequent evaporation are repeated another 2–3 times. The residue is then weighed (sum total of alkali chlorides).

Note 1. The accuracy of determination of the mixture of alkali chlorides depends on the quality of the calcium oxide treatment of the silicate solution.

Note 2. A parallel blank determination with all the reagents involved must be run.

Separation of lithium chloride from potassium and sodium chlorides

The ignited alkali metal chlorides are ground to powder with a glass rod or a pestle, 25 ml of anhydrous acetone are added, and the mixture is thoroughly stirred. The insoluble residue is allowed to settle and the solution is decanted onto a Blue Ribbon filter paper, collecting the filtrate in a platinum dish. The acetone treatment and decantation is repeated 2–3 times more to ensure complete separation of lithium chloride.

The filtrates are combined; the acetone is evaporated on a warm water bath; the residue is cautiously ignited, and a few drops of sulfuric acid are added. The mixture is evaporated and ignited until fused. The residue is weighed as Li_2SO_4.

Gravimetric determination of lithium as phosphate [17]

Lithium ions are precipitated by a solution of disubstituted sodium phosphate at pH 9 (phenolphthalein):

$$3Li_2SO_4 + 2Na_2HPO_4 \rightleftharpoons 2Li_3PO_4 + 2Na_2SO_4 + H_2SO_4.$$

Lithium phosphate separates out slowly. The precipitant must be present at least in a 10% excess. The free acid formed during the precipitation is neutralized with a solution of sodium hydroxide.

If lithium phosphate is precipitated at pH 7 or pH 11–13, the results will be a few per cent low or a few per cent high, respectively.

Twentyfold amounts of sodium and potassium ions do not interfere. It was found spectroscopically that the precipitate of lithium phosphate contains 0.13–0.15% Na and 0.011–0.012% K.

A disadvantage of this method is that lithium phosphate must be isolated twice: from the sample solution and from the filtrate after evaporation. The experimental conditions must be rigorously observed. The results are always 0.2–0.5% low.

REAGENTS

Disubstituted sodium phosphate, saturated aqueous solution (35.3 g of $Na_2HPO_4 \cdot 12H_2O$ are dissolved in 100 ml of water at 25°C).
Sodium hydroxide, 5% solution.
Ammonia, 2.5% solution.
Phenolphthalein, 1% ethanolic solution.
Oxalic acid, crystalline.

PROCEDURE

The sample solution (10 ml), containing between 10 and 110 mg of Li as sulfate or chloride, is heated to 60–70°C and the disubstituted sodium phosphate solution, also at 60–70°C, is added drop by drop (in 30% excess), while stirring with a glass rod. The solution is heated for 5–10 minutes and then cooled; 8–10 drops of phenolphthalein solution are introduced and the sodium hydroxide solution is added to pH 9. The solution with the precipitate is placed on a water bath and evaporated to dryness. During the evaporation it is very important to check the pH of the solution with phenolphthalein; if necessary, sodium hydroxide is added to maintain the pH at 9. About 10 ml of ammonia are added to the residue, and the mixture is thoroughly stirred and left to stand for 3–4 hours. The precipitate is filtered on a Blue Ribbon filter paper, and then washed with ammonia solution until the wash waters give a negative reaction for phosphate, sulfate, and sodium ions. In this way 96–97% of lithium phosphate are separated. The filtrate is combined with the first

50 ml of the wash waters and evaporated to dryness; 5–10 ml of ammonia solution are added to the residue; the solution is left to stand for 3–4 hours, and the small precipitate of lithium phosphate is filtered off and washed.

The lithium phosphate precipitates are combined, dried, and ignited together with the filter paper in a porcelain crucible at 750–800°C until a constant weight is attained.

The conversion factor to elementary lithium is 0.1797.

Determination of lithium in ores and minerals [18]

The sample is fused with alkalis, and lithium is separated from the interfering ions by ignition with oxalic acid. This is accompanied by the formation of water-soluble alkali carbonates and water-insoluble carbonates and oxides of other elements which associate with lithium.

The determination of lithium is completed by the ferric periodate method. Lithium ions in the presence of ferric ions and periodate ions are precipitated as $LiKFe(IO)_6$, and the amount of lithium in the precipitate is found indirectly from the determination of the iron by the thiocyanate method.

The ferric periodate method permits the determination of lithium in the presence of other alkali metals, but when sodium is present in high concentrations the results are high. Accordingly, lithium must be separated from the other alkali metals by treatment with a mixture of concentrated hydrochloric acid and ethanol. Lithium is then precipitated as LiCl (solubility of LiCl strongly decreases in the presence of ethanol).

The experimental error of the method is $\pm$ 10%.

REAGENTS

Standard solution of lithium sulfate, 1 ml of solution contains 20 μg of lithium.
Potassium hydroxide, crystalline and 1 *N* solution.
Oxalic acid, ground solid.
Ammonium carbonate, 5% solution.
Hydrochloric acid, d = 1.19 g/cm^3, 1 *N* solution and 1:1 solution.
Ethanol, 96%.
Potassium periodate: 2.3 g of KIO_4 are dissolved in 50 ml of 0.5 *N* potassium hydroxide, 12 ml of 0.1 *M* ferric chloride in 0.2 *N* hydrochloric acid are added with stirring and the solution is made up to 100 ml with 2 *N* potassium hydroxide.
Potassium thiocyanate, 20% and 2% solutions.
Washing solution: 1:2 mixture of concentrated hydrochloric acid with 96% ethanol.

CONSTRUCTION OF CALIBRATION CURVE

Portions of 1, 2, 3, 4, and 5 ml of standard solution of lithium sulfate are introduced into five 20- to 25-ml beakers and evaporated on a sand bath to a residual volume of 0.1–0.2 ml. One ml of 1 *N* potassium hydroxide is added to each residue. The solution is brought to the boiling point and 2 ml of potassium periodate at 90°C are introduced. The solutions are heated for another five minutes, cooled, and the precipitates are filtered with suction through a small Blue Ribbon filter paper. The precipitates are washed four times with 0.5–0.7 ml of potassium hydroxide solution each time and are then dissolved in 10 ml of 1 *N* hydrochloric acid. The resulting filtrates are collected in 25-ml volumetric flasks and diluted to the mark with water.

Samples (2.0 ml each) of the solutions are withdrawn with a pipet into 10-ml graduated tubes; 5 ml of water and 3 ml of 20% potassium thiocyanate solution are added to each tube. The solutions are made up to 10 ml with 2% potassium thiocyanate solution and are then thoroughly stirred.

The optical densities of the solutions in this series are measured relative to water on a photoelectrocolorimeter. The calibration curve is plotted from the data thus obtained.

PROCEDURE

A 25–50 mg sample of carefully ground material, containing 0.05–3.5% Li, is fused with 100–150 mg of potassium hydroxide in a small silver crucible. Water (6–10 drops) is added to the melt. The solution with the precipitate is transferred to a 3-ml porcelain crucible and 0.5–0.6 g of oxalic acid is added. The solution is cautiously evaporated and ignited over a micro-flame until the oxalic acid is fully decomposed.

The residue is wetted with an ammonium carbonate solution. The solution is transferred to a graduated centrifuge tube, diluted to 2 ml with water, and centrifuged. A 1.0 ml portion of the centrifugate is withdrawn with a dry pipet, transferred to a 3-ml porcelain crucible, and evaporated to dryness. The residue is wetted with 2–3 drops of hydrochloric acid ($d = 1.19$ g/cm^3), evaporated, and gently ignited. To the residue 0.2 ml of hydrochloric acid and 0.5 ml of ethanol are added. The mixture is left to stand for five minutes with occasional stirring and is then filtered through a small Blue Ribbon filter paper into another 3-ml porcelain crucible. The residue in the porcelain crucible and on the filter is washed with a mixture of ethanol and hydrochloric acid. The filtrate is evaporated to dryness, and the residue is gently ignited and dissolved in one or two drops of water. One ml of potassium hydroxide solution is then added to the crucible, the solution is brought almost to the boil and 2 ml of boiling potassium periodate solution are added. The mixture is heated for another five minutes and is then cooled. The precipitate is filtered with suction on a small Blue Ribbon filter paper, washed 4 times with 0.75-ml portions of potassium hydroxide solution, and dissolved in 10 ml of 1 *N* hydrochloric acid. The solution is collected in a 25-ml volumetric flask and diluted to mark with water.

A 5.0 ml sample of the resulting solution is withdrawn with a pipet into a 10-ml graduated tube. Then 3 ml of a 20% potassium thiocyanate solution are added and the volume is made up to 10 ml with a 2% solution of potassium thiocyanate. The solution is thoroughly mixed and the optical density is measured with an electrophotocolorimeter under conditions identical with those employed in the construction of the calibration curve.

The amount of lithium in the aliquot is found from the calibration curve. Thus, the content of lithium in the sample can be calculated, taking into account that the ratio Li/Fe in ferric periodate is 0.124.

Note 1. The following aliquots are recommended for the different contents of lithium in the sample.

Li, %	Sample weight, mg	Aliquot, ml
0.05–0.3	50	5.0
0.3–1.0	25	3.0
1.0–3.5	25	1.0–2.0

Note 2. The method is suitable for the analysis of silicate rocks. In the analysis of lithium phosphate minerals the sample is fused with potassium hydroxide as described above, the flux is dissolved in 8–10 drops of water, hydrochloric acid (1:1) is added to a strongly acid reaction and 0.5 ml of 5% solution of zirconium oxychloride is introduced. The mixture is evaporated to dryness and gently ignited. The residue is wetted with 3–5 drops of water, oxalic acid added and the determination continued as described for silicate rocks.

BIBLIOGRAPHY

1. GORSHKOV, V.I., I.A. KUZNETSOV, and G.M. PANCHENKOV.—*ZhAKh*, **14**, 417. 1959.
2. HOLZAPFEL, H., H. EHRHARDT, and W. TISCHER.—*J. prakt. Chem.* [4], **18**, 62. 1962.
3. NAZARENKO, V.A., G.I. BYK, S.YA. VINKOVETSKAYA, and M.B. SHUSTOVA.—*Trudy GIREDMET, 2*, 177. Metallurgizdat. 1959.
4. KUZNETSOV, V.I.—*ZhAKh*, **3**, 295. 1948.
5. ADAMOVICH, L.P. and T.T. ALEKSEEVA.—*Uchenye Zapiski Khar'kovskogo Universiteta*, **54**, *Trudy Khimicheskogo Fakul'teta i Nauchno-Issledovatel'skogo Instituta Khimmi, KhGU*, **12**, 209. 1954.
6. THOMSON, P.F.—*Anal. Chem.*, **28**, 1527. 1956.
7. NIKOLAEV, A.V. and A.A. SOROKINA.—*Doklady AN SSSR*, **77**, 427. 1951.
8. GUTER, G.A. and G.S. HAMMOND.—*J. Am. Chem. Soc.*, **78**, 5166. 1956.
9. WHITE, C.E., M. FLETCHER, and J. PARKS.—*Anal. Chem.*, **23**, 479. 1951.
10. POLUEKTOV, N.S. *Metody analiza po fotometrii plameni (Flame-photometric Methods of Analysis)*, p. 131.—Goskhimizdat. 1959.
11. KRYUKOVA, T.A., S.I. SINYAKOVA, and T.V. AREF'EVA. *Polyarograficheskii analiz (Polarographic Analysis)*, pp. 187, 188.—Goskhimizdat. 1959.
12. PANCHENKOV, G.M., E.M. KUZNETSOVA, and N.V. AKSHINSKAYA.—*ZhAKh*, **15**, 424. 1960.
13. POLUEKTOV, N.S. and V.T. MISHCHENKO.—In: *Metody opredeleniya i analiza redkikh elementov*, p. 37. Izd. AN SSSR. 1961.
14. PANCHENKOV, G.M., V.I. GORSHKOV, and M.V. KUKLANOVA.—*ZhFKh*, **32**, 361, 616. 1958.
15. VASIL'EV, P.I. *Metody uskorennogo analiza silikatov (Methods of Rapid Analysis of Silicates)*, p. 25.—Gosgeolizdat. 1951.
16. KNIPOVICH, YU.N. and YU.V. MORACHEVSKII (Editors). *Analiz mineral'nogo syr'ya (Analysis of Mineral Raw Materials)*, p. 114.—Goskhimizdat, 1959.
17. KINDYAKOV, P.S. and A.V. KHOKHLOVA.—*Trudy Moskovskogo Instituta Tonkoi Khimicheskoi Tekhnologii im. M.V. Lomonosova*, No. 6, 9. 1956.
18. NAZARENKO, V.A. and V.YA. FILATOVA.—*ZhAKh*, **5**, 234. 1950.

Rubidium Rb and Cesium Cs

Rubidium and cesium have a valency of +1. The standard electrode potentials in aqueous media at 25°C relative to standard hydrogen electrode have the following values:

$$Rb \rightleftharpoons Rb^+ + e \qquad -2.925\ V$$

$$Cs \rightleftharpoons Cs^+ + e \qquad -2.923\ V$$

Metallic rubidium and cesium are very strong reducing agents. The ions of these metals can be reduced in solution only by electrolysis.

The hydroxides RbOH and CsOH are readily soluble in water giving strong bases.

Few chemical methods are available for the determination of Rb^+ and Cs^+ ions in the presence of each other and other alkali metal ions. The ions Rb^+ and Cs^+ form relatively few sparingly soluble compounds. They do not form stable complexes (double salts only are known) and do not participate in redox reactions.

Rubidium and cesium ions are colorless and their chemical properties resemble those of potassium ions: they are precipitated by perchloric acid, chloroplatinic acid hexanitrocobalt (III) sodium, sodium tetraphenylborate, hexanitrodiphenylamine, and other reagents. The solubility of rubidium compounds is intermediate between the corresponding compounds of potassium and cesium. Cesium in the presence of rubidium and potassium can be detected by certain reagents with a fair degree of reliability. A number of rubidium and cesium salts form well-developed crystals which can be identified microscopically.

The separation of potassium, rubidium, and cesium is difficult owing to the great similarity in the properties of their compounds.

The most satisfactory separation of rubidium and cesium can be obtained by chromatographic methods.

Organic and inorganic cation exchangers, such as zirconium phosphate or ammonium phosphomolybdate, are used for this purpose. The ions Fr^+, Cs^+, and Rb^+ are separated [1] on a KU-1 cation exchanger; the ions of Li^+, Na^+, K^+, Rb^+, and Cs^+ can also be separated on different cation exchangers [2,3].

The precipitates $Me_2[PtCl_6]$ of potassium, rubidium, and cesium chloroplatinates are yellow or orange colored. Rubidium and cesium chloroplatinates are less soluble than potassium chloroplatinates. Rubidium and cesium ions can be detected in the presence of potassium ions by a saturated aqueous solution of potassium chloroplatinate.

Rubidium and cesium form triiodides MeI_3. These dissociate according to the equation:

$$MeI_3 \rightleftharpoons MeI + I_2.$$

Cesium triiodide is much more stable than rubidium triiodide and this fact allows the separate titrimetric determination of cesium and rubidium.

Potassium tetraiodobismuthite precipitates cesium ions Cs^+ from aqueous solutions and from solutions in concentrated acetic acid as $Cs_3Bi_2I_9$. The formation of this compound is the basis of gravimetric, titrimetric, and photometric determinations of cesium. Rubidium ions (if present in not more than threefold amount) and the ions of Li, Na, K, Mg, Ca, and Al do not interfere.

Rubidium and cesium ions are precipitated by solutions of stannic halide complexes as the compounds Me_2SnX_6, where Me = Rb or Cs, while X = Cl, Br, I. These complexes are used to separate rubidium and cesium from potassium.

Cesium and rubidium ions form typically shaped crystals with halides of other elements (gold and palladium chlorides, gold and silver chlorides).

Hexachlorotellurous acid H_2TeCl_6 precipitates cesium ions from solutions in 11–12 N hydrochloric acid. Small amounts of potassium ions do not interfere [4] in the process.

Silicotungstic acid precipitates only cesium ions as $Cs_4[Si(W_3O_{10})_4] \cdot xH_2O$ [5] from 6 N hydrochloric acid solutions, and the corresponding rubidium salt is more readily soluble. In this way 0.01–0.02 g of CsCl can be separated from 0.02–0.03 g of RbCl. Ammonium ions interfere while lithium and alkaline-earth metals do not.

Silicomolybdic acid forms microscopically identifiable crystals with rubidium and cesium ions and also with NH_4^+, K^+, and Tl^+ ions and under certain conditions; it can be used for their detection.

Cesium ions in sulfuric acid solutions are coprecipitated with ammonium silicomolybdate almost quantitatively [6], and, therefore, cesium can be separated from solutions of potassium salts since potassium ions do not precipitate under these conditions.

Rubidium and cesium phosphomolybdates are less soluble than potassium phosphomolybdates. Rubidium and cesium are sometimes separated as phosphomolybdate in the course of analysis.

If a cesium salt is added to $Cd_2[Fe(CN)_6]$ in 5% potassium iodide solution, the white precipitate of $Cs_2Cd[Fe(CN)_6]$ separates out [7]. Cesium in concentrations of 0.02–0.2 mg in 10 ml final solution can be detected and nephelometrically determined in the form of this compound. Lithium and sodium ions and 100-fold amounts of potassium ions do not interfere; tenfold amounts of ammonium and rubidium ions and heavy metal ions which form precipitates with $Fe(CN)_6^{4-}$ ions do interfere (see p. 35).

Silver sodium hexanitrobismuthite $Na_2AgBi(NO_2)_6$ quantitatively precipitates rubidium and cesium ions. The resulting yellow crystalline precipitates of variable composition are sparingly soluble. The reagent is used to separate rubidium and cesium from potassium.

Rubidium and cesium in dilute aqueous solutions can be concentrated by coprecipitation with potassium hexanitrocobaltiate. To do this, $Na_3[CO(NO_2)_6]$ is added to the sample solution below 10°C. The method will isolate 1–10 mg rubidium and cesium from 10 liters of solution. It has been employed in the determination of rubidium in sea water.

Rubidium and cesium hexanitrocobaltiates are less soluble than potassium hexanitrocobaltiates.

Sodium tetraphenylborate $Na[B(C_6H_5)_4]$ quantitatively precipitates potassium, rubidium, and cesium ions under certain conditions and is used in gravimetric and titrimetric determination of these elements.

Sodium triphenylcyanoborate $Na[(C_6H_5)_3B(CN)]$ precipitates cesium ions almost quantitatively [8]. The white, relatively coarse crystalline precipitate separates out immediately after a filtered 3% solution of the reagent has been added to a 0.005 M solution of Cs_2SO_4. The solubility product of cesium triphenylcyanoborate is $3 \cdot 10^{-6}$. The salt $Cs[(C_6H_5)_3B(CN)]$ is approximately 100 times more soluble than the salt $Cs[B(C_6H_5)_4]$. The reagent does not precipitate rubidium ions from $1.4 \cdot 10^{-2}$ M solution of Rb_2SO_4, but Rb^+ ions are coprecipitated with $Cs[(C_6H_5)_3B(CN)]$. Potassium ions are not precipitated from a 5% solution of KNO_3. The ions of Ca, Sr, Ba, Mg, Ni, Co, Mn(II), Zn, Fe(II, III), Cd, Pb, Cu(II), U(VI), Al, Ce(III), Cr(III), and Bi are not precipitated. The ions of Tl(I), Ag(I), and Cu(I) are precipitated.

The precipitate of $Cs[(C_6H_5)_3B(CN)]$ is readily soluble in acetone and in 1:1 mixture of acetone with water; it is practically insoluble in benzene and sparingly soluble in dioxane.

Sodium triphenylcyanoborate is used in the detection and gravimetric determination of cesium; the results are invariably low. The determination of cesium may be terminated by argentometric or thallometric titrations.

The sodium salt of hexanitrodiphenylamine (dipicrylamine) gives characteristic precipitates with potassium, rubidium, cesium, and thallous ions. Sodium, lithium, magnesium, and alkaline-earth metal ions are not precipitated. The ions of Al, Fe(III), Cr(III), Ni, Co, Bi, and Hg give amorphous precipitates. The reagent is employed in the detection and photometric determination of rubidium and cesium.

The cesium ions are isomorphously coprecipitated [9] with the sparingly soluble dipicrylamine salts of NH_4^+, K^+, Rb^+, and Tl^+. Dipicrylamine salts of thallium and ammonium may be used as carriers in the isolation of trace amounts of cesium (12 μg) from very dilute solutions.

Cesium dipicrylamine salt is extractable from aqueous alkaline solutions with nitrobenzene [10]. In this way small amounts of radioactive cesium can be separated from numerous long-lived products of fission and major quantities of uranium. The extractability of cesium decreases when large amounts of sodium ions are present in the solution.

The monosodium salt of hexanitrohydrazobenzene (saturated aqueous solution) precipitates rubidium, cesium, and potassium ions and is used in the photometric determination of these elements [11]. The filtered precipitates are dissolved in acetone and the optical density of the resulting solutions is measured. Between 10 and 130 μg rubidium or cesium can be determined. Small amounts of sodium and larger amounts of Li, Ca, Sr, Ba, and Mg ions do not interfere.

Picric acid and picrolonic acid in the form of ethanolic solutions are used in the detection of rubidium and cesium by microscopic crystal identification. However, sodium, ammonium, and potassium form similar crystals.

5-Nitrobarbituric acid, 2,4-dinitrophenol, 2,4,6-trinitro-m-cresol, 2,4,6-trinitro-resorcinol, and dinitrobenzofuroxane are used for microscopic and spot identification of rubidium and cesium ions. None of these reagents is very selective, and many extraneous ions interfere.

Tartaric acid or sodium bitartrate precipitate white crystalline bitartrates $RbHC_4H_4O_6$ or $CsHC_4H_4O_6$ from neutral solutions of rubidium or cesium salts. These precipitates are readily soluble in mineral acids, but sparingly soluble in water and acetic acid. Potassium and ammonium ions are precipitated in a similar manner. Bitartrates of all these elements readily form supersaturated solutions.

Rubidium and cesium ions are reduced on a dropping mercury electrode in aqueous solutions at very low potentials. Rubidium and cesium cannot be determined polarographically in aqueous solutions.

In a medium of 80% isopropanol, with 0.1 *M* lithium hydroxide as supporting electrolyte, the half-wave potential of cesium is -2.03 V, while that of rubidum is -1.97 V (20–30°C) relative to saturated calomel electrode [12]; thus rubidium and cesium can be polarographically determined. If both elements are present their overall content is determined. Rubidium salts color the flame reddish violet; the color is visually indistinguishable from that produced by potassium salts. The characteristic spectral lines of rubidium are 421.6 and 420.2 mμ.

Cesium salts color the flame violet-blue; their characteristic spectral lines are 459.3 and 455.5 mμ.

Flame-photometric [13] and spectroscopic determinations of rubidium and cesium are the most reliable analytical techniques and are widely employed in the analysis of different materials. There are no reliable chemical methods for the detection or determination of rubidium in the presence of other alkali metals. Chemical methods for the detection and determination of cesium are of secondary importance.

Flame photometry will determine rubidium and cesium in the presence of other alkali and alkaline-earth metals. The method was successfully employed to determine small amounts of cesium in rocks [14, 15].

Rubidium and cesium in rocks, minerals, and meteorites can be determined by neutron activation methods [16].

An experimental check of the methods of detection of rubidium and cesium ions was carried out by Geilmann and Gebauhr [17].

Titrimetric determination of cesium in pollucite [18]

This method is based on the ability of cadmium hexacyanoferrate $Cd[CdFe(CN)_6]$ to quantitatively exchange the cadmium ions in the outer sphere of the complex against cesium ions with the formation of a less soluble salt:

$$Cd[CdFe(CN)_6] + 2Cs^+ \rightleftharpoons Cs_2[CdFe(CN)_6] + Cd^{2+}.$$

The amount of cadmium ions passing into solution is equivalent to the amount of cesium ions. The released cadmium ions are titrated against the solution of $K_4[Fe(CN)_6]$. Alternatively, cadmium may be precipitated by hydrogen sulfide, the CdS precipitate dissolved in hydrochloric acid and the cadmium determined by complexometric titration.

The method is suitable for the determination of cesium in pollucite—aluminum cesium silicate $H_2Cs_4Al_4(SiO_3)_9$. The interfering ions include rubidium, fourfold amounts of potassium, copper, zinc, iron, and other ions which form insoluble ferrocyanides. Sodium, magnesium, calcium, and aluminum ions do not interfere.

The ion-exchange reaction is conducted in a neutral or sulfuric acid medium, by passing the sample solution through powdered cadmium hexacyanoferrate (II).

REAGENTS

Hydrofluoric acid, concentrated (38–40% H_2F_2).

Sulfuric acid, dilute (1:3).

Cadmiun hexacyanoferrate (II), crystalline. In a two-liter beaker are placed 300 ml of solution containing 40.69g of cadmium sulfate and 48ml of hydrochloric acid, $d = 1.19$ g/cm^3. The solution is mechanically stirred and 300 ml of solution containing 38.50 g of sodium ferrocyanide are added in small portions. The mixture is diluted with water to one liter, thoroughly stirred and left to stand overnight. The precipitate is filtered through a Blue Ribbon filter paper on a Buchner funnel, washed with distilled water to a negative reaction for sulfate ions and dried in the air. The preparation is stored in a bottle with a ground glass stopper; it is not hygrogscopic and does not deteriorate on storage.

Ammonium sulfate, 50% solution.

Hydrochloric acid, $d = 1.19$ g/cm^3.

Potassium ferrocyanide, $K_4[Fe(CN)_6] \cdot 3H_2O$, 0.05 *M* solution.

Potassium ferricyanide, $K_3[Fe(CN)_6]$, 1% solution.

Diphenylamine, 1% solution in concentrated sulfuric acid.

Zinc sulfate, 0.075 *M* solution.

PROCEDURE

Isolation of cesium from pollucite

A 0.2–0.3 g sample of pollucite, which has been finely ground and dried to constant weight at 105–110°C, is treated in a platinum dish on a sand bath with four 5-ml portions of hydrofluoric acid in the heat; 3 ml of sulfuric acid are added and the contents of the dish are evaporated until the appearance of sulfuric acid fumes. The residue is cooled, and dissolved in 10 ml of water; the insoluble matter is separated on Blue Ribbon filter paper and washed with water. The overall volume of the filtrate and wash waters should be 50 ml.

Ten grams of cadmium hexacyanoferrate (II) are placed into No. 3 or No. 4 sintered glass filter in a suction flask, wetted with a small amount of water and the salt on the bottom of the filter is compacted by gentle suction.

The sample solution is heated to 60–70°C and is passed through the layer of cadmium hexacyanoferrate (II) on the filter at the rate of 1 drop in 4–5 seconds, the percolation rate being adjusted with a water pump. The layer is washed with 3–4 portions of hot water (5–6 ml each). The wash waters are combined with the filtrate; the total volume of the solution should be 70–80 ml.

Titration of cadmium

To the resulting solution collected in a 300-ml conical flask are added 10 ml of ammonium sulfate solution and 10 ml hydrochloric acid solution. About 20 ml of $K_4[Fe(CN)_6]$ solution are introduced from a buret with vigorous stirring. The solution with the precipitate of $5Cd_2[Fe(CN)_6] \cdot (NH_4)_4[Fe(CN)_6]$ is left to stand for 5–10 minutes, 0.5 ml of $K_3[Fe(CN)_6]$ solution and 3–4 drops of diphenylamine solution are added and the excess $[Fe(CN)_6]^{4-}$ ions are titrated against a solution of zinc sulfate until a lilac coloration appears, after which the reagent is added in an excess of 4–5 drops. The solution of $K_4[Fe(CN)_6]$ is then added again from a buret until the lilac-colored solution turns pale green and then again lilac colored on the addition of one drop of zinc sulfate solution.

The volume ratio of potassium ferrocyanide and zinc sulfate at this point is determined. The concentration of potassium ferrocyanide solution is determined by titration against a standard solution of cadmium sulfate.

Calculation

The content of cesium in pollucite is calculated from the formula

$$\%Cs = (V_1 - V_2 f)\,TF \cdot 100/g.$$

where V_1 is the volume (ml) of 0.05 M $K_4[Fe(CN)_6]$ solution consumed in the titration;

V_2 is the volume (ml) of 0.075 M zinc sulfate solution consumed in the titration;

f is the conversion factor from 1 ml of zinc sulfate solution to 1 ml of $K_4[Fe(CN)_6]$ solution (1 ml of 0.075 M $ZnSO_4$ solution is equivalent to 1 ml of 0.05 M solution of $K_4[Fe(CN)_6]$, so that in this case $f = 1$);

T is the conversion titer of $K_4[Fe(CN)_6]$ to cadmium, in g of Cd/ml;

F is the conversion factor from cadmium to cesium equal to 2.365;

g is the sample weight of pollucite, in grams.

BIBLIOGRAPHY

1. POZDNYAKOV, A.A.—*ZhAKh,* **16**, 647. 1961.
2. STAROBINETS, G.A. and G.S. MARTINCHIK.—*ZhAKh,* **17**, 538. 1961.
3. HOLZAPFEL, H. and H. EHRHARDT.—*J. prakt. Chem.,* [4], **21**, 92. 1963.
4. MONTGOMERY, H.A.C.—*Analyst,* **85**, 687. 1960.
5. O'LEARY, W.J. and J. PAPISH.—*Ind. Eng. Chem., Anal. Ed.,* **6**, 107. 1934.
6. FABRIKOVA, E.A.—*ZhAKh,* **17**, 22. 1961.
7. KOZLOV, A.S.—In: *Redkie shchelochnye elementy,* p. 79, Novosibirsk, AN SSSR, Sibirskoe otdelenie. Khimiko-Metallurgicheskii Institut. 1960.
8. HAVIR, J.—*Coll. Czech. Chem. Comm.,* **26**, 1775. 1961.
9. KORENMAN, I.M. and G.A. SHATALINA.—*ZhAKh,* **13**, 299. 1958.
10. KYRS, M., J. PELCIK, and P. POLONSKY.—*Coll. Czech. Chem. Comm.,* **25**, 2642. 1960.
11. CHERKESOV, A.I.—*Uchenye Zapiski Saratovskogo Universiteta,* **42**, 85. 1955.
12. SCHÖBER, G. and V. GUTMANN.—*Michrochim. acta,* No. 3, 319. 1958.
13. POLUEKTOV, N.S. *Metody analiza po fotometrii plameni (Flame Photometric Analysis),* p. 156.—Goskhimizdat. 1959.
14. FABRIKOVA, E.A.—*ZhAKh,* **14**, 41. 1959; **16**, 22. 1961.
15. LEBEDEV, V.I.—*ZhAKh,* **16**, 272. 1961.
16. CABELL, M.J. and A.A. SMALES.—*Analyst,* **82**, 390. 1957.
17. GEILAMNN, W. and W. GEBAUHR.—*Z. anal. Chem.,* **142**, 241. 1954.
18. MIZHIDIIN, YU. *Issledovanie kompleksnykh ferrotsianidov kadmiya i ikh ispol'zovanie u analiticheskoi khimii (Complex Cadmium Ferrocyanides and their Application in Analytical Chemistry),* p. 112.—MGU. 1964.

Beryllium Be

Beryllium has a valency of +2. The standard electrode potentials in aqueous solutions at 25°C relative to standard hydrogen electrode have the following values:

$Be \rightleftharpoons Be^{2+} + 2e$	$-1.847V$
$Be + 2H_2O \rightleftharpoons Be(OH)_2 + 2H^+ + 2e$	$-1.820V$
$2Be + 3H_2O \rightleftharpoons Be_2O_3^{2-} + 6H^+ + 4e$	$-1.387V$
$Be + 2H_2O \rightleftharpoons BeO_3^{2-} + 4H^+ + 2e$	$-0.909V$

Acid solutions contain the ions of Be^{2+}, while weakly acid solutions contain $BeOH^+$, Be_2OH^{3+}, Be_2O^{2+}, and other ions. In alkaline solutions the colorless BeO_2^{2-}, BeO_3^{2-}, etc. anions are formed (Figure 1).

Beryllium hydroxide $Be(OH)_2$ is amphoteric. It begins to precipitate at pH ~6 from 0.01 *M* beryllium salt solutions and redissolves at pH ~13.5.

Beryllium ions closely resemble aluminum ions, having many reactions in common with them. For a long time the separation of beryllium from aluminum was a difficult analytical task. It has become greatly simplified since the introduction into analytical practice of Complexone III, which forms a stable complexonate with aluminum and a labile complexonate with beryllium. During the detection of beryllium and its determination, Complexone III masks Al^{3+} ions as well as the ions of many other elements.

Beryllium hydroxide is quantitatively precipitated [1] by ammonia at pH 8.5 from solutions containing an excess of Complexone III. Ions of bivalent, trivalent,

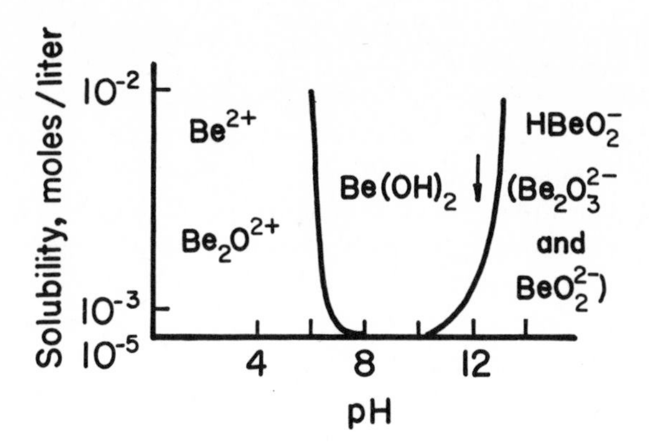

Figure 1
Solubility of beryllium hydroxide $Be(OH)_2$ and beryllium species existing in solution as a function of pH

and tetravalent elements form stable complexonates and remain in solution. The ions of Ti(IV), Sn(IV), and Cr(III) are not masked by Complexone III. The washed precipitate of beryllium hydroxide is ignited and weighed as BeO.

On the addition of excess NaF beryllium hydroxide forms the practically undissociated fluoride, NaOH being formed in equivalent amount:

$$BeSO_4 + 2NaOH \xrightarrow{pH\ 8.5} Be(OH)_2 + Na_2SO_4;$$

$$Be(OH)_2 + 2NaF \longrightarrow BeF_2 + 2NaOH.$$

The liberated NaOH is potentiometrically titrated against a solution of sulfuric acid. Al, Zr, U, Th, and REE interfere.

These reactions are not perfectly stoichiometric and so the method is empirical. However, satisfactory results are obtained if the procedure is rigidly adhered to.

The solution of sulfuric acid is standardized with metallic beryllium which readily dissolves in acids and alkalis:

$$Be + 2KOH \rightleftharpoons K_2BeO_2 + H_2.$$

The amount of beryllium is found by measuring the volume of the liberated hydrogen or the increase in pressure in a closed vessel of a known volume. Metallic beryllium is determined in slags and sediments in this way. Metallic aluminum interferes. A correction must be introduced if the sample material contains beryllium carbide or beryllium nitride (Be_2C, Be_3N_2).

Beryllium is quantitatively precipitated [2] at pH 5.5 as the crystalline phosphate $BeNH_4PO_4$ from solutions containing Complexone III. The Mg, Ca, Fe, Al, Cu, Ni, and other ions remain in solution. The filtered and washed precipitate is ignited and weighed as $Be_2P_2O_7$.

The gravimetric phosphate method is used in the determination of beryllium in concentrates of any composition [3].

Beryllium is quantitatively precipitated as the crystalline beryllium ammonium arsenate $BeNH_4AsO_4$ of constant composition from acetate solutions at pH 5.2 in the presence of Complexone III. The content of beryllium is found by titrating the arsenate iodometrically in the usual manner (in 1:3 HCl) [4] (p. 46).

Beryllium forms various complex compounds with inorganic and organic ligands. The citrate and tartrate complexes are quite stable. Freshly precipitated beryllium hydroxide dissolves in alkali carbonate solutions and is reprecipitated when the solutions are boiled (unlike aluminum hydroxide).

Beryllium ions form fluoride complexes:

$$BeF_2 \rightleftharpoons Be^{2+} + 2F^- \qquad pK \sim 2$$

$$BeF_4^{2-} \rightleftharpoons BeF_3^- + F^- \qquad pK_1 = 2; pK_2 = 3.6;$$

$$pK_3 = 5.0; pK_4 = 4.3\text{–}5.9$$

Beryllium carbonate complex is precipitated by $[CO(NH_3)_6]^{3+}$ ions as $[CO(NH_3)_6]\,[Be_2(CO_3)_2(OH)_3(H_2O)_2] \cdot 3H_2O$. This complex is used in the gravimetric determination of beryllium (p. 43).

Beryllium reacts with salicylic and sulfosalicylic acids, forming a number of stable complex compounds. These reagents are used in the titrimetric determination of beryllium (p. 44).

Various hydroxylated organic compounds are reagents for beryllium ions and are used under certain conditions in its detection and photometric determination. An example is H-resorcinol and Arsenazo I, which contain the following groupings.

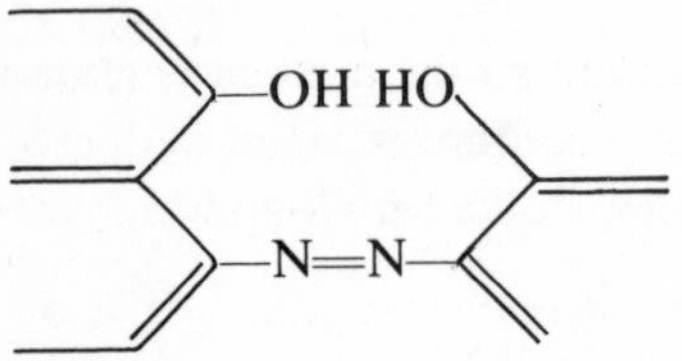

OH HO

—N=N—

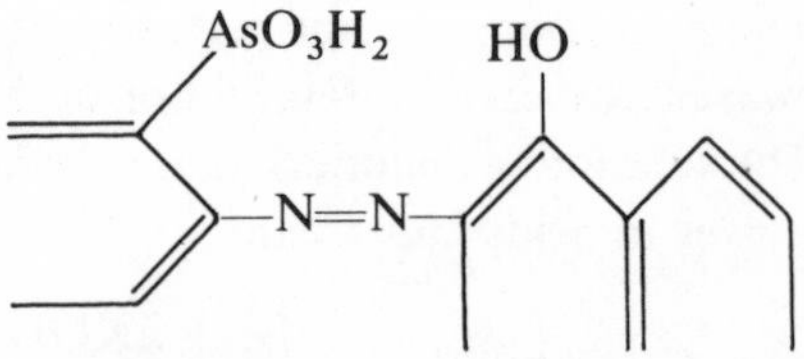

AsO_3H_2 HO

—N=N—

Arsenazo I gives a red-violet coloration with beryllium ions at pH 6–6.5 (urotropin) [5]; the reagent itself is pink. The minimum detectable amount is 0.02 μg of Be in 1 ml (dilution 1:50,000,000). Magnesium, zinc, manganese, and alkali metals do not interfere. Aluminum (III) ions are masked by tartrate or Complexone III; ferric ions, REE, and cupric ions are masked by oxalates, while uranium (VI) is masked by hydrogen peroxide. Arsenazo I is used in the detection of beryllium and its photometric determination.

Thoron reacts [6] with beryllium ions in the molar ratio of 3:2. The reacting species are $BeOH^+$ and the quintuply charged anion of the reagent. The optimum pH is approximately 12.5, and the complex formation constant is $(4.5 \pm 0.4) \cdot 10^{18}$

The violet-blue color of a 0.02% aqueous solution of Beryllone II at pH 12–13 turns light blue in the presence of beryllium ions [7]. The reagent becomes slightly more selective in the presence of Complexone III. Ferric ions, which cannot be masked by other substances, interfere. The reagent is used in the detection [7] of beryllium and its photometric determination [8].

p-Nitrobenzeneazoorcinol reacts [9] with beryllium ions (probably $Be(OH)^+$) in 0.5 N NaOH in the molar ratio of 1:1. The resulting beryllium compound has an absorption maximum at 490–500 mμ, while the solution of the reagent alone in 0.5 N NaOH has an absorption maximum at 450–460 mμ. The color develops rapidly and is stable for at least 48 hours. The molar extinction coefficient is $31{,}440 \pm$ $\pm\ 140$ at 500 mμ. The complex formation constant is $(2.87 \pm 0.06) \cdot 10^{-3}$. The reagent is used in the detection [10] of beryllium and its photometric determination [11].

Morin at pH 13 forms a fluorescent compound with beryllium ions (green fluorescence in the ultraviolet). Morin is highly sensitive and may be used to determine small amounts of beryllium in ores and silicate rocks without preliminary separation from accompanying elements, the ions of which are masked [12–14]. The intensity of the fluorescence is proportional to the concentration of beryllium. It also depends on the concentration of morin and the pH of the solution. It decreases slowly when the solution is left to stand. The sample and the standard solutions must be prepared at the same time. The introduction of ascorbic acid, citrate ions, and Complexone III makes it possible to determine microgram amounts of beryllium in the presence of Al, Ca, Mg, Mn, Mo, Cd, and Zn (up to 5 mg each), Fe (up to 0.35 mg), Ti (0.2 mg), and Cr (0.03 mg).

Aluminon reacts with beryllium ions to form a red-colored compound which is used in the photometric determination of this element (p. 47).

8-Hydroxyquinoline does not precipitate beryllium ions from acetate buffer solution (pH 5.7) but is used to separate beryllium from a number of other elements, whose ions precipitate out under these conditions (Al, Fe, Mo, W, U, Cu, Ni, Co, Zn, Cd, Hg, Bi).

A number of β-diketones form sparingly soluble compounds with beryllium ions.

2,2-Dimethylhexanedione-3,5 quantitatively precipitates beryllium ions at pH 7–8 and is used in the gravimetric determination of beryllium [15, 16] (p. 42).

Acetylacetone forms a fairly stable compound with beryllium ions. Beryllium acetylacetonate is readily extracted with chloroform in the presence of Complexone III in weakly alkaline medium (pH 9). Al, Cr, Co, Fe, Ni, Mn, Zn, Cd, Pb, Cu, Ca, and Mg ions are not extracted under these conditions. Beryllium can be separated from aluminum and other elements [17]. Satisfactory results are obtained in the separation of 4–24 mg of BeO from 24–4 mg of Al_2O_3.

Beryllium can also be determined by spectroscopic and activation methods.

For a survey of quantitative determination methods of beryllium, see [18].

Gravimetric determination of small amounts of beryllium in ores and ore dressing products by 2,2-dimethylhexanedione-3,5 [15, 16]

2,2-Dimethylhexanedione-3,5 reacts with beryllium ions at pH 7–8 to form a sparingly soluble crystalline compound with the composition

$$Be[CH_3COCHCOC(CH_3)_3]_2.$$

The precipitate is readily washed and brought to constant weight at 45–55°C; it sublimes above 70–75°C.

The method involving the use of Complexone III is suitable for the determination of beryllium in ores (0.1–3% Be).

In the presence of Complexone III 1–2 mg of beryllium can be determined without interference by 600 mg of Al, 200 mg of Fe(III), 30 mg of Ca, 20 mg of REE, 30 mg of U(VI), and 20 mg of Ti. Sulfate, chloride, nitrate, fluoride, and acetate ions may be present in 20-fold amounts, carbonate ions in 30-fold amounts and phosphate ions in 100-fold amounts. Under these conditions stannous tin is not masked by Complexone III and so interferes with the determination. This is not a serious disadvantage, however, since tin is rarely present together with beryllium.

The method has the following advantages as compared with the Complexone–phosphate method: a compound of definite composition is obtained in a single precipitation, so that the duration of the determination is considerably reduced; the crystalline structure of the precipitate is well developed, the precipitate is readily washed and can be brought to constant weight at a low temperature.

The conversion factor to beryllium is 0.03096.

REAGENTS

Potassium bifluoride, KHF_2, crystalline.
Sulfuric acid, dilute (1:1).
Hydrochloric acid, dilute.
Complexone III, 5% solution.
Ammonia, dilute (1:1) solution.
2,2-*Dimethylhexanedione*-3,5, saturated solution. The solution should be prepared 2–3 days before use in order to allow the keto-enol equilibrium to be established.

PROCEDURE

A 0.25–1.0 g sample, containing 0.05–0.5% beryllium, is fused with five times the amount of potassium bifluoride in a platinum dish in a muffle furnace at 800–900°C. Sulfuric acid (15 ml) is added to the melt and the solution is evaporated until thick white fumes are no longer evolved.

The contents of the dish are transferred to a 400-ml beaker, 10 ml of hydrochloric acid are added, and the volume made up to 100 ml with water. The mixture is heated until the salts are fully dissolved. The solution is filtered if necessary (White Ribbon filter paper), and cooled to room temperature. If stannous tin is present in the sample, it is removed by precipitating with hydrogen sulfide from the hot acid solution. Complexone III solution (35–40 ml) is added to the solution. The solution is neutralized with ammonia to pH 7–8 (Universal Indicator Paper) and 20–30 ml of 2,2-dimethylhexanedione-3,5 solution per 1 mg of beryllium is added. The solution is mixed by a jet of 15–20 ml of water.

The solution becomes turbid on the addition of the reagent; after 2–3 hours the precipitate coagulates on the bottom of the beaker in large loose flakes and the supernatant solution becomes perfectly clear. The precipitate is collected on a tared No. 4 sintered glass filter, washed several times with cold water and dried to constant weight at 45–55°C.

Gravimetric determination of beryllium in alloys by hexamminecobaltic chloride [19, 20]

The method is based on the formation of a sparingly soluble compound between the beryllium carbonate complex and hexamminecobaltic chloride.

Ions which form stable ethylenediamine tetraacetates do not interfere; in the presence of tartaric acid, Nb, Ta, Ti, and Sn which form stable tartrate complexes also do not interfere. Fluoride ions which form stable fluoride complexes with beryllium ions interfere as do major amounts of U(VI) which are coprecipitated with beryllium as the carbonate complexes.

Beryllium can be determined in beryllium concentrates, half-products of metallic beryllium production, and beryllium alloys provided that the amount of beryllium in the sample is not less than 1 mg. The experimental error is $\pm$ 0.5–2%.

REAGENTS

Ammonia, 25% solution.
Hydrochloric acid, $d = 1.19$ g/cm^3 and diluted (1:1).
Nitric acid, $d = 1.4$ g/cm^3.
Ammonium chloride, crystalline.
Complexone III, saturated solution.
Tartaric acid, 10% solution.
Ammonium carbonate, cold-saturated solution.
Hexamminecobaltic chloride, $[CO(NH_3)_6]Cl_3$. Eleven grams of $CoCl_2 \cdot 6H_2O$ are dissolved in minimum amount of water, and 23 ml of 25% ammonia solution, 7.4 g of ammonium chloride and 1 g finely ground activated carbon are added. The mixture is cooled in ice. Four ml of 30% hydrogen peroxide are added in small portions. The mixture is cooled to room temperature and then heated for five minutes on a water bath. It is acidified with 1:1 hydrochloric acid and cooled. The precipitate is filtered off and dissolved on the filter with hot water. The solution is heated to 80°C. Hydrochloric

acid (d = 1.19 g/cm^3) is added until the solution, is strongly acid. The solution is cooled, and the $[Co(NH_3)_6]Cl_3$ precipitate is filtered off, washed with ethanol and dried in the air. A saturated aqueous solution of $[Co(NH_3)_6]Cl_3$ is prepared: a 0.2% solution is prepared by dilution.

Washing solution, 60% ethanol, containing 1 ml of 2% solution of hexamminecobaltic chloride per 100 ml.

Ether.

Ethanol, 96%.

PROCEDURE

The sample of the alloy, which should contain not less than 2 mg of beryllium, is dissolved in 10 ml of hydrochloric acid (1:1) with gentle heating. Nitric acid (3–5 drops) is added and the solution is evaporated to dryness. The residue is treated with hydrochloric acid (1:1) and the mixture evaporated to dryness. The dehydration of silicic acid is repeated once more.

The precipitate is filtered off and the filtrate evaporated almost to dryness. One g of ammonium chloride, 10–15 ml of Complexone III solution and 5 ml of tartaric acid solution are added to the residue and the mixture is heated. Ammonium carbonate solution is added to the clear solution until the initially formed precipitate is fully redissolved.

Without stirring, 1–2 ml of saturated solution of hexamminecobaltic chloride and 2–5 drops of ethanol are added to the solution. After 2–3 minutes the beaker is tilted without mixing the liquid layers. Excess reagent solution is added until the interphase boundary becomes orange-yellow. The solution is stirred and diluted with water to a concentration of the salts of 2–3%. The solution and the precipitate are left to stand for $1\frac{1}{2}$–2 hours, filtered on No. 3 or No. 4 sintered glass crucible (tared to constant weight in a vacuum desiccator), and the precipitate is quantitatively transferred into the crucible with 0.2% hexamminecobaltic chloride solution.

The walls of the beaker and the crucible are flushed 2–3 times with 3–4 ml portions of the washing solution, then 3 times with 3 ml of 96% ethanol and 2–3 times with ether.

The crucible with the precipitate is transferred to a vacuum desiccator which is evacuated with a water pump for 25–30 minutes; air is readmitted into the desiccator by cautiously turning the stopcock. The crucible with the precipitate is weighed.

The conversion factor from $[Co(NH_3)_6][Be_2(CO_3)_2(OH)_3 \cdot (H_2O)_2] \cdot 3H_2O$ to beryllium is 0.0410 and to beryllium oxide 0.1138.

Titrimetric determination of beryllium by salicylic acid or sufosalicylic acid [21–23]

Beryllium reacts with salicylic and sulfosalicylic acids, forming a number of stable complex compounds.

These compounds are dissociated to a lesser extent than the compounds of beryllium with quinalizarin, aluminon, Acid Alizarin Blue BB, Acid Chrome Blue K, Chromoxane Pure Blue BLD (alberon), etc. These reagents can be used as indicators for beryllium. When beryllium ions are titrated against a salt of salicylic or sulfosalicylic acid at pH 9–10, 3 moles of titrant are consumed per 1 mole of beryllium sulfate. It is more convenient to use sulfosalicylic acid since this compound is readily soluble in water. The best metal indicators are Acid Chrome Blue K (sensitivity 0.1 μg of Be/ml) and Chromoxane Pure Blue BLD (sensitivity 0.025 μg of Be/ml).

The required pH value is best maintained by ammonium chloride buffer solution (pH 9–10) or glycocol and veronal buffer solution (pH 9.5–10). Ammonium acetate buffer solution cannot be used since the sparingly soluble beryllium oxyacetate then precipitates out and the color change at the equivalence point is not sharp.

Mg, Ca, Sr, Ba, Zn, Cd, Mn(II), Ni, Co, and Hg(II) do not interfere with the determination of beryllium in the presence of Complexone III. Permitted extraneous ions include 80-fold amounts of Al, not more than 10 mg of Fe(III) (ferric hydroxide precipitates out if more iron is present), and 40-fold amounts of Cu.

REAGENTS

Beryllium sulfate, $BeSO_4 \cdot 4H_2O$, 0.2 *M* solution.

Sulfosalicylic acid, 0.1 *M* solution.

Chromoxane Pure Blue BLD (*alberon*), 0.1% aqueous solution.

Ammonium chloride buffer solution, pH 9–10. Twenty g of NH_4Cl and 100 ml of 25% ammonia solution are diluted with water to one liter.

Complexone III, 0.1 *M* solution.

PROCEDURE

The sample solution, which should contain 4–15 mg of beryllium, is diluted with water to 50 ml in a volumetric flask. A 15.0 ml portion of the solution is withdrawn with a pipet with a rubber bulb into a 100- to 150-ml conical flask; the solution is diluted with water to 50 ml. One ml of Complexone III solution is added to bind traces of calcium, magnesium, etc., and then 6 drops of alberon and 10 ml of ammonium chloride buffer solution are added. The mixture is heated to 70–80°C and titrated against sulfosalicylic acid until the red-violet solution turns pure yellow. One ml of 0.1 *M* sulfosalicylic acid solution is equivalent to 0.300 mg of beryllium.

To titrate 1.0–5.0 mg of beryllium in the presence of aluminum, copper, or ferric iron, a solution of Complexone III is added to the solution in 10% excess, the pH is adjusted to 6, the solution heated to 60–70°C, 6 drops of alberon and the buffer solution are added and the solution is titrated with sulfosalicylic acid. In the presence of copper the solution is titrated until the color changes from blue-violet to pure green.

Titrimetric Complexone III–arsenate determination of beryllium in minerals and concentrates [4]

Beryllium ions are precipitated as beryllium ammonium arsenate in acetate buffer medium at pH 5.2, after which the arsenate ions are determined iodometrically:

$$BeCl_2 + NH_4Cl + Na_3AsO_4 \rightleftharpoons BeNH_4AsO_4 + 3NaCl$$

$$3BeNH_4AsO_4 + 6KI + 6HCl \rightleftharpoons 3I_2 + 3BeCl_2 + (NH_4)_3AsO_3 + \\ + 2K_3AsO_3 + 3H_2O$$

$$I_2 + 2Na_2S_2O_3 \rightleftharpoons Na_2S_4O_6 + 2NaI$$

A single precipitation of beryllium as $BeNH_4AsO_4$ in the presence of Complexone III results in a practically complete separation of beryllium from aluminum (Be:Al = 1:8), copper (1:50), nickel (1:8), and iron (1:16) ions. Hydrogen peroxide is also added to keep titanium ions in solution. The method is suitable for the determination of major amounts (more than 2%) of beryllium in beryllim concentrates, industrial half-products, and alloys.

The experimental error is ± 1.5%; one determination takes 6–8 hours.

REAGENTS

Potassium bifluoride, KHF_2, crystalline.
Sulfuric acid, dilute (1:1).
Hydrogen peroxide, 3% solution.
Complexone III, saturated solution.
Ammonium nitrate, crystalline.
Sodium arsenate, crystalline.
Ammonia, dilute (1:1).
Ammonium acetate, 15% solution.
Silver nitrate, 1% solution.
Hydrochloric acid, dilute (1:1 and 1:3).
Potassium iodide, crystalline.
Sodium thiosulfate, 0.1 *N* solution.
Starch, freshly prepared 0.5% solution.
Washing solution, mixture of 3% solutions of ammonium acetate and ammonium nitrate.

PROCEDURE

To 0.5 g of finely ground sample material are added 4–5 g of potassium bifluoride. The mixture is heated in a platinum dish until solidified and then in a muffle furnace, first at dull-red heat and then at 700–800°C until a clear melt is obtained.

To the cooled melt are added 15 ml of sulfuric acid. The solution is heated on a sand bath until sulfuric acid fumes are no longer evolved and is then heated over a burner flame. The residue is again fused in the muffle furnace at 500°C until a clear melt is obtained. The melt is cooled, leached with hot water in a 300-ml beaker, the solution diluted with water to 70–80 ml, 10 ml of hydrochloric acid (1:1) added and the mixture is heated until the solid particles are completely dissolved.

To the solution are added 2 ml of hydrogen peroxide, 7 ml of Complexone III solution, 10 g of ammonium nitrate, and 5 g of sodium arsenate, and the mixture is stirred until the salts are dissolved. A solution of ammonia is added until the resulting turbidity no longer disappears on stirring and 20 ml of ammonium acetate solution are added with constant stirring. The solution with the precipitate is boiled for 2–3 minutes over a weak burner flame and is then left for 1–2 hours on a boiling water bath.

After cooling, the precipitate is filtered through Blue Ribbon filter paper, washed a few times with the washing solution to a negative reaction for arsenate ions (test with $AgNO_3$ in neutral medium for the chocolate-colored silver arsenate). The precipitate is dissolved on the filter in hot hydrochloric acid (1:3). The filtrate is collected in a 300-ml conical flask and the filter paper washed with the hydrochloric acid. The solution obtained by dissolving the ammonium beryllium arsenate precipitate is combined with the wash waters; the total volume should be approximately 150 ml.

The resulting solution is cooled and 5 g of potassium iodide are introduced. The flask is stoppered and placed in the dark. After 5 minutes the liberated iodine is titrated against sodium thiosulfate until pale yellow; a few drops of starch solution are added and the titration with sodium thiosulfate is continued until the solution is colorless.

One ml of 0.1 *N* sodium thiosulfate solution is equivalent to 0.45 mg of beryllium or to 1.25 mg of beryllium oxide.

Photometric determination of beryllium by aluminon [24–28]

Aluminon (ammonium salt of aurintricarboxylic acid) reacts with beryllium ions to form a red-colored complex compound, the solutions of which have an absorption maximum at 530 mμ. The colored compound is best obtained at pH 4.6–5.4; the molar extinction coefficient is 9,200 ± 200.

At a concentration of 2–50 μg of beryllium in 50 ml of solution, 2 ml of 0.4% aluminon solution must be added to form the color compound.

In the presence of 100 mg of Complexone III beryllium may be determined without interference by 1,000 μg each of Cu, Ni, Co, and W, 5,000 μg of Pb, 500 μg each of Mn and Mo, 200 μg of Cr(III) and Cd, 100 μg of Bi, and 50 μg each of Al, Fe(III), Ti, Zr, and Sb. Niobium is masked with tartaric acid.

DETERMINATION OF BERYLLIUM IN NIOBIUM-BASED ALLOYS [24]

REAGENTS

Standard solution of beryllium chloride, 5 μg of Be/ml.
Complexone III, 5% solution.
Aluminon, 0.4% solution. One gram of aluminon is dissolved in 50–70 ml of water. Two g of benzoic acid dissolved in 25 ml of ethanol are added to the solution and then 100 ml of acetate buffer solution (pH 5.1–5.3). The solution is filtered into a 250 ml volumetric flask and made up to mark with water; it is stored in a dark glass bottle. The reagent is stable for 7–10 days.
Acetate buffer solution, pH 5.1–5.3. Acetic acid (34 ml, $d = 1.05$ g/cm^3) is diluted to 100 ml with water, 32 ml of 25% ammonia solution are added and the solution is diluted with water to one liter.
Potassium pyrosulfate, crystalline.
Ammonia, dilute (1:3) solution.
Tartaric acid, 10% solution.
Hydrochloric acid, dilute (1:1).
Nitric acid, dilute (1:5).

CONSTRUCTION OF CALIBRATION CURVE

Two ml of aluminon solution are introduced into each of five 50-ml volumetric flasks containing standard solutions of beryllium chloride corresponding to 5, 10, 20, and 25 μg of beryllium. Then 2 ml of Complexone III solution and 10 ml of acetate buffer solution are added to each flask and the solutions are made up to mark with water. The optical density of the solutions is measured, using a green filter, with respect to a blank solution containing all the components, except beryllium, in equal amounts.

PROCEDURE

The alloy (0.1 g) in a platinum or quartz crucible is heated at 700–800°C in a muffle furnace. The resulting oxides are fused with 1.5–2.0 g of potassium pyrosulfate until a transparent melt is obtained. The cooled melt is leached in a 100- to 150-ml beaker with 10 ml of solution of tartaric acid. The crucible is carefully extracted from the beaker with a glass rod, rinsed with 10–15 ml of hot water. Ammonia solution is added until the pH is 6 (Universal Indicator Paper). The solution is filtered into a 100-ml volumetric flask and diluted to mark with water.

In a 50-ml volumetric flask are placed 2 ml of aluminon solution, 2 ml of Complexone III solution and then an aliquot of the alloy solution which should contain between 5 and 25 μg of Be. Ten ml of acetate buffer solution are then added to the flask, the solution made up to mark with water and the optical density measured as described for the calibration curve.

The amount of beryllium in the aliquot is read off the calibration curve.

DETERMINATION OF BERYLLIUM IN THE CALCIUM, STRONTIUM AND BARIUM CARBONATES EMPLOYED IN VACUUM TUBE INDUSTRY [28]

Ions of alkaline-earth elements do not react with aluminon and thus do not interfere with the photometric determination of beryllium. Aluminum ions which are almost always present in alkaline-earth carbonates are masked by Complexone III, which is introduced prior to aluminon.

REAGENTS (see p. 48)

Standard solution of beryllium chloride, 2 μg of Be/ml.

CONSTRUCTION OF CALIBRATION CURVE

Into five 50 ml volumetric flasks are introduced 1, 2, 3, 4, and 5 ml of standard solution of beryllium chloride (2 μg of Be/ml); 1 ml of aluminon solution, 1 ml of Complexone III solution and 10 ml each of acetate buffer solution are introduced into each flask and the solutions are diluted to 25 ml with water. They are then heated at 70–80°C on a sand bath, cooled, diluted to mark with water and mixed. The optical density of each solution is measured at 530 mμ on a universal photometer or photoelectrocolorimeter relative to a solution prepared in a similar manner but not containing beryllium.

PROCEDURE

The carbonate sample (0.05–0.1 g) is cautiously treated in a 50-ml beaker with small portions of hydrochloric acid until CO_2 is no longer evolved. The solution is evaporated to dryness on a sandbath, the residue is dissolved in water, the solution transferred to a 100-ml volumetric flask and made up to mark with water. A 0.5 ml portion of the solution is withdrawn with a pipet, transferred to a 50 ml volumetric flask and all the operations described for the construction of the calibration curve are performed.

The amount of beryllium is read off the calibration curve.

DETERMINATION OF BERYLLIUM IN BRONZE [27]

REAGENTS (see p. 48)

Reagent solution. Into a 250-ml flask are introduced 100 ml of buffer solution, pH 4.4 (75 ml of 0.1 *N* CH_3COOH + 25 ml of 0.1 *N* NH_4OH), and 1.25 g of Complexone III and 0.1055 g of aurintricarboxylic acid or 0.1182 g of aluminon are added. When the dissolution is complete, the volume is made up to 250 ml with the buffer solution. When stored in a closed vessel in the dark, the solution is stable for more than two months.

CONSTRUCTION OF CALIBRATION CURVE

The calibration curve is constructed for beryllium concentrations of 10, 20, 30, 49, 50, and 60 μg/100 ml.

PROCEDURE

A 0.10 g sample of bronze is dissolved in a little nitric acid, the solution is transferred to a 100-ml volumetric flask and diluted to mark with water. Ten ml of the resulting solution are transferred to a 100-ml volumetric flask. Ammonia solution is added to a faint turbidity, which is dissolved by adding one drop of nitric acid. The solution is made up to mark with water. A part of the solution containing 20–50 μg of Be is mixed with an equal volume of the reagent solution.

The mixture is heated for 5 minutes on a boiling water bath or else is left to stand for 30 minutes at room temperature. It is cooled and the optical density is measured at 520 mμ in a 1-cm cell. The comparison solution is prepared by mixing equal volumes of the reagent solution and a solution containing copper ions in a concentration equal to the copper content in the sample aliquot.

BIBLIOGRAPHY

1. PŘIBIL, R. *Komplexony v chemicke analyse (Complexones in Chemical Analysis)*, pp. 473.—Prague Nakl. Ceskosl. Akad. Ved. 1957.
2. HURE, J., M. KREMER, and F. BERQUIER.—*Anal. chim. acta*, **7**, 37. 1952.
3. CHERNIKHOV, YU.A. and V.G. GORYUSHINA.—*Trudy Giredmet*, **2**, 97. Metallurgizdat. 1959,
4. GORYUSHINA, V.G. and T.A. ARCHAKOVA.—*Zav. Lab.*, **22**, 532. 1956; *Trudy Giredmet*, **2**, 106. Metallurgizdat. 1959.
5. KUZNETSOV, V.I.—*ZhAKh*, **10**, 276. 1955.
6. ADAMOVICH, L.P. and R.S. DIDENKO.—*Uchenye Zapiski Khar'kovskogo Gosudarstvennogo Universiteta*, **44**. *Trudy Khimicheskogo Fakul'teta i Nauchno-Issledovatel'skogo Instituta Khimii KhGU*, **12**, 195. 1954.
7. LUKIN, A.M. and G.B. ZAVARIKHINA.—*ZhAKh*, **11**, 392. 1956.
8. KARANOVICH, G.G.—*ZhAKh*, **11**, 400. 1956.
9. ADAMOVICH, L.P.—*Uchenye Zapiski Khar'kovskogo Gosudarstvennogo Universiteta*, **44**. *Trudy Khimicheskogo Fakul'teta i Nauchno-Issledovatel'skogo Instituta Khimii KhGU*, **12**, 167. 1954.
10. KOMAROVSKII, A.S. and N.S. POLUEKTOV.—*ZhPKh*, **7**, 839. 1934.
11. STROSS, W. and G.H. OSBORN.—*J. Soc. Chem. Ind.*, **63**, 249. 1944.
12. SANDELL, E. *Colorimetric Determination of Traces of Metals*.—Interscience, N.Y. 1944.
13. PLOTNIKOVA, R.N.—*Trudy Kazakhskogo Instituta Mineral'nogo Syr'ya* No. 3, 318. 1960.
14. SCHERBOV, D.P. and R.N. PLOTNIKOVA.—*Zav. Lab.*, **27**, 1058. 1961.
15. PRZHEVAL'SKII, E.S. and L.M. MOISEEVA.—*ZhAKh*, **15**, 117. 1960.
16. MOISEEVA, L.M., N.M. KUZNETSOVA, and I.I. PAL'SHINA.—*ZhAKh*, **15**, 561. 1960.
17. ALIMARIN, I.P. and I.M. GIBALO.—*ZhAKh*, **11**, 389. 1956.
18. GORYUSHINA, V.G.—In: *Metody opredeleniya i analiza redkikh elementov*, p. 79. Izd. AN SSSR.
19. PIRTEA, T.I. and V. CONSTANTINESCU.—*Z. anal. Chem.*, **165**, 183. 1959.

20. Vinogradov, A.V. and R.M. Apirina.—In: *Metody opredeleniya i analiza redkikh elementov,* p. 99. Izd. AN SSSR. 1961.
21. Adamovich, L.P. and T.U. Kravchenko.—*Zav. Lab.,* **23**, 416. 1957.
22. Mustafin, I.S. and L.O. Matveev.—*Zav. Lab.,* **24**, 259. 1958.
23. Kalinichenko, L.P. and I.I. Kalinichenko.—*ZhAKh,* **17**, 840. 1962.
24. Tsyvina, B.S. and M.B. Ogareva.—*Zav. Lab.,* **28**, 917. 1962.
25. Luke, C. and M. Campbell.—*Anal. Chem.,* **24**, 1056. 1952.
26. Tsyvina, V.S., M.B. Ogareva, and S.I. Plyushchikova.—In: *Metody opredeleniya i analiza redkikh elementov,* p. 97. Izd. AN SSSR. 1961.
27. Adamovich, L.P. and B.V. Yutsis.—*Zav. Lab.,* **28**, 920. 1962.
28. Molot, L.A. and N.S. Frumina.—*Uchenye Zapiski Saratovskogo Universiteta,* **75**, 90. 1962.

Scandium Sc

Scandium has a valency of $+3$. The standard electrode potentials in aqueous medium at 25°C relative to standard potential of hydrogen electrode have the following values:

$$Sc \rightleftharpoons Sc^{3+} + 3e \qquad -2.077V$$

$$Sc + 3H_2O \rightleftharpoons Sc(OH)_3 + 3H^+ + 3e \qquad -1.787V$$

The ions of Sc^{3+} are colorless. Their properties more closely resemble those of aluminum ions than those of yttrium and the lanthanides.

Scandium hydroxide, like aluminum hydroxide, is amphoteric and in aqueous medium can dissociate both as acid and base.

The acid dissociation of hydrated scandium ion can be expressed by the following equation:

$$[Sc(H_2O)_6]^{3+} + H_2O \rightleftharpoons [Sc(H_2O)_5OH]^{2+} + H_3O^+.$$

The dissociation constant at 20°C is $1.17 \cdot 10^{-5}$.

The protonic dissociation process is complicated by the formation of dimers and possibly also of higher polymers:

$$2[Sc(H_2O)_5OH]^{2+} \rightleftharpoons [Sc(H_2O)_5OH]_2^{4+}$$

The dimerization constant is $(13.8–2.17) \cdot 10^3$.

The degree of polymerization increases with increasing concentration of hydroxyl ions.

The values of the acidic dissociation constant and the dimerization constant vary greatly with the temperature and the ionic strength of the solution.

Freshly precipitated $Sc(OH)_3$ is readily soluble in dilute acids with the formation of scandium salts and much less so in concentrated solutions of alkalis, ammonia, solutions of ammonium chloride and ammonium nitrate, with the formation of scandiate ions and the complex ammines $[Sc(NH_3)_6]^{3+}$, respectively.

Scandium ions begin to precipitate out of 0.005 *M* chloride solutions or 0.0025 *M* sulfate solutions at pH 4.8–4.9 as basic salts [1]. The precipitation is completed at pH 5.45 and 5.10, respectively. The approximate composition of the precipitates are $Sc(OH)_2Cl \cdot xH_2O$ and $Sc(OH)SO_4 \cdot xH_2O$. On the addition of NaOH in an amount equivalent to that of scandium, basic scandium chloride and sulfate are converted to the hydroxide [1].

Ammonium carbonate and alkali carbonates precipitate a voluminous, white, amorphous basic carbonate $Sc(OH)CO_3 \cdot H_2O$ out of solutions of scandium salts. This precipitate is moderately soluble in cold solutions of ammonium and sodium carbonate; on heating, the solubility increases considerably and this is accompanied by the formation of double salts. The compounds $NH_4Sc(CO_3)_2 \cdot 2H_2O$, $2Sc_2(CO_3)_3 \cdot (NH_4)_2CO_3 \cdot 6H_2O$, and $Sc_2(CO_3)_3 \cdot 4Na_2CO_3 \cdot 6H_2O$ have been isolated in the crystalline form. They are soluble in concentrated solutions of ammonium and sodium carbonate. When dissolved in water, the double carbonates are hydrolyzed.

Scandium oxide Sc_2O_3 is a fine white powder. It is very sparingly soluble in cold dilute acids, much more soluble in the heat, and very readily soluble in concentrated acids.

Scandium oxide is obtained on igniting scandium hydroxide, carbonate, nitrate, or oxalate.

Scandium salts are hydrolyzed in aqueous solutions to a much greater extent than the salts of REE but less than the corresponding aluminum salts.

Scandium has a tendency to form complex compounds with a coordination number of 6. Double salts (fluorides, oxalates, carbonates) and complex compounds with various organic ligands of scandium are known. Scandium is a stronger complex former than REE or aluminum. In this respect scandium resembles zirconium and particularly thorium.

The aqueous solution of scandium sulfate $Sc_2(SO_4)_3$ contains the scandium salt of the complex acid $Sc[Sc(SO_4)_3]$.

No sufficiently selective reagents are known for the detection and quantitative determination of scandium. This is due to the fact that the properties of scandium are intermediate between those of aluminum and the elements of the yttrium subgroup and are also similar in many respects to those of other trivalent and tetravalent elements such as iron and thorium. Thus, scandium is usually isolated by precipitation as the basic ammonium tartrate, after which it is extracted with ether as the thiocyanate complex.

Scandium chloride is 1,000–10,000 times more soluble in concentrated hydrochloric acid (in the presence of ether) than are the chlorides of yttrium, lanthanides, or aluminum, and therefore scandium can be quantitatively separated from these elements. The crystalline precipitate which appears when aluminum, yttrium, and REE chlorides are precipitated from aqueous ether solution does not occlude scandium chloride.

Scandium chloride is extracted with tributyl phosphate. Scandium may be separated from lanthanides by a single extraction from solutions in 5–8 *M* hydrochloric acid.

During the extraction of ferric iron from 6 *N* HCl with ether, practically all the scandium ions remain in the aqueous phase.

Scandium fluoride ScF_3 resembles ThF_4 as regards its solubility in water and in mineral acids. However, scandium fluoride is readily soluble in solutions of KF, NaF, and $(NH_4)F$ with the formation of hexafluoroscandates K_3ScF_6, Na_3ScF_6, and $(NH_4)_3ScF_6$. Thorium and REE fluorides are insoluble in excess solution of alkali fluoride. Nevertheless, scandium cannot be separated from REE owing to various difficulties.

Scandium thiocyanate, obtained by adding ammonium thiocyanate to a $ScCl_3$ solution, is extracted with ether. Under optimum conditions (20–30 g of NH_4SCN in 100 ml of 0.5–1 *M* HCl solution and in the absence of sulfate and phosphate ions), the distribution coefficient reaches 16–17.4, when 94.1–94.6% scandium pass into the organic phase. Scandium can be readily separated from lanthanides, Zr, Hf, Th, Fe(II), Mn, Be, Mg, Ti, and U in this manner.

Scandium nitrate is readily extracted with ether from solutions saturated in $LiNO_3$. The distribution coefficient of $Sc(NO_3)_3$ extracted with ether from 1 *N* nitric acid solution saturated with $LiNO_3$ at 35°C is 4.97. Under these conditions 83.3% of scandium nitrate is extracted. In the absence of lithium nitrate little scandium nitrate is extracted.

Scandium nitrate is also readily extracted with tributyl phosphate from nitric acid solutions. The distribution coefficient for the extraction from 15.6 *N* HNO_3 is more than 1000. Scandium can be separated from lanthanides in this way.

Sodium pyrophosphate precipitates $Sc_4(P_2O_7)_3$. The precipitate is insoluble in mineral acids and only very sparingly soluble in hot sulfuric acid. Scandium pyrophosphate is very slowly dissolved by a large excess of sodium pyrophosphate in the heat, while heavy metal pyrophosphates are readily soluble in the presence of excess sodium pyrophosphate. Scandium pyrophosphate is dissolved only in the presence of sodium fluoride and in ammoniacal solution of ammonium carbonate and is reprecipitated on the addition of acetic acid or ammonia.

When oxalic acid or ammonium oxalate is added to solutions of scandium salts, a white crystalline precipitate is formed, which is readily soluble in excess ammonium oxalate. Scandium cannot be quantitatively precipitated as the oxalate $Sc_2(C_2O_4)_3$. The solubility of the oxalate is 156 mg in one liter of water at 25°C; it is much more soluble than REE oxalates.

Ammonium tartrate in the presence of excess ammonia reacts with scandium ions to form ammonium scandium double tartrate, whose approximate composition is $NH_4OOCCH(OH)CH(OH)COOSc(OH)_2 \cdot nH_2O$. The completeness of the precipitation depends on the pH of the solution. At pH 7–13 the solubility of ammonium scandium tartrate is least, 0.25 mg of Sc_2O_3 per liter. It increases both in more acidic and in more alkaline media. The solubility of ammonium scandium tartrate is very different from the solubilities of similar compounds of Zr, Hf, Th, Mn, Fe(III), Al, and Ti and of elements of the cerium subgroup.

Ammonium tartrate is used in the separation of scandium (p. 58).

Inositolhexaphosphoric acid $C_6H_6O_6[P(OH)_3OP(OH)_3]_3$ reacts with scandium ions to form the very sparingly soluble compound $C_6H_6O_6[P(O_3Sc)OP(O_3Sc)]_3$; the resulting white flocculent precipitate is insoluble even in concentrated hydrochloric acid. It does not react with nitric acid, mixtures of nitric and hydrochloric acids, or with hydrogen peroxide. The reagent also precipitates the ions of Ti, Zr, Hf, and Th; the precipitates are insoluble in acids. Unlike scandium and thorium compounds, zirconium and hafnium compounds are readily soluble in oxalic acid, while the titanium compound is soluble in the presence of alkali fluorides.

The thorium salt of inositolhexaphosphoric acid is readily soluble in the presence of nitrilotriacetic acid. Inositolhexaphosphoric acid is used in the detection and separation of scandium.

Benzeneseleninic acid quantitatively precipitates scandium ions and is used in the gravimetric determination of this element (p. 59).

Mandelic acid $C_6H_5CH(OH)COOH$ quantitatively precipitates [2] scandium ions from solutions at pH 1.8–3.2. The composition of the precipitate is $H_3[Sc(C_8H_6O_3)_3] \cdot nH_2O$; it is soluble in ammonia with the formation of $(NH_4)_3[Sc(C_8H_6O_3)_3]$. Mandelic acid is employed to separate scandium from lanthanides and from small amounts of thorium. The precipitate usually contains mandelic acid as impurity and is therefore ignited to the oxide Sc_2O_3 at 800°C.

8-Hydroxyquinoline at pH 7.5 precipitates the lemon-yellow compound $Sc(C_9H_6ON)_3 \cdot C_9H_7ON$ from the aqueous solution of scandium perchlorate $Sc(ClO_4)_3$. In precipitating scandium ions an excess of 8-hydroxyquinoline is first added to the solution and is followed by ammonia–ammonium acetate buffer solution. The filtered and washed precipitate is dried at 100–110°C and weighed.

All ions precipitated by 8-hydroxyquinoline interfere with the determination of scandium.

The compound of scandium with 8-hydroxyquinoline is soluble in benzene, chloroform, and other organic solvents.

In the presence of iodide ions and diantipyrylmethane, scandium forms a complex compound extractable with chloroform. In this way scandium can be separated from the accompanying elements (p. 57).

Alizarin, morin, and aurintricarboxylic acid react with scandium and aluminum ions in a similar manner.

Quinalizarin forms a blue compound with scandium and also with beryllium, magnesium, and REE. Scandium quinalizarinate differs from the corresponding compounds of beryllium in magnesium in that it can be quantitatively extracted from the water phase with ethyl acetate or with isoamyl alcohol. Quinalizarin is used in the detection of scandium.

Schiff's base, which is readily formed by condensation of o-aminophenylarsonic acid with salicylic aldehyde, gives an intense yellow coloration or a precipitate with scandium ions in a neutral or weakly acetic acid solution [3]. In the detection of scandium ions it is advisable not to use the ready-made preparation of the sparingly soluble reagent, but rather to prepare the reagent in situ by adding first salicylic aldehyde and then o-aminophenylarsonic acid to the sample solution. The detectable minimum is 1 μg of Sc in 1 ml of solution (limiting dilution 1:1,000,000).

Aluminum, yttrium, REE, indium, thallous thallium, and also ions of bivalent elements do not interfere with the detection of scandium. Gallium ions form a yellowish-colored precipitate.

The following ions interfere: Ti, Th, Nb, U(IV), Zr, Hf, Sn(IV), Ta, Sb, Bi, Fe(III), V(V), and Pd(III).

Arsenazo III is used in photometric determination of 0.001–0.1% of scandium in aluminosilicates, coal ash, and cassiterites after isolation and concentration [4].

Derivatives of 2,3,7-trihydroxy-6-fluorone form colored compounds with scandium ions. The most sensitive reagent is 9-propyl-2,3,7-trihydroxy-6-fluorone [5]. This reagent forms a red-colored compound with scandium at pH 3–7. If the concentration of scandium ions is sufficiently high, a red-colored precipitate separates out. Colloidal solutions of this compound can be stabilized with gelatine. Weakly acid solutions of the reagent are yellow with yellow-green fluorescence. The optimum pH of formation of the colored scandium compound is at 5.6. At pH 6 the optical density of the solutions of the reagent decreases. The absorption maximum of the reagent solution is at 480 mμ while that of the scandium compound is at 510 mμ. The reagent reacts with scandium ions in the molar ratio of 1:1.

Xylenol Orange, sulfonazo, Chlorophosphonazo III, and 1-(2-pyridylazo)-resorcinol form colored compounds with scandium ions and are used in photometric determination of the element (pp. 61, 62, 63, and 64).

Scandium can be quantitatively determined by emission and X-ray spectroscopy.

Isolation of scandium by extraction as iodide complex with diantipyrylmethane and complexometric titration of scandium [6]

In moderately acid medium scandium ions react with diantipyrylmethane in the presence of iodide ions to form a mixed complex which is readily soluble in chloroform and dichloroethane. Under these conditions REE, Fe, Cr, Mn, Mg, Ba, Ca,

Co, Ni, and As do not form compounds soluble in chloroform or dichloroethane and do not interfere with the separation of scandium.

The ions of Hf(IV) and Zr(IV) react similarly to scandium, but are easily masked with Complexone III.

Satisfactory results were obtained in the isolation of 8 mg of scandium from a solution containing 1 g of Al, Cr, Co, Mn, Fe, Y, Ln, Be, Ni or Mg; 0.1 g of Ce, Ca, Ba, Ti, and REE and 0.2 g of Hf or Zr.

Cd, In, Hg, Bi, and Sb, which form stable complex iodide anions, interfere with the separation.

Scandium is reextracted with ammoniacal solution of Complexone III and the determination is terminated by complexometric titration.

REAGENTS

Potassium iodide, crystalline.
Ascorbic acid, crystalline.
Chloroform.
Diantipyrylmethane.
Complexone III, 0.02 and 0.025 M solutions.
Zinc sulfate, 0.025 M solution.
Ammonia, 25% solution.
Indicator mixture (Eriochrome Black ET–00 and sodium chloride in the ratio of 1:100).
Buffer solution, pH 10, prepared by mixing solutions of ammonia and ammonium chloride.

PROCEDURES

Separation of scandium by extraction

The sample solution, which should be 0.75–1.25 N in HCl and should contain between 2 and 30 mg of Sc, is quantitatively transferred to a 60-ml separating funnel, and 2 g of potassium iodide, 0.2 g of ascorbic acid (to prevent oxidation of iodide ions), 15 ml of chloroform and 2.5 g of diantipyrylmethane are added. The total volume of the aqueous phase should be 15–30 ml. The mixture is mechanically shaken for 15–20 minutes. The phases are allowed to separate. The chloroform layer is decanted into another separating funnel, and the extraction process with 15 ml of chloroform and 1–2.5 g of diantipyrylmethane in the first funnel is repeated. After the second extraction the water phase is washed with 10 ml of chloroform. All chloroform extracts are combined and filtered through a dry filter paper collecting the filtrate in a separating funnel.

Note. If the sample solution contains not more than 50 mg of zirconium or hafnium, they are masked by adding 3–4 ml of 0.2 M solution of Complexone III and shaking briefly, after which scandium is extracted as above.

Complexometric determination of scandium

To the chloroform extract in a separating funnel obtained as above are added 40 ml of 0.025 *M* solution of Complexone III and 3 ml of ammonia solution and the mixture is shaken for 15–20 minutes. The chloroform layer is separated and washed with water. The wash waters are added to the filtrate, 15 ml of buffer solution (pH 10) are added, a little indicator mixture on the tip of a spatula is introduced and the excess Complexone III is titrated against the solution of zinc sulfate until the blue solution turns wine-red.

One ml of 0.025 *M* Complexone III is equivalent to 1.124 mg of scandium.

Isolation of scandium by the tartrate method [7, 8]

The method is based on the precipitation of scandium as the basic tartrate in the presence of yttrium as coprecipitant. The precipitate is ignited and dissolved in hydrochloric acid, and scandium is determined photometrically or by sulfonazo.

The method is suitable for the analysis of coal ash, fluorite, wolframite concentrate, and other objects.

REAGENTS

Solution of yttrium chloride or yttrium nitrate, containing 2 mg of Y in 1 ml of solution.
Ammonium tartrate, 3.5% and 40% solutions.
Sodium hydroxide, 2% and 1 *N* solutions.
Hydrochloric acid, $d = 1.19$ g/cm^3.
Ammonia, 25% solution.
Hydrofluoric acid, concentrated.
Sulfuric acid, $d = 1.84$ g/cm^3.
Solutions of methyl orange and neutral red indicators.
Sodium chloride, 1% solution.

PROCEDURE

The weighed sample (0.5 g if the concentration of scandium is of the order of 0.01%) is treated in a platinum dish with 10 ml of hydrofluoric and 10 ml of sulfuric acids, and the solution is evaporated on a sand bath to remove most of the sulfuric acid. The residue is wetted with 5 ml of hydrochloric acid and water is added to a volume of 10 ml. The solution is neutralized to methyl orange with the solution of sodium hydroxide, diluted with double the volume of 2% sodium hydroxide solution, heated to boiling and held for 30 minutes on a boiling water bath until the precipitate is fully coagulated. The precipitate is filtered through White Ribbon filter paper, washed with sodium chloride solution,and dissolved in a small amount of hydrochloric acid. The solution is evaporated to 5 ml. The residue is diluted with water to a volume of 40 ml in a 150-ml beaker, and 5 ml of yttrium salt solution and 50 ml of 40% ammonium tartrate solution are added. The mixture is neutralized to Neutral Red with the solution of ammonia. It is heated to boiling and again neutralized with solution of ammonia. Two ml of ammonia are added in excess, and the solution is placed on a water bath until the precipitate begins to separate out. The solution with the precipitate is then left overnight.

The precipitate is filtered through White Ribbon filter paper, washed with 3.5% ammonium tartrate solution, dried, and the filter paper ashed and ignited at 500–600°C in a platinum crucible. The residue is dissolved by heating with 2–3 ml of hydrochloric acid, it is evaporated almost to dryness and dissolved in 5 ml of water. The solution is filtered into a 25-ml volumetric flask, diluted with water to the mark and thoroughly mixed.

Scandium is determined photometrically by sulfonazo, by withdrawing 10 ml of the resulting solution with a pipet into a 25-ml volumetric flask, after which the procedure is continued as described below (p. 60).

Note. If the sample material contains large amounts of iron (100–150 mg), most of it is separated as chloride after the solubilization by extraction with ether from 6 *N* HCl.

Gravimetric determination of scandium in the presence of zirconium by benzeneseleninic acid [9, 10]

Benzeneseleninic acid and its ammonium salt precipitate scandium from neutral and weakly acid solutions as a white crystalline compound, readily soluble in mineral acids.

The precipitate when dried at 105–120°C has the composition $Sc(C_6H_5SeO_2)_3$, and contains 7.38% scandium; the conversion factor to scandium is 0.0738. Ammonia decomposes the compound with the formation of scandium hydroxide. On being heated to 600°C the compound is quantitatively converted to scandium oxide.

Zirconium forms a similar compound with benzeneseleninic acid, but in more acid medium. In the determination zirconium is precipitated first, after which scandium is precipitated from the filtrate.

At a ratio of Sc:Zr $\leqq$ 1:10 satisfactory results are obtained. The sensitivity is 5 μg of Sc/ml.

REAGENTS

Ammonium benzeneseleninate, 20% and 0.5% aqueous solutions.
Hydrochloric acid, 1 *N* solution.
Ammonia, dilute (1:1) solution.
Methyl orange, indicator solution.

PROCEDURES

Determination of scandium in solution of the pure salt

The sample solution, which should contain 0.1–3 mg of scandium, is diluted with water to 50 ml in a 100-ml beaker and heated on a water bath. One drop of methyl orange indicator is added and the solution is neutralized with ammonia solution to the first color change. To the hot solution are added 5 ml of 20% ammonium benzeneseleninate solution and the mixture is cooled to room temperature.

The precipitate is filtered on a No. 4 porous glass microcrucible and washed with 10–15 ml of double-distilled water and with an equal volume of ethanol. It is dried at 105°C for 20 minutes and then weighed.

Determination of scandium in the presence of zirconium

The sample solution, containing 0.1–3 mg of scandium and not more than tenfold amount of zirconium, is diluted to 50 ml with hydrochloric acid. It is heated and zirconium is precipitated by adding 5 ml of 20% ammonium benzeneseleninate solution. It is then cooled. The precipitate is filtered on Blue Ribbon filter paper and thoroughly washed with 0.5% ammonium benzeneseleninate solution. The scandium-containing filtrate is evaporated to 50 ml on a water bath and scandium is determined as described above.

Photometric determination of scandium in the presence of REE by sulfonazo [8]

Sulfonazo (sulfone-bis-(4-hydroxyphenyl)-(3-azo-2′)-1′-hydroxy-8′-aminonaphthalene-3′,6′-disulfonic acid) is a highly sensitive and selective reagent for scandium ions. The violet-pink solutions of the reagent become violet or blue-violet on the addition of scandium salts; the coloration is stable for many hours. The light absorption by the solutions obeys Beer's law in the concentration range of 5–80 μg of Sc in 25 ml of solution. The maximum absorption is at 610–620 mμ. The reaction is carried out at pH 4.0–5.0 (urotropin buffer solution). The complex is formed somewhat more slowly if acetate buffer solution (pH 5.0–5.5) is employed, but the reagent becomes more selective.

The sensitivity of the reagent is 0.04 μg of Sc/ml.

Alkali metals, alkaline-earth metals, REE, and Mn, Tl, Be, Cr(III), Cd, Pb, Ge, Mo, and Re do not interfere.

Ferric ions and titanium, zirconium, and certain other ions which are hydrolyzed under these experimental conditions must be removed. V(V), Co^{2+} and Ga^{3+} ions which form intensely colored compounds with sulfonazo and also Au(III) ions which decolorize the reagent interfere. The ions of indium, copper, hexavalent uranium, nickel, aluminum, zinc, and phosphate, citrate, and tartrate ions must not be present.

The method is suitable for the determination of scandium in silicate rocks and coal ash after the isolation of scandium by the tartrate method (p. 58).

REAGENTS

Standard solution of scandium salt, 10 μg of Sc/ml solution.

Sulfonazo, 0.03% aqueous solution.

Buffer mixture. Sodium acetate (8.2 g) is dissolved in water, 100 g of urotropin are added; the solution is stirred, filtered if necessary, acidified to pH 5.0–5.2 with hydrochloric acid and diluted to one liter with water.

CONSTRUCTION OF CALIBRATION CURVE

Volumes of 1, 2, 3, 4, and 5 ml of a standard solution of scandium salt are introduced into five 25-ml volumetric flasks; 3 ml of sulfonazo solution and 5 ml of buffer mixture are added to each flask and the solutions are made up to mark with water. After 10–15 minutes the optical density is measured relative to a solution containing the same reagents except the scandium salt. The calibration curve is then plotted.

PROCEDURE

The reagents are added to the sample solution as in the plotting of the calibration curve, the optical density is measured and the scandium content is read off the calibration curve.

Photometric determination of scandium in wolframite by Chlorophosphonazo III [7–11]

Chlorophosphonazo III (2,7-bis-(4-chloro-2-phosphinic acid phenyl-azo)-1,8-dihydroxynaphthalene-3,6-disulfonic acid) reacts with scandium ions to give a characteristically colored compound. If the concentration of scandium is sufficiently high, a precipitate separates out. The coloration develops practically instantaneously and is stable for several days. The absorption maximum is at 640 and 690 mμ. The optimum pH range is 2–4. The sensitivity is 0.1–0.2 μg of Sc/ml.

The reagent reacts with Th, Zr, Ti, Nb, Ta, U, and REE. The effect of Ti, Zr, Fe, and Al may be greatly reduced by the introduction of tartaric acid. Fluoride, oxalate, phosphate, and ethylenediaminetetraacetate ions interfere. Nitrate and sulfate ions as well as 20,000-fold amounts (pH 2–2.5) of tartrate ions may be present in the sample solution.

The method is suitable for the determination of scandium in wolframite after solubilization of the sample by sodium carbonate fusion, treatment of the residue with sulfuric acid and ether extraction of scandium as thiocyanate. Other elements which are partly or fully extracted by this procedure do not interfere with the subsequent determination.

REAGENTS

Sodium carbonate, crystalline and 1% solution.
Sulfuric acid, $d = 1.84$ g/cm^3.
Formalin.
Ammonia, 25% solution.
Hydrochloric acid, 0.1 and 1 *N* solutions and 1:1 solution.
Ammonium thiocyanate, crystalline.
Ascorbic acid, crystalline.
Ether.
Nitric acid, $d = 1.4$ g/cm^3.
Acetate buffer solution (pH 2.5).
Tartaric acid, 10% solution (neutralized to pH 2.5 with sodium hydroxide).
Chlorophosphonazo III, 0.05% solution.
Standard solution of scandium chloride, 5 μg of Sc/ml.

CONSTRUCTION OF CALIBRATION CURVE

Into five 25-ml volumetric flasks is introduced standard scandium solution in amounts of 5, 10, 15, 20, and 25 μg of Sc, and 5 ml of tartaric acid solution, 1 ml of Chlorophosphonazo III solution and 5 ml of acetate buffer solution are added to each flask. The solutions are made up to mark with water and mixed. Their optical density is measured with respect to a solution containing the same reagents except the scandium salt. The results are used to plot the calibration curve.

PROCEDURE

In order to separate the scandium, 0.05–0.10 g of a finely ground sample containing 150–300 μg of scandium are fused in a platinum or nickel crucible with 1 g of sodium carbonate and the melt is leached with hot water in a 100-ml beaker. The solution is decanted onto Blue Ribbon filter paper. The residue is washed twice with small portions of sodium carbonate solution, washed back into the beaker with a jet of hot water from a washbottle and evaporated to dryness on a sand bath. The dry residue is treated with 3–5 ml of sulfuric acid, heated on a water bath, and cooled. Water (10 ml) is added with stirring, then 3–5 drops of formalin are added to reduce manganese and the solution is heated until the excess formalin is removed and the liquid becomes colorless.

The solution is diluted with 25–30 ml of water and filtered. Ammonia solution is added to the filtrate to a faint permanent odor. The hydroxide residue is filtered onto White Ribbon filter paper and dissolved in hot hydrochloric acid (1:1); the filter paper is washed with 1 *N* hydrochloric acid. The hydrochloric acid solution is evaporated almost to dryness on the sand bath. The residue is dissolved in 5 ml of 0.1 *N* hydrochloric acid and transferred to a 25- to 30-ml separating funnel. Three g of ammonium thiocyanate and a few grains of ascorbic acid are introduced into the funnel to reduce the ferric iron and the mixture is shaken with 10 ml of ether. The extraction is repeated twice, adding 1 ml of 1 *N* hydrochloric acid each time. The ether extracts are combined, acidified with 2 ml of 1 *N* hydrochloric acid and evaporated to dryness on a water bath in a 100-ml beaker.

The dry residue is cautiously treated with 1 ml of nitric acid, covered with a watch glass and heated on a water bath to decompose the orange-red products of decomposition of thiocyanic acid. The solution is evaporated almost to dryness. The residue is dissolved in 5 ml of 0.1 *N* hydrochloric acid. The solution is transferred to a 50-ml volumetric flask, diluted to mark with water and thoroughly mixed. A 2.0–4.0 ml aliquot is then withdrawn with a pipet, transferred to a 25-ml flask and the reagents are added in the same amounts and in the same sequence as in the construction of the calibration curve.

The amount of scandium in the solution is read off the calibration curve.

Photometric determination of scandium by 1-(2-pyridylazo)-resorcinol [12, 13]

Scandium ions form two complex compounds with 1-(2-pyridylazo)-resorcinol. One of them is formed at pH $\geqq 2$ (molar ratio 1:1) with an absorption maximum at 505 mμ; the equilibrium constant of formation is $10^{-2.17}$. Molar extinction coefficients are $1.47 \cdot 10^4$ at 520 mμ and $4.80 \cdot 10^3$ at 560 mμ. The other complex compound is formed at pH $\geqq 3.92$; it has an absorption maximum at 510 mμ; the molar ratio scandium : reagent = 1:2.

The optical density of the solutions is proportional to the concentration of scandium between 0.05 and 2.0 μg of Sc/ml.

Be, Mo, Ca, and Mg do not interfere with the determination of scandium; Zn, Co, Ni, Cd, Cu, Ga, In, Tl, Zr, Th, U, Fe, and Al ions interfere by forming colored compounds with the reagents and must be removed. Fluoride, phosphate, thiocyanate, sulfate, and thiosulfate ions decompose the compounds of scandium with the reagent and must not be present in the solution. The method is suitable for the determination of scandium in solutions of its pure salts or in different objects after careful separation from interfering ions.

REAGENTS

Standard scandium nitrate solution, 10 μg of Sc/ml.
1-(2-*pyridylazo*)-*rescorcinol*, sodium salt, 0.05% aqueous solution.
Ammonium acetate, 20% solution.

CONSTRUCTION OF CALIBRATION CURVE

Into five 50-ml volumetric flasks are introduced 1, 2, 3, 4, and 5 ml of standard scandium solution; 2 ml of 1-(2-pyridylazo)-resorcinol solution and 10 ml of ammonium acetate solution are then added to each flask. The solutions are diluted to mark with water and the optical density is measured at 515 mμ (l = 1 cm) with respect to a solution containing all the reagents except for the scandium salt. The calibration curve is then plotted.

PROCEDURE

Reagents are added as described above to the sample solution. The optical density is determined and the scandium content is found from the calibration curve.

Photometric determination of scandium in magnesium alloys by Xylenol Orange [14, 15]

Xylenol Orange reacts with scandium ions in acid medium (pH 1.5–5) to form a soluble red-violet compound; the reagent itself in acid medium has a yellow color, which turns red-violet at pH above 5.5. The color is fully developed within ten minutes after mixing of the solutions, and is stable for two days. The light absorption of the solution obeys Beer's law within a wide concentration range of scandium.

The sensitivity of the reagent is 0.1 μg of Sc/ml.

The determination is carried out at pH 1.5. Ions of REE, La, Pr, Nd, Sm, Ce(III), Y, Zn, Cd, Al, Mn, and Fe(II) do not interfere. Zr, Th, In, Ga, and Bi, which form colored compounds with the reagent, interfere. Ferric and ceric ions are previously reduced by ascorbic acid.

The method is applicable to the determination of scandium in metallic magnesium and magnesium alloys without separation of the components of the alloy.

REAGENTS

Standard solution of scandium nitrate, 10 μg of Sc/ml.
Xylenol Orange, 0.05% aqueous solution.
Buffer solution, pH 1.5. Hydrochloric acid (0.2 *N*, 263 ml) is mixed with 500 ml of 0.2 *N* solution of potassium chloride.
Sodium acetate, 50% solution.
Ascorbic acid, 2% solution, freshly prepared.
Hydrochloric acid, dilute (1:1).

CONSTRUCTION OF CALIBRATION CURVE

Into five 50-ml volumetric flasks is introduced a standard scandium salt solution in amounts of 10, 20, 30, 40, and 50 μg of Sc; 5 ml of buffer solution and 5 ml of Xylenol Orange solution are introduced into each flask and the solutions are diluted to mark with water. After 20 minutes the optical densities of the solutions are measured (relative to a solution containing the same reagents except the scandium salt) using a photoelectrocolorimeter with a green filter. The calibration curve is then plotted.

PROCEDURE

A one-gram sample of the alloy containing 0.002–0.005% scandium is dissolved in 10–20 ml of hydrochloric acid in a 100-ml beaker. The solution is evaporated to a residual volume of 10 ml, quantitatively transferred to a 50-ml volumetric flask, rinsing the walls with small portions of water. Five ml ascorbic acid solution is added to the flask, then sodium acetate solution until lilac to Congo Red paper, 5 ml of buffer solution and 5 ml of Xylenol Orange solution, and the mixture is diluted to mark with water. After 20 minutes the optical density of the solution is determined as described for the construction of the calibration curve. The amount of scandium is found from the calibration curve.

Note. If the scandium content exceeds 0.005%, the sample weight is correspondingly reduced; scandium is determined on an aliquot of alloy solution containing between 20 and 40 μg of Sc.

BIBLIOGRAPHY

1. IVANOV-EMIN, B.N. and E.A. OSTROUMOV.—*ZhOKh*, **14**, 772. 1944.
2. ALIMARIN, I.P. and SHÊN HAN-HSI.—*ZhAKh*, **15**, 31. 1960.
3. KUZNETSOV, V.I.—*ZhOKh*, **14**, 897. 1944.
4. BELOPOL'SKII, M.P. and N.P. POPOV.—In: *Metody khimicheskogo analiza mineral'nogo syr'ya*, No. 7, 48 (Vsesoyuznyi nauchno-issledovatel'skii institut mineral'nogo syr'ya). Gosgeoltekhizdat. 1963.
5. NAZARENKO, V.A. and E.A. BIRYUK.—In: *Redkozemel'nye elementy*, p. 313, D.I. RYABCHIKOV, Editor.—Izd. AN SSSR. 1963.
6. ZHIVOPISTSEV, V.P. and I.S. KALMYKOVA.—*ZhAKh*, **19**, 69. 1964; Uchenye Zapiski Permskogo Universiteta, **25**, 120. 1963.
7. FISCHER, W. and R. BOCK.—*Z. anorg. Chem.*, **249**, 146. 1942.
8. BRUDZ', V.G., V.I. TITOV, E.P. OSIKO, D.A. DRAPKINA, and K.A. SMIRNOVA.—*ZhAKh*, **17**, 568. 1962.
9. ALIMARIN, I.P. and V.S. SOTNIKOV.—*Doklady AN SSSR*, **113**, 105. 1957; *ZhAKh*, **13**, 332. 1958.
10. ALIMARIN, I.P. and N.V. SHAKHOVA.—*ZhAKh*, **16**, 412. 1961.
11. ALIMARIN, I.P. and V.I. FADEEVA.—*ZhAKh*, **17**, 1020. 1962; *Vestnik MGU*, seriya II, khimiya, No. 4, 67. 1963.
12. BUSEV, A.I. and CHANG FAN.—*Vestnik MGU*, chem. ser., II, No. 6, 46. 1960.
13. SOMMER, L. and M. HNILIČKOVA.—*Anal. chim. acta*, **27**, 241. 1962.
14. VOLODARSKAYA, R.S. and G.N. DEREVYANKO.—*Zav. Lab.*, **29**, 148. 1963.
15. KON'KOVA, O.V.—*ZhAKh*, **19**, 73. 1964.

Yttrium Y

Yttrium has a valency of +3. In solution it forms stable colorless ions Y^{3+} which are difficult to separate from the ions of scandium (III) and rare-earth elements by chemical methods [1].

Reactions of yttrium ions in many respects resemble those of scandium ions.

The sparingly soluble hydroxide $Y(OH)_3$ is converted to the oxide Y_2O_3 on ignition.

The standard electrode potentials in aqueous medium at 25°C, relative to standared hydrogen electrode, have the following values:

$$Y \rightarrow Y^{3+} + 3e \qquad -2.372V$$

$$2Y + 3H_2O \rightarrow Y_2O_3 + 6H^+ + 6e \qquad -1.676V$$

Hydrofluoric acid precipitates the white YF_3.

Oxalic acid reacts with yttrium ions to form a white precipitate which is soluble in excess oxalate.

The red Erioviolet B (4-aminobenzene-azo-chromotropic acid) colors the solution violet in the presence of Y^{3+} ions.

Yttrium does not give any specific reactions and thus is detected by spectroscopic and, mainly, X-ray methods.

Yttrium is determined photometrically as pyrocatechol borate complex (see below) and with some other reagents.

Photometric determination of yttrium in the presence of a number of REE as pyrocatechol borate complex [1]

Pyrocatechol Violet reacts with yttrium ions in alkaline medium to form a blue complex compound; ions of other elements interfere. It is advisable to replace Pyrocatechol Violet with its red boric acid complex.

Yttrium ions react with pyrocatechol borate complex (BPCC) in the molar ratio 1:1. The solutions obey Beer's law in the concentration range between $1\cdot10^{-5}$–$4\cdot10^{-5}$ g-ion/liter (0.9–4.5 μg/ml). The absorption maximum is at 590 mμ, and the molar extinction coefficients are 18,500 at 590 mμ and 19,000 at 607 mμ.

Ce(III) and La(III) ions also react with BPCC but the resulting complexes are less stable. The effect of La is eliminated by increasing the concentration of H_3BO_3 to 0.05 *M*, and that of Ce by adding H_2O_2 to oxidize Ce(III) to Ce(IV). Under these conditions about 77% of yttrium is bound as complex with BPCC. Equal amounts of Pr, Nd, double amounts of La, and 0.5-fold amounts of Ce and Sm do not interfere.

REAGENTS

Standard solution of yttrium chloride, 20 μg of Y/ml.
Ammonium acetate buffer solution, pH 8.7. Acetic acid (26 ml, d = 1.05 g/cm^3) is diluted with water to 200 ml; 32 ml of 25% ammonia solution are added. The mixture is diluted to one liter with water.
Pyrocatechol Violet, 0.1% solution.
Boric acid, 4% solution.
Hydrogen peroxide, 3% solution.

CONSTRUCTION OF CALIBRATION CURVE

One ml of Pyrocatechol Violet solution and 2 ml of boric acid solution are introduced into each of five 25-ml volumetric flasks; the flasks are shaken and portions of 1, 2, 3, 4, and 5 ml of the standard yttrium chloride solution and 10 ml of ammonium acetate buffer solution are added to each flask. The solutions are mixed, 3–4 drops of hydrogen peroxide are introduced into each flask and the flasks are made up to the mark with ammonium acetate buffer solution. After 10–15 minutes the optical densitites are measured relative to the blank solution, on an electrophotocolorimeter and the calibration curve is plotted.

PROCEDURE

Reagents as above are added to the sample solution, the optical density is determined, and the yttrium content read off the calibration curve.

BIBLIOGRAPHY

1. SERDYUK, L.S. and U.F. SILICH.—*ZhAKh*, **17**, 802. 1962; **18**, 166. 1963.

Lanthanum La and the Lanthanides Ce, Pr, Nd, Pm, Sm, Eu, Gd, Tb, Dy, Ho, Er, Tm, Yb, Lu

The lanthanides constitute a special group of elements with highly similar properties which are difficult to separate from one another·

The lanthanides, together with lanthanum, scandium, and yttrium, form the group of rare-earth elements (REE).

As a rule, lanthanum and all lanthanides have a valency of + 3. Cerium also forms a number of compounds in which it is tetravalent. Tetravalent praseodymium and terbium oxides and salts of bivalent samarium, europium, and ytterbium are also known.

The standard electrode potentials in aqueous medium at 25°C, relative to standard hydrogen electrode, have the following values:

$La \rightleftharpoons La^{3+} + 3e$	– 2.522 V
$2La + 3H_2O \rightleftharpoons La_2O_3 + 6H^+ + 6e$	– 1.856 V
$Ce \rightleftharpoons Ce^{3+} + 3e$	– 2.483 V
$Ce^{3+} + 2H_2O \rightleftharpoons Ce(OH)_2^{2+} + 2H^+ + e$	+ 1.731 V
$Pr \rightleftharpoons Pr^{3+} + 3e$	– 2.462 V
$2Pr + 3H_2O \rightleftharpoons Pr_2O_3 + 6H^+ + 6e$	– 1.829 V
$Pr_2O_3 + H_2O \rightleftharpoons 2PrO_2 + 2H^+ + 2e$	+ 0.863 V
$Pr(OH)_3 \rightleftharpoons PrO_2 + H_2O + H^+ + e$	+ 1.431 V
$Nd \rightleftharpoons Nd^{3+} + 3e$	– 2.431 V
$2Nd + 3H_2O \rightleftharpoons Nd_2O_3 + 6H^+ + 6e$	– 1.811 V
$Pm \rightleftharpoons Pm^{3+} + 3e$	– 2.423 V
$Sm \rightleftharpoons Sm^{3+} + 3e$	– 3.121 V
$Sm^{2+} \rightleftharpoons Sm^{3+} + e$	– 1.000 V
$Eu \rightleftharpoons Eu^{2+} + 2e$	– 3.395 V

$Eu + 3H_2O \rightleftharpoons Eu(OH)_3 + 3H^+ + 3e$	– 2.002 V
$Eu^{2+} \rightleftharpoons Eu^{3+} + e$	– 0.429 V
$Gd \rightleftharpoons Gd^{3+} + 3e$	– 2.397 V
$Gd + 3H_2O \rightleftharpoons Gd(OH)_3 + 3H^+ + 3e$	– 1.994 V
$Tb \rightleftharpoons Tb^{3+} + 3e$	– 2.391 V
$Tb + 3H_2O \rightleftharpoons Tb(OH)_3 + 3H^+ + 3e$	– 1.999 V
$Dy \rightleftharpoons Dy^{3+} + 3e$	– 2.353 V
$Dy + 3H_2O \rightleftharpoons Dy(OH)_3 + 3H^+ + 3e$	– 1.956 V
$Ho \rightleftharpoons Ho^{3+} + 3e$	– 2.319 V
$Ho + 3H_2O \rightleftharpoons Ho(OH)_3 + 3H^+ + 3e$	– 1.937 V
$Er \rightleftharpoons Er^{3+} + 3e$	– 2.296 V
$Tm \rightleftharpoons Tm^{3+} + 3e$	– 2.278 V
$Tm + 3H_2O \rightleftharpoons Tm(OH)_3 + 3H^+ + 3e$	– 1.913 V
$Yb^{2+} \rightleftharpoons Yb^{3+} + e$	– 1.205 V
$Lu \rightleftharpoons Lu^{3+} + 3e$	– 2.255 V
$2Lu + 3H_2O \rightleftharpoons Lu_2O_3 + 6H^+ + 6e$	– 1.892 V

Metallic lanthanum and lanthanides are strong reducing agents.

Lanthanum and lanthanides are present in solution as the stable ions Ln^{3+}. Except for cerium, none of the elements is present in solutions as Ln^{4+} ions. Ce^{4+} ions are a strong oxidizing agent in acid medium. The ions of Eu^{2+}, and especially Yb^{2+} and Sm^{2+}, reduce hydrogen ions to elementary hydrogen in aqueous solutions.

The ions La^{3+}, Ce^{3+}, Gd^{3+}, Yb^{3+}, Lu^{3+}, and Tb^{3+} and also Y^{3+} and Sc^{3+} are colorless; Pr^{3+} ions are green, Nd^{3+} ions violet-pink, and Er^{3+} ions pink.

With the exception of lanthanum and lutetium which do not absorb light, the ions of trivalent lanthanides have absorption spectra with very sharp bands in the ultraviolet, visible, and infrared [1]. These bands are characteristic for each ion and have the same wavelengths in the solid state and as aqueous solution.

If a high-resolution spectrophotometer is available, the absorption spectra may be used for a quantitative analysis of a mixture of lanthanides.

The molar extinction coefficients are small, but milligram amounts of many lanthanides can nevertheless be detected.

Separate spectrophotometric determinations of praseodymium, neodymium, holmium, erbium, and thulium in the presence of the other lanthanides can be effected using ordinary spectrophotometers [2].

Emission spectroscopy, X-ray spectroscopy, and certain other physical methods are used in the detection and determination of more than 0.01% of individual lanthanides in their mixture. The emission spectra of lanthanides have a very large number of lines.

The trivalent ions of lanthanum and lanthanides, including yttrium and scandium, are very similar. They give almost identical reactions and therefore cannot be separated in solution by analytical reagents.

Tetravalent cerium can be readily determined in the presence of other lanthanides by chemical methods. Praseodymium is detected in the presence of other lanthanides by taking advantage of its ability to form oxides in which it is tetravalent. Europium may be separated and determined by reducing it to the bivalent state in solution. Bivalent ytterbium and samarium ions in solution are so unstable that they cannot be determined by any chemical method.

Lanthanides can be successfully separated by various chromatographic techniques (ion-exchange chromatography with the use of complex formers, paper chromatography). Extraction methods are also employed.

After separation by paper chromatography or ion-exchange chromatography it is usual to determine the individual lanthanides by photometric methods.

The hydroxides of lanthanum and lanthanides $Ln(OH)_3$ are practically insoluble in water and act as bases only. In the sequence from lanthanum to lutetium the hydroxides become progressively less basic. They are soluble in acids and insoluble in excess ammonia and solutions of alkali hydroxides.

Lanthanide hydroxides or basic salts are precipitated at pH 6.8–8.5. The hydroxides of individual elements are precipitated at the following pH values:

Element	La	Ce	Pr and Nd	Sm	Gd	Dy	Yb	Lu
pH	7.3–8.4	7.1–7.4	7.0–7.4	6.8	6.2	7.0	6.2–7.1	6.0

Lanthanum and lanthanide ions are not precipitated by ammonia from solutions containing organic hydroxy acids (tartaric, citric, malic, etc. acids) owing to complex formation.

The ignition of lanthanum hydroxide $La(OH)_3$ yields the white trioxide La_2O_3, which in the presence of traces of lanthanides has a grayish tinge.

Cerium hydroxide $Ce(OH)_3$ yields CeO_2 when ignited in air.

Ignition of praseodymium and terbium oxalates or nitrates yields oxides of variable composition. Praseodymium and terbium are partly trivalent and partly tetravalent in these oxides. The formation of such oxides is used to detect praseodymium in the presence of other REE.

Lanthanum and lanthanide fluorides, phosphates, and carbonates are sparingly soluble in water. The fluorides are also sparingly soluble in acids. The lanthanides of the cerium subgroup form relatively sparingly soluble double sulfates with alkali metal sulfates. The chlorides and nitrates are soluble in water.

Oxalic acid precipitates lanthanum and lanthanides from weakly acid solutions and is used for their separation. Alkali oxalates and ammonium oxalate must not be used, as the resulting oxalates will be partly solubilized owing to the formation of soluble complexes. The amorphous precipitates of oxalates become crystalline at 60°C. The ions Zr^{4+}, Th^{4+}, Bi^{3+}, and Sb^{3+} are precipitated together with lanthanides.

The composition of the white crystalline precipitate of lanthanum oxalate is $La_2(C_2O_4)_3 \cdot 10H_2O$.

The solubility product of cerium (III) oxalate is approximately 10^{-19}.

Dihydroxytartaric acid precipitates the ions of all lanthanides from weakly acid or neutral solutions [3], the initially amorphous precipitates eventually becoming crystalline. The composition of the lanthanum precipitate is approximately $4La_2O_3 \cdot 5C_4H_4O_8 \cdot 24H_2O$. The reagent is used to detect lanthanum, yttrium, gadolinium, erbium, etc. The ions of Be, Mg, Sr, Ba, Hg(II), Al, In, Tl(I), Ti(IV), V(V), Bi(III), U(VI), Mn(II), Re(VII), Fe(II), Fe(III), and Cd are not precipitated.

Alizarin, quinalizarin, morin, and a number of other organic reagents form colored compounds with the lanthanum and lanthanide ions. The selectivity of these reagents is low; ions of Al^{3+}, Sc^{3+}, Mg^{2+}, and many other elements react in a similar manner.

Xylenol Orange is used in the differential spectrophotometric determination of large amounts of total REE [4] at pH 5.6 (570 mμ). This method is as accurate as the gravimetric method.

8-Hydroxyquinoline quantitatively precipitates Ce^{3+} ions from ammoniacal tartrate-containing solution [5] as $Ce(C_9H_6ON)_3$ (yellow crystalline precipitate). The precipitate is not formed in weakly acetic acid solution, and this makes it possible to separate Ce^{3+} from Th^{4+}, the latter being precipitated. The reagent is employed in the gravimetric and titrimetric determination of cerium.

8-Hydroxyquinoline, 5,7-dichloro-8-hydroxyquinoline, and other 5,7-dihalogeno-8-hydroxyquinolines form with lanthanide ions sparingly soluble compounds, which are used for quantitative determinations. The compounds with 8-hydroxyquinoline are not extracted with organic solvents. The compounds with 5,7-dichloro-8-hydroxyquinoline are extracted with chloroform.

9-(o-Hydroxyphenyl)-2,3,7-trihydroxyfluorone (salicylfluorone) reacts [6] with lanthanide ions at pH $\sim$6.7. The resulting colloidal solutions of lanthanide compounds have a maximum absorption at 530 mμ. They are stabilized by the addition of gelatine. Salicylfluorone is used in the photometric determination of the sum total of lanthanides after their separation. Calcium and magnesium ions are masked with sulfosalicylic acid. Thorium, zirconium, uranium, ferric and ferrous iron, and tetravalent titanium give colored compounds.

Arsenazo I is a group reagent for lanthanide ions [7]. The aqueous solution of the reagent is pink. The reagent gives a red-violet coloration with lanthanide ions in neutral medium (pH $\sim$7.2, urotropin) and will detect these ions at a dilution of 1:3,000,000. Large amounts of ammonium ions, Li, K, Na, Ag, Ca, Sr, Ba, Mg, Zn, Cd, Hg(II), Sn(IV), Pb, V(V), Mo, Co, and Ni and chlorides do not interfere. The interfering ions are phosphates, pyrophosphates, fluorides, and tungstates which mask the lanthanide ions. REE can also be detected in the presence of Cu(II) and Al(III) ions after adding salicylates, and U(VI) ions after the addition of hydrogen peroxide. If Ti(IV), Zr(IV), Th(IV), and Fe(III) are not present in excessive amounts, they can be removed by coprecipitation with stannic acid formed by hydrolysis of $SnCl_4$. Arsenazo I is used in the photometric determination of total lanthanides [8, 9].

Carboxylic hydroxy acids (tartaric, citric, etc. acids) form anionic complexes with the ions of trivalent lanthanides. They are used during the separation of lanthanides by ion-exchange chromatography.

Aminopolycarboxylic acids (iminodiacetic, nitrilotriacetic, ethylenediaminetetra-acetic acids) react with trivalent lanthanides to form complex compounds of varying stabilities which are used in chromatographic and other separations. The stability of the complex compounds with ethylenediaminetetraacetic acid [10] increases with the decreasing ionic radius of the lanthanide (ionic strength 0.1, temperature 20°C):

	La	Ce	Pr	Nd	Sm	Eu	Gd	Tb	Dy	Ho	Er	Tm	Yb	Lu
p*K*	14.72	15.39	15.75	16.06	16.55	16.69	16.70	17.25	17.57	17.67	17.98	18.59	16.68	19.06

The mixture of REE is determined by complexometry (p. 75).

Lanthanide ions are precipitated from solutions containing pyrocatechol in excess on the addition of ethylenediamine, pyridine, or quinoline [11]. In the case of ethylenediamine (En) the composition of the precipitate corresponds to the formula $En_3[Ln(C_6H_4O_2)_3]_2$, where Ln is the lanthanide ion. The precipitates are insoluble in excess ammonia, unlike the corresponding precipitates of many heavy metals. La, Y, Gd, and Er may be determined in this way at concentrations of 100, 5, 10, and 100 μg/ml, respectively. The ions Cu(II), Be, Mg, Ba, Hg(II), Ti(IV), Zr(IV), Pb(II), V(V), As(III), Sb(III), Nb(V), Ta(V), Bi(III), Mo(VI), W(VI), U(VI), Se(IV), Te(IV), Mn(II), Re(VII), Ni, Pd, Pt(IV) and BO_2^- are not precipitated. The ions Al(III), Ga(III), In(III), Sn(IV), Th(IV), Zn, Cd, Cr(III), Fe(III), and Co(II) form ammonia-soluble precipitates with ethylenediamine and piperidine.

β-Diketones react with trivalent lanthanide ions as follows:

$$3\;R{-}C(=O){-}CH_2{-}C(=O){-}R' + Ln^{3+} \longrightarrow \left[\left(R{-}C(=O){-}CH{=}C({-}O){-}R'\right)_3 Ln\right] + 3H^+$$

Compounds in which $R = R' = CH_3$ (acetylacetonates), $R = R' = C_6H_5$ (dibenzoylmethanates), and $R = CF_3$, $R' = C_4H_3S$ (thenoyltrifluoroacetonates) have been most thoroughly studied. These compounds are usually obtained by the addition of the corresponding β-diketone to the acidified solution of the lanthanide salt, while carefully controlling the rise in pH (to prevent formation of basic compounds). The separated compounds are crystalline materials, which are readily soluble in organic solvents.

All the above compounds are group reagents and react with individual trivalent lanthanide ions in approximately the same manner. However, for tetravalent Ce and bivalent Eu relatively selective chemical analytical methods have been developed.

The compounds of cerium (III) are readily oxidized in alkaline medium to the yellow cerium(IV) compounds, unlike the compounds of lanthanum and other lanthanides. Thus, under certain conditions cerium (III) behaves as a reducing agent. For example, it reduces phosphomolybdic acid to molybdenum blue.

The Ce(IV) ions have strong oxidizing properties in acid solutions. The formal potential Ce(IV)/Ce(III) will depend on the nature of the acid:

Solution	9 *N* $HClO_4$	1 *N* $HClO_4$	1 *N* HNO_3	1 *M* H_2SO_4	1 *N* HCl
Experimental value of formal potential, Ce(IV)/Ce(III)	1.90	1.70	1.64	1.44	1.28

In acid solutions cerium (IV) ions oxidize the ions Fe^{2+}, $Fe(CN)_6^-$ and also H_2S, SO_2, NH_3, $(COOH)_2$, H_2O_2 and other ions. In the oxidation of benzidine, N-phenyl-anthranilic acid, etc. intensely colored products are formed. If the pH of the solution is increased, the oxidizing effect of Ce(IV) ions rapidly decreases and the reducing effect of Ce(III) ions increases correspondingly.

Cerium (IV) can be titrated against a solution of hydroquinone (p. 76).

Tetravalent cerium gives the same reactions as Th(IV), Zr(IV), Ti(IV), and U(IV), and its properties are quite different from those of trivalent lanthanides.

Ce(IV) ions are stable in strongly acid solutions; at lower acidities hydrolysis takes place with the separation of basic salts. Cerium(IV) forms the yellow cations $CeOH^{3+}$ and the orange-colored cations $Ce(OH)_2^{2+}$.

Ammoniacal solutions of silver nitrate oxidize cerium (III) salts in the heat to cerium (IV) hydroxide; the finely dispersed metallic silver which is formed in the process colors the bulky precipitate black. Ce(III) ions can be detected in this way in the presence of other lanthanide ions. The ions Mn(II), Fe(II), and Co(II) interfere.

The colorless salts of trivalent cerium are oxidized to the yellow tetravalent cerium salts when heated in sulfuric acid solution with ammonium persulfate in the presence of Ag^+ ions.

Sodium bismuthate oxidizes cerium (III) ions in acid solutions.

Cerium hydroxide $Ce(OH)_4$ has a light yellow color and precipitates out at pH > 0.8. On the addition of ammonia, KOH, or salts of weak acids with strong bases to the solution of cerium salts, the yellow amorphous $Ce(OH)_4$ precipitates out. On being ignited it is converted to CeO_2. The ignited CeO_2 is very sparingly soluble in HCl, but soluble in hot concentrated sulfuric acid.

To separate cerium from lanthanides in solution, the pH is adjusted to 1.5–6.0 and the cerium is oxidized to the tetravalent state when $Ce(OH)_4$ precipitates out.

The introduction of potassium hydroxide or ammonia into solutions of tetravalent and trivalent cerium salts in the presence of hydrogen peroxide yields the precipitate of a red-brown peroxide compound. The compound is stable in neutral and alkaline media.

Cerium (IV) phosphate is a very sparingly soluble compound.

Tetravalent cerium ions are quantitatively precipitated by potassium iodate from nitric acid solutions. In this way cerium can be separated from lanthanides. The composition of the precipitate varies with the conditions of precipitation.

Tetravalent cerium forms soluble complex compounds with tartrate, citrate, oxalate, chloride, sulfate, nitrate, and perchlorate ions.

The precipitate of cerium (IV) oxalate dissolves in excess oxalic acid. The resulting oxalate complexes are usually decomposed during the separation of the $Ce_2(C_2O_4)_3$ precipitate owing to the reduction of cerium to the trivalent state.

Cerimolybdic heteropolyacid is used in the photometric determination of cerium in solutions of pure salts (p. 80).

Dibenzoylmethane $C_6H_5COCH_2COC_6H_5$ (methanolic solution) precipitates the ruby red crystalline $Ce(C_{15}H_{11}O_2)_4$ from methanolic solutions of cerium (IV) salts if the pH of the solution has been adjusted [12] to 8 by the addition of methanolic ammonia solution. Under these conditions cerium (IV) is precipitated quantitatively. Cerium (III) is not precipitated but is oxidized by atmospheric oxygen to Ce(IV) when the precipitate of $Ce(C_{15}H_{11}O_2)_4$ is again formed. The remaining lanthanides which are trivalent are precipitated from the filtrate by methanolic solution of oxalic acid after the red crystals of $Ce(C_{15}H_{11}O_2)_4$ have been filtered off.

Dibenzoylmethane also precipitates thorium (IV) ions [12] at pH 4–5 as the compound $Th(C_{15}H_{11}O_2)_4$ and is used to separate thorium from cerium (III), lanthanum, and lanthanides.

Europium may be determined by redox titration.

To determine europium in mixtures of lanthanide compounds, the solution of the lanthanide chlorides is passed through a Jones reductor with metallic zinc; europium (III) is reduced to europium (II). The solution emerging from the reductor is collected in a vessel containing an excess of $FeCl_3$ solution. The ferrous ions which are formed in an equivalent amount are titrated against a solution of potassium dichromate after adding H_3PO_4. Samarium and ytterbium, as well as other lanthanides, are not reduced by metallic zinc and thus do not interfere.

All lanthanides, except europium and ytterbium, have half-wave reduction potentials which are so close that they cannot be identified in the presence of each other [13]. The polarographic technique is only suitable for the determination of europium and ytterbium and also of cerium after conversion to the tetravalent state.

Europium can be determined in the presence of samarium and dysprosium in the presence of holmium by the neutron activation method [14]. Neutrons from a radium–beryllium source are used for irradiation. The presence of other lanthanides (except lutetium) in amounts equal to or less than those of europium and dysprosium results in an error of less than 1%. The presence of equal amounts of lutetium results in an error of 5%.

For a review of analytical methods employed for lanthanides, see [15, 16].

Complexometric determination of total REE in phosphorus-containing materials [17, 18]

REE, calcium, and iron are separated from phosphorus, silicon, and aluminum by fusing the sample material with potassium–sodium carbonate. Calcium together with REE is isolated from the solution by ammonia, and the elements are then separated as oxalates. The oxalates are decomposed by heating with nitric acid and the total REE are titrated against Complexone III in the presence of Xylenol Orange. At pH 5.2–5.4, the color of the solution abruptly changes at the equivalence point from raspberry red to yellow (the characteristic color of free Xylenol Orange). The method gives satisfactory results if the total content of REE is more than 0.1%. Smaller amounts of REE are determined photometrically.

The error of the complexometric determination is 1–2%.

REAGENTS

Potassium–sodium carbonate, crystalline.
Hydrogen peroxide, 30% solution (perhydrol).
Sodium carbonate, 10% solution.
Oxalic acid, 1 and 10% solutions.
Nitric acid, d = 1.4 g/cm^3.
Ammonia, 1% and 2 N solutions.
Complexone III, 0.01 M solution. The solution is best standardized against the total of REE after their separation from the sample.
Ascorbic acid, 1% solution.
Ammonium acetate buffer solution, pH 5.2–5.4. Thirty-two ml of acetic acid (d = 1.05 g/cm^3) are diluted with water to 200 ml, 34 ml of 25% solution of ammonia are added. The solution is diluted to 500 ml with water.
Xylenol Orange, 0.5% solution.

PROCEDURE

Isolation of total REE

The sample material (0.5–1 g) is ignited in a porcelain crucible, mixed with a 15-fold amount of potassium–sodium carbonate and transferred to a platinum crucible with the aid of potassium–sodium carbonate. The crucible is placed in a muffle furnace at 600–700°C and the temperature of the furnace is gradually raised to 950°C. The sample is kept at this temperature for 40–50 minutes. The crucible is immediately immersed into a dish of water, cooled, and placed into a 500-ml beaker. The water from the dish is poured over the crucible and more water is added to a total volume of 200–250 ml.

The beaker is heated on a sand bath and boiled for 30–40 minutes, periodically adding a few drops of hydrogen peroxide to accelerate the leaching of the melt.

The hot solution is filtered through Blue Ribbon filter paper. The beaker and the crucible are rinsed with a hot solution of sodium carbonate and the precipitate on the filter is washed 4–5 times with this solution. The crucible is removed from the beaker and rinsed with a hot 10% oxalic acid solution. The precipitate is returned to the beaker which was used for the leaching with the aid of small portions of hot water and then with hot 10% oxalic acid. Another 100 ml of 10% oxalic acid are introduced into the beaker. The mixture is brought to the boil while stirring, boiled for 2–3 minutes and left to stand overnight.

The precipitate is separated on Blue Ribbon filter paper, washed with 1% oxalic acid solution. The filter paper with the precipitate is dried, transferred to a porcelain crucible, and the filter paper ashed. The precipitate is ignited for 30–40 minutes. The residue is wetted with water; 10–15 ml of nitric acid and 1 ml of hydrogen peroxide are added. The solution is transferred to a 200-ml beaker. It is evaporated on a sand bath to a residual volume of 5 ml, diluted with water to 70–80 ml, and heated to the appearance of water vapors. A 2 *N* solution of ammonia is added while stirring until a strong odor appears.

The solution with the precipitate is heated for 5–10 minutes, and cooled and the precipitate is separated on Blue Ribbon filter paper and washed with 1% solution of ammonia.

The precipitate is washed into a 100-ml beaker with the aid of 20–30 ml of 10% oxalic acid. The mixture is brought to the boil and then left on the water bath for 2 hours. The residue is filtered onto Blue Ribbon filter paper and washed with 1% solution of oxalic acid into the cone of the filter paper. The filter paper is punctured, the precipitated oxalates are rinsed back into the beaker in which the precipitation was performed with the aid of 10 ml of nitric acid and the filter paper is washed. The solution is evaporated on a sand bath in the presence of 1 ml of hydrogen peroxide to a residual volume of 1 ml and the residue is rinsed with hot water into a 250- to 300-ml conical flask.

Complexometric titration of total REE

To the solution, which should contain not less than 1 mg of total REE, are added 2–3 drops of ascorbic acid solution, 20 ml of ammonium acetate buffer solution, and 3–5 drops of Xylenol Orange solution. The solution is diluted with water to 100 ml and titrated against Complexone III, added from a microburet, until the raspberry-red color turns pure yellow.

Determination of cerium in cast iron by titration against hydroquinone [19]

This method is based on the titration of Ce(IV) against a solution of hydroquinone in the presence of ferroin as indicator. Ce(IV) may also be reduced by solutions of ferrous sulfate ($E_0 = +0.76$ V), ascorbic acid ($E_0 = +0.43$ V), stannous chloride ($E_0 = +0.2$ V). Hydroquinone ($E_0 = +0.68$ V) is more convenient, since it instantaneously and quantitatively reduces cerium (IV) at room temperature. Solutions of hydroquinone in 1–3% sulfuric acid are stable on storage.

Oxidizing agents such as iodine, bromine, chlorine, and also bromates, chromates, and vanadates interfere. In the determination in cast iron, cerium is separated as the fluoride CeF_3 at pH 2–5 ($L_{CeF_3} \approx 10^{-15}$). The iron is reduced by ascorbic acid prior to the precipitation of CeF_3. The precipitated fluorides are filtered off and washed; the filter paper is ashed and the residue is treated with sulfuric acid until complete removal of the fluorides. Cerium is oxidized to the tetravalent state by ammonium persulfate and titrated against a solution of hydroquinone.

Lanthanum, neodymium, and praseodymium do not interfere. Cerium in concentrations of 0.05–0.1% can be determined fairly rapidly (5–6 hours) and reliably.

REAGENTS

Hydrochloric acid, dilute (1:1).
Ascorbic acid, crystalline.
Ammonia, concentrated, $d = 0.91$ g/cm^3 and 25% solution.
Sodium fluoride, crystalline.
Sulfuric acid, $d = 1.84$ g/cm^3 and dilute (1:4).
Ammonium persulfate, 25% solution.
Hydroquinone, 0.005 N solution in 1% sulfuric acid.
Ferroin, 0.05 M solutions of o-phenanthroline and ferrous sulfate are mixed in equal volumes.

PROCEDURE

A 1–3 g sample of cast iron which should contain 2–3 mg of cerium, is placed in a 250-ml conical flask, 25–30 ml hydrochloric acid are added and the mixture is heated on a sand bath until the dissolution is complete while preventing the reaction from becoming too vigorous. The graphite is filtered off. The filtrate is collected in a 100-ml conical flask and the filter paper washed with water; the volume of the filtrate together with the wash waters should not exceed 50 ml. Ascorbic acid (0.5 g) is added to the cold solution. Then, a 25% solution of ammonia is added drop by drop to the appearance of the precipitate. Sodium fluoride (0.5 g) is then added, the flask is stoppered with a rubber plug and shaken mechanically for one hour.

The precipitated fluorides are separated on Blue Ribbon filter paper, and washed 4 or 5 times with hot water. The filter paper with the precipitate is placed in a platinum crucible; the filter paper is ashed and the precipitate is gently ignited at 300–400°C in a muffle furnace.

Fifteen ml of sulfuric acid (1:4) are added to the ignited precipitate; the mixture is evaporated to dryness and the residue is ignited until thick white fumes are no longer evolved. The precipitate is dissolved in water and acidified with sulfuric acid. It is transferred to a 250-ml conical flask, diluted with water to 150 ml, and acidified with 5 ml of sulfuric acid (d = 1.84 g/cm^3); 10–25 ml of ammonium persulfate solution are added and the solution is boiled for 5–7 minutes on asbestos gauze until the excess persulfate is fully decomposed. The solution is cooled, one drop of ferroin solution is added and the solution is titrated against the hydroquinone solution to the appearance of a pink color.

Photometric determination of total REE in monazites [8, 9, 20]

Rare-earth elements react with Arsenazo I to form violet-colored compounds with a maximum absorption at 540–560 mμ.

The properties of these compounds with the ions of certain REE are shown below.

	Molar extinction coefficient		Instability constant, pK
	580 mμ	600 mμ	
Cerium	24,000	14,000	$3.1 \cdot 10^{-10}$
Lanthanum	23,500	13,500	$4.1 \cdot 10^{-10}$
Neodymium	25,000	15,000	$3.3 \cdot 10^{-10}$
Praseodymium	24,500	14,500	$3.1 \cdot 10^{-10}$
Yttrium	22,500	13,500	$2.7 \cdot 10^{-10}$

Formation of the colored compound begins at pH 2.5–3.0, and the optical density increases with increasing pH up to 6.5; between pH 7.0 and pH 9.0 the optical density is practically constant. The molar ratio of the components is 1:1.

The elements Sc, U(VI), Cu, Al, Th, V(IV), Zr, Ga, In, Pd, and Fe(III) also form colored compounds with Arsenazo I under these conditions. The separation is performed by precipitating REE together with thorium as oxalates in acid medium. The precipitation may be effected by a hot oxalic acid solution or else by making the reagent in situ with acetonedioxalic acid at pH 0.5–2.

If thorium is present, its content is determined first photometrically by Arsenazo I at pH 1.6–1.8 (REE do not interfere at this acidity value) and then the total of thorium and REE is determined by the same reagent at pH 6–7. The REE content is found by difference.

The method is applicable to the determination of the total of REE in monazites.

REAGENTS

Sulfuric acid, dilute (1:2).
Hydrogen peroxide, 3% solution.
Ammonia, dilute, 1:4 solution.
Oxalic acid, 1% and 10% solutions.

Nitric acid, $d = 1.4$ g/cm^3 and 0.1 M solution.
Arsenazo I, 0.05% solution.
Boric buffer solution, pH 7.5. 10.53 g of boric acid and 2.84 g of sodium tetraborate are dissolved in water and diluted to one liter with water.
Standard solution of cerium (III) *chloride*, 20 μg of Ce/ml.
Hydrochloric acid, dilute (1:2).
Methyl orange, 1% solution.

CONSTRUCTION OF CALIBRATION CURVE

Into five 50-ml volumetric flasks are introduced 1, 2, 3, 4, and 5 ml of standard solution of cerium (III) chloride; 3 ml of Arsenazo I solution and 10 ml of boric buffer solution are added to each flask and the solutions are made up to mark with water. The optical density of the solutions is measured on a universal photometer or electrophotocolorimeter relative to a solution containing 3 ml of Arsenazo I solution and 10 ml of borate buffer solution in 50 ml. The calibration curve is plotted from the results.

PROCEDURE

A 0.1–0.3 g sample of monazite in a 100-ml beaker is heated with occasional stirring, on a water bath with 10–15 ml of sulfuric acid for $1\frac{1}{2}$–2 hours. The solution is evaporated to a residual volume of 2–3 ml; 10–15 ml of hydrogen peroxide are added, and the solution is mixed. Water (20–25 ml) is added and the solution filtered through a White Ribbon filter paper into a 150- to 200-ml beaker. Three drops of methyl orange are added to the filtrate and the filtrate is neutralized by ammonia solution, with stirring. Two ml of hydrochloric acid are then added (the excess acid increases the solubility of REE oxalates). The solution is heated almost to the boil; 15 ml of 10% oxalic acid solution are added, and the solution is kept at the boil for 20–30 minutes. The solution together with the precipitate is left overnight.

The precipitate is filtered through Blue Ribbon filter paper and washed with 1% oxalic acid solution. The solution is allowed to drain from the filter. The filter is then punctured and the oxalates washed with the aid of nitric acid ($d = 1.4$ g/cm^3) into a 100-ml beaker. The mixture is heated until the precipitate has dissolved; it is evaporated to a residual volume of 1–2 ml, diluted with water to 25 ml, and transferred to a 100-ml volumetric flask.

An aliquot of the solution containing 20–100 μg of REE is withdrawn with a pipet and placed in a 50-ml volumetric flask. Three ml of Arsenazo I solution and 10 ml of borate buffer solution are added and the solution is made up to mark with water. The optical density of the solution is measured at 540–580 mμ (D_1).

An equal aliquot volume is withdrawn with a pipet into another 50-ml volumetric flask; 3 ml of Arsenazo I and then 5 ml of 0.1 M nitric acid are added and the solution is diluted to mark with water. The optical density is measured as before (D_2).

The difference $D = D_1 - D_2$ corresponds to the light absorption by a solution containing compounds of Arsenazo I with REE.

The content of REE in the aliquot is found from the calibration curve.

Photometric determination of cerium (IV) in solutions of pure salts as cerimolybdic heteropolyacid [21]

The method is based on the determination of the optical density of solutions of cerimolybdic heteropolyacid (yellow form) $H_8[CeMo_{12}O_{40}] \cdot xH_2O$, which is formed when a solution of a cerium (IV) salt is added to a solution of ammonium molybdate. The formation of the compound is most complete in 0.2 *N* sulfuric acid in the presence of a 60-fold amount of ammonium molybdate. The color is stable for one hour. The maximum absorption is at 380 mμ. The measurement is conducted on an electrophotocolorimeter with a violet filter with maximum light transmission at 360 mμ.

The sensitivity of the method is 1 μg of Ce/ml; the colored solutions obey Beer's law between 1 and 16 μg of Ce/ml. Cerium may be determined in the presence of REE by this method.

REAGENTS

Standard solution of cerium (IV) *sulfate* in 0.2 *N* sulfuric acid, 50 μg of Ce/ml.
Sodium molybdate or *ammonium molybdate* 5% solution, with pH adjusted to 7 by the addition of sulfuric acid.
Sulfuric acid, 0.2 *N* solution.

CONSTRUCTION OF CALIBRATION CURVE

Into five 25-ml volumetric flasks are introduced 1, 2, 3, 4, and 5 ml of standard solution of cerium (IV) sulfate; 12 ml of sodium molybdate or ammonium molybdate solution are added to each flask. The solutions are mixed and left to stand for 10 minutes. They are diluted to 25 ml with sulfuric acid and their optical densities at 380 mμ are measured on an electrophotocolorimeter.

PROCEDURE

To the sample solution (0.025–0.4 mg of Ce) are added the reagents as above, the optical density is determined and the cerium content read off the calibration curve.

BIBLIOGRAPHY

1. Moeller, Th. and J.C. Brantley.—*Anal. Chem.*, **22**, 433. 1950.
2. Shcherbov, D.P., V.A. Mirkin, and V.V. Klimov.—*Trudy Kazakhskogo Nauchno-Issledovatel'skogo Instuta Mineral'nogo Syr'ya,* No. 3, 296. 1960.
3. Beck, G.—*Mikrochim. acta,* 1495. 1956.
4. Chernikhov, Yu.A., T.M. Malyutina, and B.M. Dobkina.—In: *Redkozemel'nye elementy,* p. 302, D.I. Ryabchikov, Editor. Izd. AN SSSR. 1963.

5. BERG, R. and E. BECKER.—*Z. anal. Chem.*, **119**, 1. 1940.
6. ZAIKOVSKII, F.V. and G.F. SADOVA.—*ZhAKh,* **16**, 29. 1961.
7. KUZNETSOV, V.I.—*ZhAKh,* **7**, 226. 1952.
8. KUTEINIKOV, A.F. and G.A. LANSKII.—*ZhAKh,* **14**, 686. 1959.
9. ZAIKOVSKII, F.V. and V.S. BASHMAKOVA.—*ZhAKh,* **14**, 50. 1959.
10. WHEELWRIGHT, E.J., F.H. SPEDDING, and G. SCHWARZENBACH.—*J. Am. Chem. Soc.,* **75**, 4196. 1953.
11. BECK, G.—*Mikrochim. acta,* 337. 1954.
12. WOLF, L. and D. STATHER.—*J. prakt. Chem.*, [4], **2**, 329. 1955.
13. KRYUKOVA, T.A., S.I. SINYAKOVA, and T.V. AREF'EVA. *Polyarograficheskii analiz (Polarographic Analysis),* p. 273.—Goskhimizdat. 1959.
14. MEINKE, W.W. and R.E. ANDERSON.—*Anal. Chem.,* **26**, 907. 1954.
15. RYABCHIKOV, D.I. and V.A. RYABUKHIN.—In: *Metody opredeleniya i analiza redkikh elementov,* p. 128. Izd. AN SSSR. 1961.
16. VICKERY, R.C. *Analytical Chemistry of the Rare Earths,* p. 139.—Pergamon Press. 1961.
17. KINUNNEN, J. and B. WENNERSTRAND.—*Chemist-Analyst,* **46**, 92. 1957.
18. LUK'YANOV, V.F. and A.A. MOZZHORINA.—In: *Metody opredeleniya i analiza redkikh elementov,* p. 158. Izd. AN SSSR. 1961.
19. AMSHEEVA, A.A. and D.V. BEZUGLYI.—*ZhAKh,* **16**, 683. 1961.
20. KUTEINIKOV, A.F.—*Zav. Lab.*, **28**, 1179. 1962.
21. SHAKHOVA, Z.F. and S.A. GAVRILOVA.—*ZhAKh,* **13**, 211. 1958.

Thorium Th

Thorium has a valency of $+4$. Thorium compounds of lower valencies are unstable in aqueous solutions. The ions of Th(IV) are colorless. The standard electrode potential in aqueous medium at 25°C relative to standard potential of hydrogen electrode is -1.899 V for the reaction

$$Th \rightleftharpoons Th^{4+} + 4e.$$

The reactions of Th^{4+} greatly resemble those of Zr^{4+} and Ti^{4+}.

The white thorium hydroxide begins to precipitate out of thorium salt solutions at pH 3(0.01 M); it separates out quantitatively around pH 6; the solubility product of thorium hydroxide is about 10^{-50}.

Potassium fluoride and hydrofluoric acid precipitate a bulky, gelatinous precipitate of $ThF_4 \cdot xH_2O$ which is sparingly soluble in water in excess H_2F_2 (unlike the fluorides of Zr, Ti, Be, Al, Ta, Nb, but not Ce(IV) and U(IV) which also form sparingly soluble fluorides) and in dilute nitric acid. The solubility product of thorium fluoride is about 10^{-27} (ionic strength 0.5).

The addition of ammonium carbonate to solutions of thorium salts results in the separation of a white precipitate which is soluble in excess precipitant owing to the formation of $(NH_4)_6[Th(CO_3)_5]$. The solution of this complex forms a peroxy compound of thorium on the addition of hydrogen peroxide. The thallium salt of the carbonate complex of thorium separates out as tiny prisms, suitable for microscopic crystal identification [1]. The phosphate $Th_3(PO_4)_4$ separates out at pH 2.7. Sodium hypophosphate $Na_2H_2P_2O_6$ precipitates the amorphous white $ThP_2O_6 \cdot \cdot xH_2O$ out of strongly acid solutions (HCl); the precipitate is practically insoluble in alkali hydroxides and in concentrated hydrochloric acid [2]. A similar reaction is given by Zr(IV) and Ti(IV), but lanthanides do not interfere with the separation of thorium.

Thorium pyrophosphate ThP_2O_7 and thorium ferrocyanide $Th[Fe(CN)_6]$ are sparingly soluble in dilute acids.

Thorium can be separated from beryllium [3] by precipitation with sodium selenite Na_2SeO_3. The precipitate of thorium selenite is ignited to thorium dioxide, ThO_2.

Potassium iodate reacts with thorium ions to form a sparingly soluble white compound [4] even in strong nitric acid media and it is used in the separation of thorium and its titrimetric determination (pp. 88, 89).

From solutions of thorium salts, oxalic acid precipitates the white crystalline $Th(C_2O_4)_2 \cdot 6H_2O$, which is practically insoluble in excess precipitant. The precipitate is also formed when ammonium oxalate is added to weak hydrochloric acid or weak nitric acid solutions. The precipitate dissolves when boiled with ammonia in the presence of excess oxalate ions, and it separates out again if the solution is subsequently acidified with hydrochloric acid. In the presence of excess ammonium or alkali oxalate the complex ions $[Th(C_2O_4)_4]^{4-}$ are formed; they decompose on acidification, since the concentration of oxalate ions $C_2O_4^{2-}$ decreases owing to the formation of the undissociated $H_2C_2O_4$. The precipitation of thorium oxalate from dilute nitric acid solutions is utilized for the separation of thorium and also cerium and other lanthanides. For example, it is utilized in the analysis of minerals (p. 86).

Other dicarboxylic acids (succinic, adipic, phthalic acids) also precipitate Th(IV) ions from weakly acid solutions.

Benzoic, o-chlorobenzoic, m-nitrobenzoic, and o- and p-aminobenzoic acids precipitate Th(IV) and Zr(IV) ions. These acids make it possible to separate thorium from lanthanides in the analysis of monazite sand and also to separate thorium from uranium.

Thorium forms complex compounds with anions of tartaric, citric, and other organic hydroxy acids.

Complexone III reacts with thorium ions in the molar ratio of 1:1 and it is used in the titrimetric determination of thorium (p. 91).

N-Benzoylphenylhydroxylamine quantitatively precipitates thorium ions at pH 2 as a white crystalline precipitate. The precipitate is stable on being heated to 220°C. The reagent is used in the gravimetric determination of thorium and its separation from lanthanides (pH 4.5) and uranium (in the presence of ammonium carbonate at pH 7–8.5). The compound of thorium with N-benzoylphenylhydroxylamine is readily extracted with benzene, chloroform, isoamyl alcohol, and other organic solvents. The separation of thorium from lanthanides is based on its extraction with isoamyl alcohol at pH 4.5. This reagent will also separate thorium from cerium, which is precipitated and extracted only at pH 6.5–7.5.

Azo compounds with the grouping

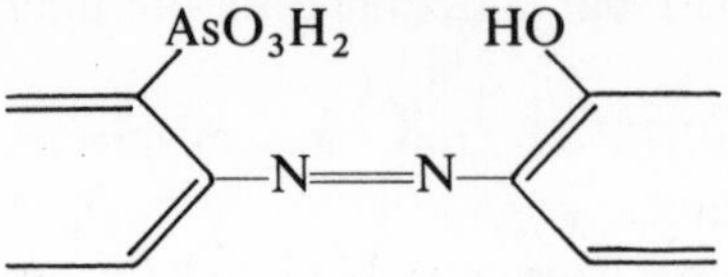

$$AsO_3H_2 \quad HO$$

$$-N{=}N-$$

form colored compounds with thorium ions [5].

Thoron, o-(2-hydroxy-3,6-disulfo-1-naphthylazo)-benzenearsonic acid, is often used in analytical practice. This reagent is orange-colored when dissolved in water; acidified solutions of the reagent become light yellow. When a thorium salt is added to the yellow solution of the reagent in 0.5 *M* HCl (pH 0.70), a red coloration appears; in more concentrated solutions of thorium salts a red-colored precipitate separates out. The absorption maximum of the solution of the reagent in 0.5 *M* HCl is at 480 mμ, while the absorption maximum of the compound of thorium with the reagent in 0.5 *M* HCl is at 510 mμ. Thorium ions interact with the reagent in the molar ratio of 1:2. The optimum pH value is 1.7.

The reaction proceeds according to the equation:

$$Th^{4+} + 2H_3R^{2-} \rightleftharpoons Th(H_3R)_2,$$

where R is the thoron anion. The complex formation constant is $(7.9 \pm 0.9) \cdot 10^9$.

Thoron is used in the detection of thorium [7] and its photometric determination in dilute hydrochloric acid medium. The interfering ions are U(IV), Fe(III), Ti(IV), Zr(IV), Sn(IV), large amounts of SO_4^{2-}, F^-, $C_2O_4^{2-}$, PO_4^{3-} ions and organic hydroxy compounds. REE, Al(III) and U(VI) do not interfere.

Microgram amounts of thorium can be separated from gram amounts of REE, aluminum, and iron [8] on AV-17 anion exchanger impregnated with a solution of thoron. The method is suitable for the determination of more than $1 \cdot 10^{-4}\%$ of thorium in oxides and salts of REE. Other anion exchangers (EDE-10, EDE-10P, AN-2F, AV-16) are unsuitable.

Arsenazo III interacts with thorium ions even in 4–10 *N* hydrochloric acid unlike the ions of most other elements [9, 10]. The only ions interfering with the photometric determination of thorium (p. 96) are Zr(IV), Ti(IV), and U(IV). The optical density of the solutions is determined at 665 mμ.

Thorium can also be photometrically determined by Arsenazo I [11, 12, 13]. In weakly acid medium Arsenazo I forms a blue-violet complex compound with thorium ions in the molar ratio 1:1; in the absence of thorium the solutions of the reagent are pink. The absorption maximum is at 580 mμ (Figure 2). The optimum pH range is 1.3–3.0 (Figure 3). The compound is formed immediately after the component solutions are mixed and it is stable for one month in diffused light. The instability constant is $1.6 \cdot 10^{-16}$.

The sensitivity of the reagent is 0.7 μg of Th/ml. The maximum color development is noted at Arsenazo I contents of 2–3 mg/50 ml.

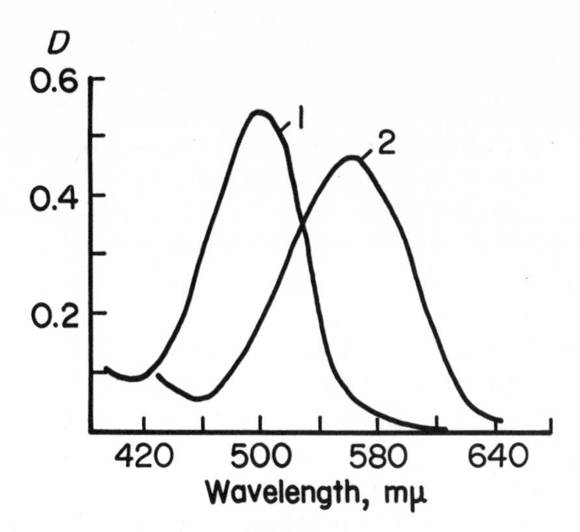

Figure 2

Curve 1: light absorption of a $2 \cdot 10^{-5}$ *M* Arsenazo I solution (at pH 1.9). Curve 2: light absorption of a thorium–Arsenazo I complex (reagent concentration, $2 \cdot 10^{-5}$ *M*; Th concentration, $4.7 \cdot 10^{-4}$ *M*).

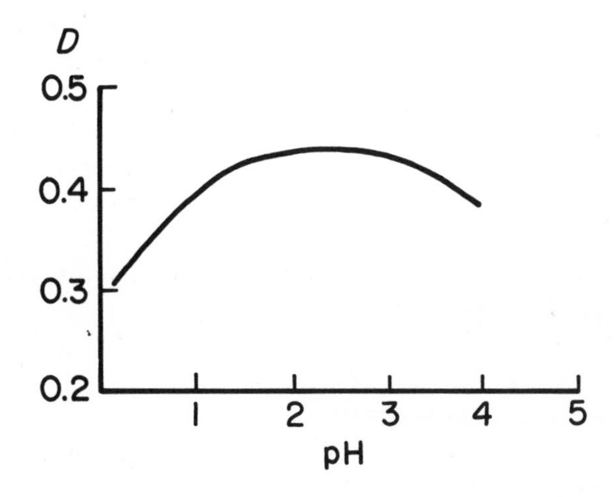

Figure 3

Absorption of light by the complex of thorium with Arsenazo I as a function of pH

The ions U(IV), Pu(IV), Ti(IV), Fe(III), Zr(IV), and Al react with Arsenazo I similarly to thorium. REE ions do not interfere with the determination at pH 1.9 even if present in 1300-fold amount. Traces of thorium in the form of oxalate can thus be separated from the interfering ions, using REE oxalates as coprecipitants.

Thorium can be determined in the presence of zirconium and titanium if these elements are masked by the addition of tartaric acid and ascorbic acid, respectively; 50 mg of tartaric acid will mask a 35-fold amount of zirconium and 50 mg of ascorbic acid in 50 ml of solution will determine thorium in the presence of the tenfold amount of titanium.

Ce(IV) ions in more than 100-fold amounts interfere by oxidizing the ascorbic acid. Fluorides, sulfates, phosphates, and other anions which form precipitates or complex compounds with thorium ions also interfere.

1-(2-Pyridylazo)-resorcinol forms a colored compound with thorium ions and is used in the photometric determination of thorium (p. 97).

Alizarin in ethanol solution colors thorium-containing solutions violet and is used in the detection of thorium [14].

8-Hydroxyquinoline [15, 16] quantitatively precipitates Th(IV) even from dilute acetic acid solutions pH 3.7–4.0; in this way thorium(IV) can be separated from cerium(III) which begins to precipitate at pH 4.4. At room temperature the yellow precipitate of $Th(C_9H_6ON)_4$ is formed, and above 70°C the orange-colored $Th(C_9H_6ON)_4 \cdot C_9H_7ON$ precipitates.

The addition of morin to 1 *N* hydrochloric acid containing thorium ions results in a yellow-green fluorescence which disappears when the concentration of the acid is increased. On the other hand, the fluorescence of zirconium is still noticeable even in 10 *N* HCl.

Quercetin reacts with thorium ions in ethanol–water medium to form a yellow compound. It is used in the photometric determination of thorium.

Xylenol Orange [17] reacts with Th(IV) ions at pH 2.7–3.3 to form a compound (molar ratio 1:1) with a maximum absorption at 570 mμ. The stability constant is $6.7 \cdot 10^{59}$ (ionic strength 10^{-3}). The reagent is used [18] as complexometric indicator during the titration of thorium(IV) ions and in the photometric determination of thorium [17].

For a survey of the analytical methods used for the detection of thorium and its determination, see [19, 20].

Isolation of thorium from monazites [21, 22]

Monazite is decomposed by fusion with sodium peroxide, and thorium and total REE are precipitated as oxalates. The oxalates are decomposed by heating with nitric acid and thorium is then determined photometrically by Arsenazo I, quercetin, or 1-(2-pyridylazo)-resorcinol.

REAGENTS

Sodium peroxide, crystalline.
Hydrochloric acid, 2 *N* solution.
Oxalic acid, 1% and 10% solutions.
Nitric acid, dilute (1:1).

PROCEDURE

A 50–100 mg monazite sample, which should contain about 250–500 μg of thorium, is placed in a porcelain crucible over a layer of about 0.5 g of sodium peroxide. it is covered by another layer of 0.5 g of sodium peroxide. The crucible is placed inside a larger crucible and heated over a burner flame, the temperature being increased regularly until a uniform melt is obtained.

The melt is kept liquid for about ten minutes; it is cooled and the crucible is placed inside a 100-ml beaker. Water (20–25 ml), acidulated with a few drops of hydrochloric acid, is added to the melt. After the melt has been leached, the crucible is taken out of the beaker and rinsed, first with water and then with hydrochloric acid, and the wash waters are combined with the solution. The solution is acidified with hydrochloric acid to pH 1; another 5 ml of hydrochloric acid are added and the solution is filtered into another 250-ml beaker using Blue Ribbon filter paper. The filtrate is heated to the appearance of water vapors; 10 ml of hot 10% oxalic acid solution are added; the solution is heated for another 1–2 minutes and left to stand overnight.

The precipitate of thorium and REE oxalates is separated on Blue Ribbon filter paper and washed with 1% oxalic acid solution. The filter paper is punctured and the precipitate is washed into a 100- to 150-ml beaker with the aid of nitric acid. The contents of the beaker are heated until the oxalates are fully dissolved. The solution is evaporated to a residual volume of 5 ml. It is then transferred to a 50-ml volumetric flask, then made up to mark with water and thoroughly mixed.

An aliquot is used to determine thorium photometrically (p. 92ff.).

Note. If less than 0.1% of thorium is present in monazites, the sample is solubilized by treatment with sulfuric acid; if more than 0.5% of thorium is present, the sample volume is increased or the aliquot decreased for the photometric determination.

Chromatographic separation of thorium from accompanying elements [23, 24]

In this method the ions of thorium and of the accompanying elements are sorbed on the strongly basic cation exchanger KU-2 in the H^+-form. Titanium is eluted with 1 *N* HCl, REE with 2 *N* HCl, zirconium with 0.5% oxalic acid solution, and thorium with saturated solution of ammonium oxalate. Thorium is determined photometrically or complexometrically, depending on its amount in the filtrate.

The method is applicable for the determination of thorium 10^{-3}–10^{-4}% in any silicate rock containing large amounts of zirconium, titanium and rare-earth elements.

REAGENTS

Sodium peroxide, crystalline.
Hydrochloric acid, dilute (1:1), 1 *N* and 2 *N* solutions.
Ammonia, 1 *N* and 25% solutions.
Oxalic acid, O.1 *N* solution.
Ammonium oxalate, saturated solution.
Nitric acid, $d = 1.4$ g/cm^3.
KU-2 *cation exchanger* in the H^+-form, particle size 0.15–0.3 mm.

PROCEDURE

A 0.1–0.2 g sample of the finely ground material is fused in a nickel crucible with 4–5 g of sodium peroxide until a homogeneous melt is obtained. The melt is cooled, and the crucible is placed in a 250–300-ml beaker and covered with 100 ml water. The beaker is covered with a watch glass and heated for 15–20 minutes on a sand bath. The crucible is taken out with a glass rod, rinsed with water and the wash waters are combined with the main solution.

The hydroxide precipitate is separated on White Ribbon filter paper. It is dissolved in 30–40 ml of hydrochloric acid (1:1) and the solution is diluted to 80–100 ml with water. The solution is stirred and 25% ammonia added until a strong odor appears. The precipitate is separated on White Ribbon filter paper, washed 5–6 times with 1 *N* solution of ammonia and dissolved on the filter in 20–25 ml of hot 1 *N* hydrochloric acid solution.

The solution is cooled and 5–6 drops per minute are passed through a 0.5 × 30 cm chromatographic column filled with KU-2 cation exchanger to a height of 24–25 cm.

The resin is washed with 20 ml of 2 *N* hydrochloric acid and 7 ml of oxalic acid solution. The eluate is discarded.

Twenty ml of ammonium oxalate solution are then passed through the column; the eluate is collected in a 100-ml beaker; it is evaporated to dryness on a sand bath, 5 ml of nitric acid are added and it is evaporated to a volume of 1–2 ml. Thorium is determined in the residue by any suitable photometric method.

Note. If thorium is determined photometrically by Arsenazo III, thorium oxalate need not be evaporated with nitric acid, as oxalate ions do not interfere.

Determination of thorium by the iodate method with iodometric titration [4, 25, 26]

Thorium ions are precipitated by potassium iodate; the precipitate is dissolved in sulfuric acid in the presence of added potassium iodide, and the liberated iodine is titrated against thiosulfate in the presence of starch.

In order to obtain a precipitate of the formula $4Th(IO)_3 \cdot KIO_3 \cdot 18H_2O$, the experimental procedure must be rigidly adhered to.

Zr, Ti, U(IV), and Ce(IV) interfere. Titanium and zirconium ions can be masked by the addition of oxalic acid; U(IV) ions are oxidized to uranyl ions; Ce(IV) is reduced to Ce(III) by hydrogen peroxide.

Between 4 and 16 mg of thorium in the presence of Ce(III) and REE can be determined with an error of not more than 3–4% by this method.

REAGENTS

Nitric acid, $d = 1.4\ g/cm^3$.
Potassium iodate, 1% solution in 1:9 nitric acid and 15% solution in 1:1 nitric acid.
Methanol.
Sulfuric acid, dilute (1:1).
Potassium iodide, 5% solution.
Starch, 1% solution.
Sodium thiosulfate, 0.05 *N* solution.

PROCEDURE

To 10 ml of solution containing 4–16 mg of thorium are added 5 ml of nitric acid and 10 ml of 15% potassium iodate solution. The mixture is left for ½ to 1 hour, with occasional stirring, until the precipitate is fully coagulated.

The precipitate is filtered on Blue Ribbon filter paper, the walls of the beaker being washed with 1% potassium iodate solution, and it is washed with methanol until the reaction for iodates (test with potassium iodide in sulfuric acid) is negative. The filter paper with the precipitate is returned to the beaker which was used for the precipitation; 10 ml of sulfuric acid are added and after the precipitate has fully dissolved the solution is filtered into a 100-ml volumetric flask, diluted to mark with water and thoroughly mixed.

A 20 ml aliquot is transferred with a pipet to a 300-ml conical flask; 20 ml of potassium iodide solution is added and the mixture is left to stand in the dark for 5 minutes. Water (100–150 ml) is added and the solution is titrated against sodium thiosulfate until it becomes straw-yellow; 1–2 ml of starch solution is then added and the titration continued until the last drop produces complete decoloration of the solution. It is carried out several times until three agreeing results are obtained.

One ml of 0.0500 *N* solution of thiosulfate is equivalent to 0.11 mg of thorium.

Iodate–complexometric determination of thorium [27]

Thorium in the form of iodate is separated from the accompanying elements. The precipitate is dissolved in a known amount of a standard solution of Complexone III and the excess of the latter is titrated against a standard solution of copper sulfate in the presence of 1-(2-pyridylazo)-2-naphthol (PAN). At the end point the greenish solution becomes red-violet on the addition of 1 drop of copper sulfate in excess; the color change is very sharp.

The precipitate of thorium iodate may have any composition, and there is no need to remove the excess precipitant as in the method described above. Thorium in oxalic acid solution may be separated from Ti, Zr, Ce(III), and U(IV).

The percentage error in the determination of 4–20 mg of thorium does not exceed 3–4%.

REAGENTS

Nitric acid, $d = 1.4\ g/cm^3$.
Potassium iodate, 0.5% solution in 1 *M* nitric acid and 15% solution in 1:1 nitric acid.
Complexone III, 0.01 *M* solution.
Copper sulfate, 0.01 *M* solution.
1-(2-*Pyridylazo*)-2-*naphthol*, 0.1% solution in methanol.
Ammonia, 1 *M* solution.

PROCEDURE

To 10 ml of sample solution, which should contain 4–20 mg of thorium, are added 6–7 ml of nitric acid. The solution is diluted with water to 25 ml and 5 ml of 15% potassium iodate solution are added. The mixture is heated on a sand bath until the precipitate is fully coagulated. The precipitate is separated on Blue Ribbon filter paper. It is washed 3–4 times with 0.5% potassium iodate solution and placed together with the filter paper in the beaker where the precipitation was performed. Fifteen ml of Complexone III solution are added to the beaker. The solution is neutralized to pH 3–4 (Universal Indicator Paper) with the ammonia solution and is heated on the sand bath until the precipitate is fully dissolved (10–15 minutes at 80–90°C). The paper fibers are filtered off, washed several times with water and the filtrate together with the wash waters are collected into a 100-ml conical flask; 3–4 drops of 1-(2-pyridylazo)-2-naphthol solution are added and the excess Complexone III is titrated against copper sulfate solution until the yellow-green solution turns red-violet.

One ml of the 0.0100 *M* Complexone III solution is equivalent to 2.32 mg of thorium.

Note. The pH of the solution should be kept at 3 or above during the titration or the solution will turn red-violet on the addition of the first drop of copper sulfate. This is because at pH < 3 the copper compound of 1-(2-pyridylazo)-2-naphthol is more stable than copper complexonate.

Complexometric titration of thorium in the presence of 1-(2-pyridylazo)-2-naphthol [28, 29]

Complexone III forms with thorium a compound stable in acid medium with a molar ratio of 1:1 (instability constant $10^{-23.2}$). The complexometric titration may be effected even at pH 1.8, when many other ions do not yet form stable complexomates.

A number of different indicators have been proposed to determine the end point of the titration. One of the most selective is 1-(2-pyridylazo)-2-naphthol, in the presence of which the titration is performed at pH 2.0–3.5. The compound of thorium with 1-(2-pyridylazo)-2-naphthol is orange-red. On the addition of Complexone III, the orange-red solution turns yellow-orange and then yellow-green, when about one-half of the thorium present has reacted. On further titration the color change is slow. However, at the end point the color change to greenish yellow is sharp and takes place on the addition of a single drop of Complexone III solution.

Large amounts of alkali and alkaline-earth metals, U, La, Ce(III), and traces of Ce(IV), Sn(II), and Pb do not interfere with the determination of thorium. Sulfates may be present in large amounts. Nickel, bismuth, indium, and vanadium which form colored compounds with 1-(2-pyridylazo)-2-naphthol, stable in acid solution, interfere. Fluorides, phosphates, oxalates, and other anions which form precipitates or complexes with thorium should also be absent. The tinge of the solution is different in the presence of major amounts (2.5 mg/ml) of Zn, Cd, Mn, and Cr(III) and more than 0.6 mg/ml of Cr(VI), but the color change at the end point is still sharp. In the presence of mercury and tin the solution becomes turbid towards the end, and this interferes with the determination of the end point. Titanium and zirconium ions are hydrolyzed while adsorbing the indicator, and also interfere with the determination.

In the presence of aluminum ions the titration is performed at pH 2.5; ferric ions are previously reduced by ascorbic acid.

The method is suitable for the determination of thorium in monazite sands and thorium-containing materials (after removal of phosphates).

If thorium is determined in solutions of pure salts, the mixed indicator: copper complexonate, 1-(2-pyridylazo)-2-naphthol, can be employed and the titration effected at pH 3–4. Under these conditions a more stable thorium complexonate is formed, whereas copper reacts with 1-(2-pyridylazo)-2-naphthol to form a red-violet compound which is decomposed by Complexone III. When Complexone III is added to the solution, thorium complexonate is formed first and copper complexonate next. There is a sharp color change from red-violet to yellow-green on the addition of one drop of Complexone III in excess. The color change will be even sharper if the solution is heated to a temperature of 70–80°C prior to titration.

REAGENTS

Complexone III, 0.01 *M* solution.
Ammonia, 1 *N* solution.
Hydrochloric acid, 1 *N* solution.
Ammonium acetate, 1 *N* solution.
1-(2-*Pyridylazo*)-2-*naphthol*, 0.1% solution in methanol or ethanol.
Copper complexonate, 0.01 *M* solution.

PROCEDURE

One ml of 1 *N* hydrochloric acid is added to the sample solution, which should contain 50–100 mg of thorium, in a 50-ml volumetric flask, and the solution is diluted to mark with water. To determine thorium, 15 ml of solution are withdrawn with a pipet into a 100- to 150-ml conical flask; 3–5 drops of 1-(2-pyridylazo)-2-naphthol solution added and then a few drops of ammonia, until the solution turns orange-red. The solution is then titrated against a solution of Complexone III until the color changes from orange-red to greenish yellow. One ml of 0.0100 *M* Complexone III solution is equivalent to 2.32 mg of thorium.

In the presence of foreign ions 3–5 drops of 1-(2-pyridylazo)-2-naphthol solution are added to the thorium-containing solution; if a pink or red coloration appears, 1–2 drops of hydrochloric acid are added until the solution becomes orange-colored and the solution is then titrated with Complexone III.

If thorium ions are titrated in the presence of a mixed indicator, 3–5 drops of copper complexonate solution and 3–5 drops of 1-(2-pyridylazo)-2-naphthol solution are added to the aliquot sample solution containing 15–30 mg of thorium. The solution is diluted with water to 25–30 ml; 1–2 ml of ammonium acetate solution are added (to pH 3–4, Universal Indicator Paper). The solution heated to 70–80°C (appearance of water vapor) and is then titrated with Complexone III solution until the last drop of the titrant turns the red-violet solution yellow-green.

Photometric determination of thorium by quercetin [30, 31]

Quercetin reacts with thorium ions in water–alcohol medium to form a yellow-colored compound which is used in the photometric determination.

Quercetin has two absorption maxima at 255 and 365 mμ; solutions of its thorium compound have one maximum at 440 mμ, but the determination of optical density is best effected at 455 mμ, when the absorption by the reagent is insignificant.

The colored solutions of the quercetin compound of thorium are stable for 20–30 hours at alcohol concentrations of 40% by volume; the stability increases with increasing concentration of alcohol. Beer's law is obeyed between 1 and 10 μg of Th/ml.

The optimum pH value of the solution is 3.4–6.5, but it is preferable to work at pH 3.4–4.0 (ammonium acetate buffer solution) in order to prevent hydrolysis of thorium ions.

REE ions do not interfere with the determination of thorium. The method may be employed in the analysis of monazite sand (after separation of thorium and REE as oxalates and their decomposition).

REAGENTS

Standard solution of thorium nitrate, 25 μg of Th/ml.
Quercetin, 0.05% solution in methanol, ethanol or acetone.
Ammonium acetate buffer solution, pH 3.4–4.0. Glacial acetic acid (102 ml) is mixed with 22 ml of 25% ammonia solution and the mixture diluted to two liters with water.
Methanol, ethanol or acetone.

CONSTRUCTION OF CALIBRATION CURVE

Fifteen ml of alcohol or acetone, 2 ml of quercetin solution and 1, 2, 3, 4, and 5 ml of standard thorium nitrate solution are introduced into each of five 25-ml volumetric flasks. To each flask 5 ml of buffer solution are slowly added with constant stirring; if the buffer solution is added rapidly, thorium may become hydrolyzed and fail to form a colored compound. After 20 minutes, the solutions are diluted to the mark with water and their optical densities relative to water are measured on a universal photometer with a suitable filter.

The content of thorium in the control solution is read off the calibration curve.

PROCEDURE

All the reagents as above are added to the sample solution and the thorium content is found from the calibration curve.

Photometric determination of thorium in monazites by Arsenazo II [32, 33]

Arsenazo II reacts with thorium ions in acid medium to form a blue-violet compound with a molar ratio of 1:1.

The absorption maximum by the solution of the compound is at 560 mμ. The color develops almost immediately and is stable for a long time. The absorption of light by the solutions of the complex shows only insignificant variations in the acidity range between 0.01 and 0.6 *N* HCl. Beer's law is obeyed from 10 μg of Th/25 ml and above. Sulfates mask 5–7% of the thorium when its content is 1–5 mg/ml, the acidity 0.2 *N* HCl and the reagent concentration 10^{-4} *M*. However, if the same amount of sulfate is introduced into all solutions, the experimental error can be practically eliminated. The effect of phosphates is minimized if thorium is determined in 0.1–0.4 *N* HCl. Fluoride ions, which mask thorium, are sufficiently removed when monazite is decomposed by fusion with pyrosulfate.

Zirconium and titanium interfere, but this may be neglected since their content in monazites does not exceed 0.02%. Cations which become hydrolyzed at pH > 2 (alkali metals, alkaline-earth metals, REE, and Al, Cu, and Fe(II) even when present in 1000-fold amounts) do not interfere.

The method will determine 0.5–7% of thorium in monazite in 25–30 minutes; the error is 3–5%. If thorium is present in smaller amounts, it is separated from the other elements by double precipitation as fluoride or fluoride–oxalate.

It is best to use Arsenazo II in the analysis of monazites containing 20–30% total REE.

REAGENTS

Standard thorium nitrate solution, 20 μg of Th/ml.
Arsenazo II, 0.1 aqueous solution.
Potassium pyrosulfate, crystalline.
Potassium pyrosulfate, 2% solution in 1:5 hydrochloric acid.
Monosubstituted sodium phosphate, $NaH_2PO_4 \cdot 2H_2O$, 0.5% solution.
Hydrochloric acid, dilute (1:1).
Ascorbic acid, crystalline.

CONSTRUCTION OF CALIBRATION CURVE

Into five 25-ml volumetric flasks are introduced 1, 2, 3, 4, and 5 ml portions of standard thorium nitrate solution and 2.5 ml of potassium pyrosulfate solution, 2.0 ml of sodium phosphate solution and 5 ml of Arsenazo II solution are added to each flask. The mixtures are made up to mark with water and mixed The optical density is measured after five minutes and the calibration curve is plotted.

PROCEDURE

A 10–200 mg monazite sample containing 0.5% or more of thorium is fused for a few minutes, first over a gentle burner flame and then at dull-red heat with 2 g of potassium pyrosulfate in a quartz tube or porcelain crucible. The cooled melt is dissolved in 40 ml of hydrochloric acid (the solution should be practically clear and should not contain heavy particles, whose presence indicates that monazite was not fully decomposed). The solution is transferred to a 100-ml volumetric flask, diluted to mark with water and some of it is filtered into a dry flask.

A 2.5 ml aliquot (25–100 μg of Th) is withdrawn with a pipet into a 25-ml volumetric flask; 10–20 mg of ascorbic acid is added (if the monazite contains iron), and then 2.0 ml of NaH_2PO_4 solution, and 5.0 ml of Arsenazo II solution. The solution is made up to the mark with water and mixed. The optical density is determined after five minutes.

Twenty-five ml of comparison solution should contain 5.0 ml of Arsenazo II solution and 2.5 ml of potassium pyrosulfate solution.

Note. If the content of thorium in monazite is more than 10–15%, a correspondingly smaller aliquot volume is taken and the difference between 2.5 ml and the aliquot volume is made up by adding potassium pyrosulfate.

Photometric determination of thorium by Arsenazo III [10, 34–37]

Arsenazo III reacts with thorium ions to form a chelate compound which is more stable than the compounds of thorium with Arsenazo I and Arsenazo II. In this way thorium may be determined in strongly acid solutions without preliminary separation of sulfate, phosphate, fluoride, oxalate, and other anions. The reagent is very sensitive. Some of the properties of the compounds of thorium with Arsenazo III are shown in Table 3.

Table 3
PROPERTIES OF COMPOUNDS OF THORIUM WITH ARSENAZO III

Thorium: Arsenazo III ratio in compound	pH 3.0–1.5		Concentration 3.5 M in HCl or HNO_3	
	absorption maximum, mμ	molar extinction coefficient	absorption maximum, mμ	molar extinction coefficient
1:1	620	32,300	612	41,500
	670	24,500	665	65,300
1:2	565	47,000	Not determined	
	675	40,000		
1:3	Complex formation not noted		615	94,000
			665	127,000
1:4	Complex formation not noted		570	64,100

The optimum conditions for the determination of thorium are: concentration of hydrochloric or nitric acid 3.5–6 M, molar ratio Arsenazo III:thorium = 7.5:1. In nitric acid solutions 1 g of urea is introduced for each mole of acid to prevent the oxidation of the reagent. Under these conditions the molar ratio of the complex is 1:3, its maximum absorption is at 665 mμ and its molar extinction coefficient is 127,000.

The reagent will determine thorium in zircon, and in niobium-containing substances and other materials.

REAGENTS

Standard thorium nitrate solution, 5 μg of Th/ml.
Arsenazo III, 0.05% aqueous solution.
Hydrochloric acid, $d = 1.18$ g/cm^3.
Oxalic acid, 0.5% solution.

CONSTRUCTION OF CALIBRATION CURVE

Into five 25-ml volumetric flasks are introduced 1, 2, 3, 4, and 5 ml of standard thorium nitrate solution; 1 ml of Arsenazo III solution, and 8 ml of hydrochloric acid are added to each flask. The solutions are diluted to the mark with water, mixed thoroughly, and the optical density is determined on an electrophotocolorimeter or a universal photometer. The blank solution is prepared in exactly the same manner, except that the thorium salt is not introduced.

PROCEDURE

All the reagents as above are added to the sample solution. The optical density is determined and the thorium content found from the calibration curve.

In the presence of titanium (not more than 100 μg) or zirconium (not more than 1 mg) 1 ml of oxalic acid solution is introduced prior to dilution.

Extraction-photometric determination of microgram amounts of thorium by Arsenazo III [37]

In the determination of microgram amounts of thorium the colored compound of thorium with Arsenazo III is extracted with isoamyl alcohol in the presence of diphenylguanidine hydrochloride and monochloroacetic acid. Thorium is determined following its concentration by extraction; it may be determined at dilutions of $1:5 \cdot 10^8$ (0.002 μg of Th/ml).

REAGENTS

Standard thorium nitrate solution, 1 μg of Th/ml.
Arsenazo III, 0.025% solution.
Diphenylguanidine hydrochloride, 20% solution.
Isoamyl alcohol.
Monochloroacetic acid, crystalline.
Nitric acid, 0.1 *N* solution.

CONSTRUCTION OF CALIBRATION CURVE

Into 25–30-ml separating funnels are introduced 1, 2, 3, 4, and 5 ml portions of standard thorium nitrate solution; 1 ml of Arsenazo III solution, 4.0 ml of nitric acid solution, 1 g of monochloroacetic acid, 10 ml of diphenylguanidine hydrochloride solution and 10.0 ml of isoamyl alcohol are then added to each flask. The mixtures are shaken for two minutes; the phases are allowed to separate, and the organic phase is filtered through filter paper into a dry cell and its optical density is determined with respect to a solution obtained in the same manner, but not containing thorium.

PROCEDURE

To the sample solution are added all the reagents used in the construction of the calibration curve. The optical density is measured, and the thorium content determined from the calibration curve.

Photometric determination of thorium by 1-(2-pyridylazo)-resorcinol [42]

Thorium ions react with 1-(2-pyridylazo)-resorcinol at pH 3–9 in the molar ratio of 1:4, forming a water-soluble orange-red compound with absorption maximum at 500 mμ (Figure 4). At this pH value the solutions of the reagent are yellow and have the absorption maximum at 415 mμ. The analysis is best performed at pH 6.4–6.7. The molar extinction coefficient of the solutions of the compound is 38,900; its formation reaction constant is 10^{-8} at pH 6.4–6.7. The solutions obey Beer's law between 10 and 200 μg of Th/25 ml of solution.

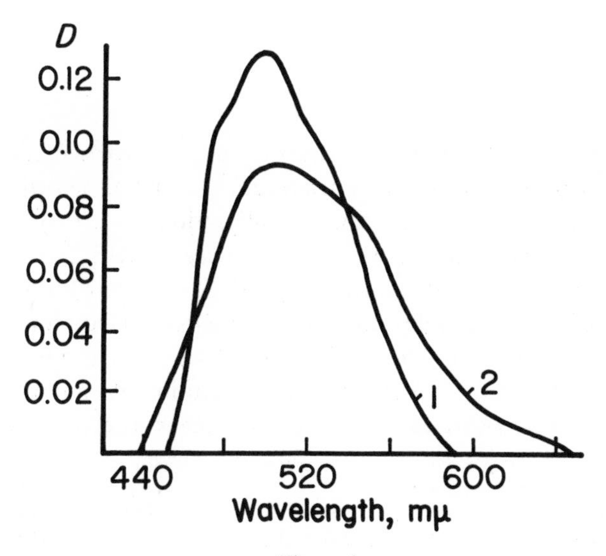

Figure 4

Light absorption by solutions of the complex of thorium with 1-(2-pyridylazo)-resorcinol at different pH values:

1) pH 5.50; 2) pH 9.95 (SF-4 spectrophotometer).

The thorium compound of 1-(2-pyridylazo)-resorcinol is more stable than thorium complexonate. On the other hand, REE complexonates are more stable. For this reason, thorium may be determined in the presence of REE ions by first adding 1-(2-pyridylazo)-resorcinol to the sample solution (this results in the formation of the colored compound of the reagent with thorium and REE ions), and then Complexone III (complexes of the reagent with REE ions are decomposed and the colorless complexonates of REE are formed).

The method is suitable for the determination of thorium in monazites and other objects, after the separation of thorium as oxalate and its decomposition.

REAGENTS

Standard thorium nitrate solution, 20 μg of Th/ml.
1-(2-*Pyridylazo*)-*resorcinol*, 0.05% aqueous solution.
Ammonium acetate, 20% solution.
Complexone III, 0.01 *M* solution.

CONSTRUCTION OF CALIBRATION CURVE

To each of five 50-ml volumetric flasks are added 2ml of 1-(2-pyridylazo)-resorcinol solution; 8 ml of water 1, 2, 3, 4, and 5 ml of standard thorium nitrate solution are added. The solutions are mixed and 15 ml of ammonium acetate solution are added to each flask. The solutions are made up to the mark with water, and thoroughly mixed. Their optical densities are measured with respect to the reagent prepared under similar conditions.

PROCEDURE

The determination of thorium in the sample solution is carried out in a similar manner. All the reagents as above are introduced; the optical density is determined and the thorium content is found from the calibration curve.

To determine thorium in monazites in the presence of REE 2 ml of 1-(2-pyridylazo)-resorcinol solution are added to an aliquot of the solution (30–70 μg of thorium in the presence of 50- to 100-fold amounts of REE); 15 ml of ammonium acetate solution is slowly introduced. The mixture is left to stand for ten minutes with occasional shaking, and 5 ml of Complexone III solution are added drop by drop with constant stirring. The solution is diluted to the mark with water and the optical density is determined as given for the calibration curve.

Note. Five ml of 0.01 *M* Complexone III solution are enough to bind 5 mg of REE (calculated as metal). If the REE content in the aliquot is known to be smaller, the amount of Complexone III is correspondingly reduced.

BIBLIOGRAPHY

1. BEHRENS, H.—*Z. anal. Chem.*, **30**, 157. 1891.
2. KOSS, M.—*Chem. Ztg.*, **36**, 686. 1912.
3. KOTA, J.—*Chem. Listy*, **27**, 100, 128, 150, 194. 1933.
4. CHERNIKOV, YU.A. and T.A. USPENSKAYA.—*Zav. Lab.*, **9**, 276. 1940.
5. KUZNETSOV, V.I.—*ZhOKh*, **14**, 914. 1944.
6. ADAMOVICH, L.P. and V.M. RUTMAN.—*Uchenye Zapiski Khar'kovskogo Universiteta*, **54**; *Trudy Khimicheskogo Fakul'teta i Nauchno-Issledovatel'skogo Instituta KhGU*, **12**, 203. 1954.
7. THOMASON, P.F., M.A. PERRY, and W.M. BYERLY.—*Anal. Chem.*, **21**, 1239. 1949.
8. NAZARENKO, V.A., E.A. BIRYUK, and E.N. POLUEKTOVA.—*Radiokhimiya*, **5**, 497. 1963.
9. SAVVIN, S.B.—*Doklady AN SSSR*, **127**, 1231. 1959.
10. LUK'YANOV, V.F., S.B. SAVVIN, and I.V. NIKOL'SKAYA.—*Zav. Lab.*, **25**, 1155. 1959.
11. KUTEINIKOV, A.F.—*Byulleten Vesoyuznogo Instituta Mineral'nogo Syr'ya*, No. 7, 10. 1957.
12. ZAIKOVSKII, F.F. and L.I. GERKHARDT.—*ZhAKh*, **13**, 274. 1958.
13. KUZNETSOV, V.I. and I.V. NIKOL'SKAYA.—*ZhAKh*, **15**, 299. 1960.
14. PAVELKA, F. *Mikrochem.*, **4**, 199. 1926.
15. HECHT, F. and W. REICH-ROHRWIG.—*Monatsh. Chem.*, **53/54**, 596. 1929.
16. ERERE, F.J.—*J. Am. Chem. Soc.*, **27**, 226. 1962.
17. BUDĚŠINSKÝ, B.—*Coll. Czech. Chem. Comm.*, **27**, 226. 1962.
18. KÖRBL, J., R. PŘIBIL, and A. EMR.—*Chem. Listy*, **50**, 1440. 1956.
19. MÖLLER, TH., G.K. SCHWEITZER, and D.D. STARR.—*Chem. Rev.*, **42**, 63. 1948.
20. RYABCHIKOV, D.I. and E.K. KORCHEMNAYA.—In: *Metody opredeleniya i analiza redkikh elementov*, p. 374. Izd. AN SSSR. 1961.
21. *Analytical Chemistry of Uranium and Thorium.*—[Russian translations of Papers, edited by P.N. PALEI. 1956.]
22. RYABCHIKOV, D.I. and E.K. GOL'BRAIKH. *Analiticheskaya khimiya toriya (Analytical Chemistry of Thorium).*—Izd. AN SSSR. 1960. [English translation available as part of the *Analytical Chemistry of the Elements* series, Ann Arbor–Humphrey Sci. Pub. 1969.]
23. VOLYNETS, M.P.—In: *Metody opredeleniya i analiza redkikh elementov*, p. 392. Izd. AN SSSR. 1961.
24. POLYAKOV, A.I. and M.P. VOLYNETS.—*Geokhimiya*, No. 5, 426. 1961.
25. MEYER, R.J. and M. SPETER.—*Chem. Ztg.*, **34**, 306. 1910.
26. MEYER, R.J.—*Ztg. anorg. chem.*, **71**, 67. 1911.
27. BUSEV, A.I., V.M. IVANOV, and V.G. TIPTSOVA.—*Zav. Lab.*, **27**, 799. 1961.
28. BUSEV, A.I., L.V. KISELEVA, and A.I. CHERKESOV.—*Zav. Lab.*, **24**, 13. 1958.
29. FLASCHKA, H. and H. ABDINE.—*Chemist-Analyst*, **45**, 58. 1956.
30. ALIMARIN, I.P., A.P. GOLOVINA, A.F. KUTEINIKOV, and N.F. STEPANOV.—*Vestnik MGU*, seriya II, Khimiya, No. 2, 203. 1958.
31. MENIS, O., D.L. MANNING, and G. GOLDSTEIN.—*Anal. Chem.*, **29**, 1426. 1957.
32. KUZNETSOV, V.I. and S.B. SAVVIN.—*ZhAKh*, **15**, 175. 1960.
33. SAVVIN, S.B., M.P. VOLYNETS, YU.A. BALASHOV, and V.V. BAGREEV.—*ZhAKh*, **15**, 446. 1960.
34. VLADIMIROVA, V.M. and N.K. DAVIDOVICH.—*Zav. Lab.*, **26**, 1210. 1960.
35. NEMODRUK, A.A. and N.E. KOCHETKOVA.—*ZhAKh*, **17**, 330. 1962.
36. SAVVIN, S.B.—*Talanta*, **8**, 673. 1961.
37. KUZNETSOV, V.A. and S.B. SAVVIN.—*Radiokhimiya*, **3**, 79. 1961.
38. BUSEV, A.I. and V.M. IVANOV.—*Izvestiya Vuzov, Khimiya i Khimicheskaya Tekhnologiya*, **4**, 914. 1961.

Uranium U

The valence states of uranium in its compounds are +3, +4, +5, and +6. The compounds of tetravalent and hexavalent uranium are the most important.

The standard electrode potentials of uranium in aqueous medium at 25°C (relative to standard hydrogen electrode), have the following values:

$U \rightleftharpoons U^{3+} + 3e$	−1.798 V
$U^{3+} \rightleftharpoons U^{4+} + e$	−0.607 V
$U^{3+} + H_2O \rightleftharpoons UOH^{3+} + H^+ + e$	−0.538 V
$U^{3+} + 4H_2O \rightleftharpoons U(OH)_4 + 4H^+ + e$	−0.019 V
$UO_2^+ \rightleftharpoons UO_2^{2+} + e$	−0.052 V
$UOH^{3+} + H_2O \rightleftharpoons UO_2^{2+} + 3H^+ + 2e$	−0.299 V
$U^{4+} + 2H_2O \rightleftharpoons UO_2^{2+} + 4H^+ + 2e$	−0.333 V

The reactions of U(III) ions resemble those of ferric ions. U(III) ions are pink-purple. Uranium (III) compounds in solutions are readily oxidized to U(IV) compounds by atmospheric oxygen.

The compound $U(OH)_3$ is brown and reduces hydrogen ions in solution.

U(III) ions are obtained by reducing U(VI) ions with very strong reducing agents such as zinc amalgam. The U(IV) cations are green. In weakly acid solutions the $U(OH)^{3+}$ ions are formed, which are slowly polymerized. The equilibrium constant of the reaction

$$U^{4+} + H_2O \rightleftharpoons U(OH)^{3+} + H^+$$

is p$K = 1.5$ at 0.5 ionic strength and 0.7 at zero ionic strength. Compounds of U(IV) are fully stable in the air. The compound $U(OH)_4$ is green. The sparingly soluble compounds and complexes of U(IV) resemble the corresponding compounds of Th(IV).

Uranium tetrafluoride UF_4 is sparingly soluble in dilute mineral acids; this is also true for the oxalate and phosphate.

U(IV) ions are precipitated by cupferron from weakly acid solutions.

Xylenol Orange reacts with U(IV) ions at pH 1.2–1.6 in the molar ratio 1:1, the resulting compound having the absorption maximum at 550 mμ [1]. The stability constant is $1.4 \cdot 10^{69}$ at 10^{-1} ionic strength. The reagent will determine 2–70 μg of U(IV) by the photometric method. U(VI) ions do not interfere even when present in 2500-fold amounts.

Tetravalent uranium forms colored compounds with Arsenazo III.

Complexone III reacts with U(IV) (and U(VI)) ions in the molar ratio of 1:1 and is used in its titrimetric determination (pp. 110 and 111).

Uranium (V) compounds are unstable and disproportionate to compounds of tetravalent and hexavalent uranium:

$$2UO_2^+ + 4H^+ \rightleftharpoons UO_2^{2+} + U^{4+} + 2H_2O.$$

Uranium (V) forms anions in alkaline solutions.

The most stable compounds of uranium are uranyl and uranate salts, in which uranium is hexavalent.

The hexachloride UCl_6 forms uranyl ions UO_2^{2+} when dissolved in water:

$$UCl_6 + 2H_2O \rightleftharpoons UO_2^{2+} + 4H^+ + 6Cl^-.$$

Uranyl ions are yellowish and are stable in acid solutions. At pH 2–3 various condensed basic ions are formed.

The reactions of U(VI) ions resemble those of W(VI) and Mo(VI).

The hydroxide $UO_2(OH)_2$ is formed at pH >3.8. It is practically quantitatively precipitated at pH 5.3 (but may remain colloidal up to pH 7) and is strongly amphoteric, forming uranates UO_4^{2-} and polyuranates $U_2O_7^{2-}$, $U_3O_{10}^{2-}$, etc., when the pH is further raised (in analogy to polytungstate and polymolybdate ions). Neutralization of uranyl salt solutions usually yields the yellow precipitate of an alkali uranate such as Na_2UO_4, $Na_2U_2O_7$, etc.

Uranyl salts give an intense yellow fluorescence in the solid state and also in melts with borax or with fluoride. This fact is taken advantage of in highly sensitive detections and semiquantitative determinations of uranium. The fluorescence of the salts disappears in aqueous solution, but is noticeable in concentrated sulfuric acid and in syrupy phosphoric acid.

Uranyl sulfide UO_2S is brown; it is soluble in dilute mineral acids.

Potassium ferrocyanide reacts with uranyl ions in acetic acid or neutral medium to yield a reddish brown precipitate of varying composition. The reagent is used for the detection of uranyl ions and their photometric determination.

Sodium phosphate precipitates uranyl ions as $UO_2HPO_4 \cdot 4H_2O$; the compound is light yellow. In the presence of ammonium salts, the precipitate of $NH_4UO_2PO_4 \cdot 6H_2O$ is formed.

Selenious acid quantitatively precipitates [2] uranyl ions as UO_2SeO_3 at pH 4–5 from solutions in 50% alcohol. The precipitate is soluble in concentrated hydrochloric acid. Uranium can be determined by the iodometric titration of H_2SeO_3 after the dissolution of the precipitate [2].

U^{4+} ions form unstable oxalate complexes and stable tartrate and citrate complexes. Unstable fluoride and carbonate complexes decompose at pH <7 and pH >12.

Hydrogen peroxide in neutral concentrated solution of U(VI) salts gives a yellow precipitate of the hydrated peroxide $UO_4 \cdot 2H_2O$, soluble in ammonium carbonate solution; the resulting solution is yellow, and solutions in aqueous alkali carbonates are orange-yellow.

Thiocyanates in the presence of U(VI) ions give an intense yellow coloration.

8-Hydroxyquinoline reacts with UO_2^{2+} to form a brown precipitate which has the composition $UO_2(C_9H_6ON)_2 \cdot C_9H_7ON$. The reagent is used in the detection of U(VI) ions and their gravimetric and photometric determination.

Ferron (7-iodo-8-hydroxyquinoline-5-sulfonic acid) gives an intense brown coloration [3] with U(VI) ions at pH 5 (urotropin buffer solution). The sensitivity of detection is 10 μg of U/ml (limiting dilution $1:10^5$). The reagent reacts with uranium in the molar ratio of 2:1. The absorption maximum of the solution of uranium compound is at 360 mμ. The selectivity of the reagent is low. Fe(III), Al(III), and many other ions interfere; acetate anions also interfere.

Borax beads obtained in the oxidizing flame are yellow and have a green fluorescence; beads obtained in the reducing part of the flame are green; NaF beads display intense fluorescence.

U(VI) ions are not precipiated by cupferron and they can thus be separated from V(V), Fe(III), and Ti(IV). After separating the cupferronates of these metals the UO_2^{2+} in the filtrate can be reduced and U(IV) ions precipitated by cupferron.

The U(VI) ions are separated from Fe(III), Ti(IV), Bi(III), Mg, Sr, Cd, Zn, Pb, and Mn ions as readily soluble carbonate complexes. The ions V(V), Mo(VI), Al(III), and Th(IV) remain in solution together with the U(VI) ions.

Uranium is separated from other elements by extraction methods: extraction of $UO_2(NO_3)_2 \cdot 2H_2O$ with ether or ethyl acetate, extraction with amines from sulfuric acid or phosphoric acid solutions, extraction with tributyl phosphate $(C_4H_9)_3PO_4$, with trialkylphosphine oxides R_3PO, etc. Uranium is extracted as complexes with SCN^- ions, compounds with acetylacetone, 8-hydroxyquinoline, cupferron, diethyldithiocarbamate, xanthate, etc.

Uranium is separated from other elements by ion-exchange and partition chromatography. Both cation exchangers and anion exchangers are employed.

In the analysis of the different materials uranium is isolated by precipitation as sparingly soluble compounds:

ammonium diuranate $(NH_4)_2\ U_2O_7$;
peruranic acid;
uranium (IV) cupferronate $U(C_6H_5N_2O_2)_4$;
ammonium uranyl cupferronate $NH_4[UO_2(C_6H_5N_2O_2)_3]$;
uranyl 8-hydroxyquinolate $UO_2(C_9H_6ON)_2 \cdot C_9H_6NOH$;
uranyltricarbonatohexamminocobaltinitrate $[UO_2(CO_3)_3(H_2O)_3]\,[CO(NH_3)_6(NO_3)_2]$;
ammonium uranyl phosphate $NH_4UO_2PO_4 \cdot 3H_2O$;
uranium (IV) phosphate $U(HPO_4)_2 \cdot 4H_2O$;
uranium (IV) fluoride $UF_4 \cdot 2.5H_2O$, etc.

The titrimetric methods for the determination of large and small amounts of uranium employed in analytical practice are based on redox reactions and on complex formation reactions.

Uranium (VI) is titrated with a solution of $CrCl_2$ or $TiCl_3$ in an atmosphere of carbon dioxide, the end point being established potentiometrically or by indicators.

The methods employed in most cases are based on the oxidation of U(IV) to U(VI) ions. Many variants of such methods have been developed. First, U(VI) ions are reduced to U(IV) with oxalic acid in daylight, electrochemically, by metallic cadmium or bismuth, by zinc, cadmium or bismuth amalgam, by sodium dithionite $Na_2S_2O_4$ and rongalite $NaHSO_2 \cdot CH_2O \cdot 2H_2O$. The U(IV) ions are then titrated against solutions of $KMnO_4$, $K_2Cr_2O_7$, NH_4VO_3, $Ce(SO_4)_2$, $Fe_2(SO_4)_3$, etc. In several methods the resulting U(IV) ions are oxidized with ferric salt in excess:

$$U^{4+} + 2Fe^{3+} + 2H_2O \rightleftharpoons UO_2^{2+} + 2Fe^{2+} + 4H^+.$$

The ferrous ions which are formed in equivalent amounts are then titrated.

Compounds which may undergo oxidation or reduction interfere with redox methods for the determination of uranium.

An important method is the complexometric determination of uranium (VI) after its reduction to uranium (IV). The titration is effected at pH 1.7 with a solution of Complexone III in the presence of Arsenazo I.

Many photometric methods for the determination of uranium are known; these involve the use of inorganic and mainly organic reagents. The reagents for uranium with the groupings

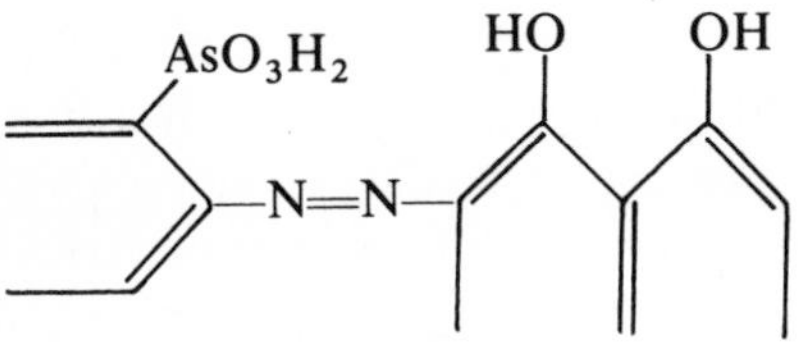

AsO_3H_2 HO OH
—N=N—

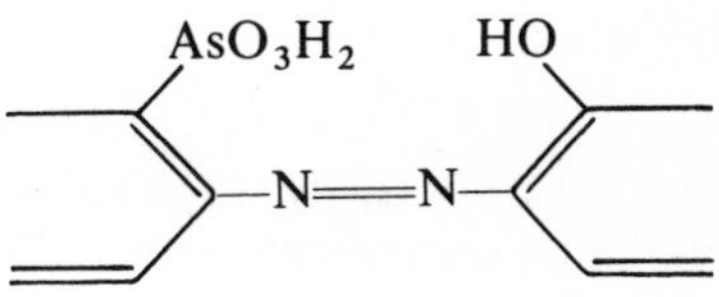

AsO_3H_2 HO
—N=N—

are highly sensitive and sufficiently selective. Examples of such reagents are Arsenazo I, Arsenazo III, and thoron. The first two reagents form intensely colored compounds with U(VI) and U(IV) ions. Thoron reacts with U(IV) ions.

Other reagents for the photometric determination of uranium include Chlorophosphonazo I and Chlorophosphonazo III which contain $—PO_3H$ and —OH groups, pyridylazo compounds such as 1-(2-pyridylazo)-2-naphthol and 1-(2-pyridylazo)-2-resorcinol, solochrome azo compounds which contain two OH groups ortho to the azo group, and also hydroxyflavone dyes such as flavonol, morin, and quercetin.

Other reagents which are also used include Alizarin S, Bromopyrogallol Red, sulfosalicylic acid, 8-hydroxyquinoline, chromotropic acid, and other hydroxylated reagents, sodium diethyldithiocarbamate, thioglycolic acid, and ascorbic acid.

Polarographic, coulometric, fluorimetric, and radiometric methods are used for the determination of uranium.

For a review of the detection and determination methods of uranium, see [4, 5].

Complexone–phosphate method of isolation of uranium from rocks and minerals [6]

Uranium is isolated as $UO_2HPO_4 \cdot 4H_2O$, a titanium salt being used as precipitant. In order to prevent the precipitation of the phosphates of Cu(II), Ni(II), Al(III), Fe(III), Cr(III), and REE, a solution of Complexone III is introduced. This makes it possible at the same time to separate uranyl ions from V(V), Mo(VI), and Cr(VI) ions, and so greatly simplifies analyses of complex materials. Finally, the precipitate is treated with a solution of sodium carbonate, when uranium is solubilized as the carbonate complex.

The method is suitable for the determination of 0.001–2% of uranium in rocks and minerals and is widely employed in analytical practice. The experimental error is 20–40%, 10–20% and 5–10% for uranium contents of the order of 0.001%, 0.01% and 0.1%, respectively.

REAGENTS

Hydrochloric acid, $d = 1.19$ g/cm^3 and 5% solution.
Hydrogen peroxide, 30% solution.
Sulfuric acid, dilute (1:1).
Nitric acid, $d = 1.4$ g/cm^3.
Disubstituted sodium or ammonium phosphate, 10% solution.
Ammonium nitrate, crystalline and 2% solution.
Complexone III, saturated aqueous solution.
Sodium carbonate, 10% solution.
Titanyl sulfate, solution, 5 mg of TiO_2/ml.

Activated charcoal.
Methyl red, 0.1% alcoholic solution.
Ammonia, 25% solution.

PROCEDURE

To a 0.5 g sample of the mineral, which should contain about 0.01% of uranium, in a 100-ml beaker are added 15–30 ml of hydrochloric acid (d = 1.19 g/cm^3) and 3–4 ml of hydrogen peroxide; the beaker is covered with a watch glass and the mixture is heated at gentle boil for 30–40 minutes. The solution is slightly cooled, 2–3 ml of hydrogen peroxide and then 5 ml of hydrochloric acid (d = 1.19 g/cm^3) are added, and the solution is boiled for another 15 minutes. Thirty ml of hot water are added; the beaker is covered with a watch glass and the solution is boiled until the salts are dissolved. The insoluble residue is separated on Blue Ribbon filter paper. The paper is washed several times with 5% hydrochloric acid, and then 2–3 times with hot water, added in 10 ml portions, and is then discarded.

The filtrate should be 80–100 ml in volume. Crystalline ammonium nitrate (2–3 g), and 10–30 ml of Complexone III solution (for sample weights of 0.2, 0.5 and 1–2 g of 10, 15 and 25–30 ml solution of Complexone III are added: a large excess of the reagent prevents the oxidation of phosphates). The solution is neutralized with ammonia to methyl red (external indicator). When ferric ions are present, the end of the neutralization is indicated by the appearance of a reddish coloration on the addition of the solution of ammonia. Another 1–2 drops of ammonia solution are added to the neutralized solution; the solution is heated to the boil and boiled for 2–3 minutes. Boiling the solution favorizes the formation of complexonates of chromium (III) and of other metals.

The resulting solution is cooled with cold water in a crystallizer; 10 ml of di-substituted sodium or ammonium phosphate are added, then 5 ml titanyl sulfate solution as coprecipitant. The solution is neutralized to methyl red with a solution of ammonia; 1–2 drops of the ammonia solution in excess is added at the end of the neutralization.

The solution with the precipitate is left to stand overnight, filtered through two thicknesses of White Ribbon filter paper and the precipitate is washed by decantation with 10–12 portions (5–7 ml) of hot ammonium nitrate solution.

The washed precipitate is returned with the aid of hot water from the filter to the beaker which was used for the precipitation and sodium carbonate solution is added to a final concentration of 5%. The beaker is covered with a watch glass, its contents are heated to the boil and evaporated to 30–40 ml. It is expedient during the boiling stage to introduce about 0.1 g of activated charcoal on the tip of a spatula in order to remove the foreign coloration. After cooling, the solution with the precipitate is transferred to a 25-ml volumetric flask, diluted to mark with water and mixed. The residue is separated on a White Ribbon filter paper previously wetted with sodium carbonate solution, and the filtrate is collected in a 100-ml beaker.

Nitric acid is added to the filtrate to pH 1–2 and the solution is evaporated to dryness. The residue is again treated with nitric acid and a little water is added to solubilize the salts. The solution is transferred to a 25-ml volumetric flask and diluted to mark with water.

Uranium is determined by one of the photometric methods given on pp. 112–115.

Note. 1. Zirconium-containing rocks, tantalum-niobium ores, and other rocks which are difficult to decompose with mixtures of hydrochloric acid with hydrogen peroxide, are treated with sulfuric and nitric acids. To do this 15 ml of sulfuric acid diluted with nitric acid, d = 1.4 g/cm^3, in the ratio of 1:1.5 are added; the beaker is covered with a watch glass and heated to incipient evolution of sulfuric acid fumes. The heating is then continued for another 40–60 minutes. The watch glass is removed and the solution is heated to expel the bulk of sulfuric acid (almost complete cessation of the evolution of white fumes). After cooling, 10–15 ml of hydrochloric acid, d = 1.19 g/cm^3 and 30–50 ml of water are added to the residue; the mixture is boiled and filtered, and the undissolved residue is washed as above.

Note 2. If more than 50 mg of chromium is present, the phosphate precipitate must be reprecipitated with the addition of Complexone III.

Diethyldithiocarbamate–complexonate extraction method of separation of uranium [7–9]

In this method the diethyldithiocarbamate compound of U(VI) is extracted in the presence of Complexone III with chloroform, ether, isoamyl alcohol, ethyl acetate, or amyl acetate from neutral solution (pH 6.5–7.5); this is followed by the reextraction of U(VI) ions with aqueous ammonium carbonate. The carbonate complex of U(VI) is decomposed and the determination is terminated by any suitable photometric method. The ions of Fe(III), Co, Ni, In, Ga, Zn, Pb, V(V IV), Nb(V), Sn(IV) form stable complexonates in weakly acid and neutral media and do not interfere. The ions of Be, Sb, Ti and partly Mn(II), which do not form stable complexonates, may precipitate out during the neutralization of the solution.

Cu, Ag, Bi, Hg, Tl, As, Se, and Te are extracted together with uranium as diethyldithiocarbamates, but when the extract is treated with ammonium carbonate, only U(VI) passes into the aqueous layer, since its carbonate complex is more stable than its diethyldithiocarbamate complex.

REAGENTS

Ammonia, 2% solution and dilute (1:1) solution.
Complexone III, 10% solution.
Sodium diethyldithiocarbamate, 2% solution.
Ammonium carbonate, saturated aqueous solution.
Nitric acid, 5% solution.
Chloroform (or ether, isoamyl alcohol, ethyl acetate, amyl acetate).

PROCEDURE

To the clear acid solution obtained from the solubilization of the ore sample, which should contain 10–50 μg of uranium and should be about 50 ml in volume, a 1:1 solution of ammonia is added until a faint turbidity appears (pH 2.5–3); then 10 ml of Complexone III solution are added (for iron-rich ores the amount of Complexone III added should be 1.5–2 times higher). The solution is neutralized with 1:1 ammonia to pH 6.5–7.5 (litmus paper) and transferred to a 200- to 250-ml separating funnel. Five ml of sodium diethyldithiocarbamate solution are added; the solution is shaken; 5–7 ml of chloroform are added, and the carbamates formed are extracted for 1–2 minutes. The organic phase is decanted into another separating funnel; 2 ml of sodium diethyldithiocarbamate solution are added to the aqueous phase, and then 5 ml of chloroform. The solution is shaken for 1–2 minutes. The addition of sodium diethyldithiocarbamate and the extraction is repeated until a fresh chloroform portion remains colorless. The aqueous phase is discarded.

The chloroform extracts are combined, 2 ml of 2% ammonia solution and 5–6 drops of ammonium carbonate solution are added. The solution is vigorously shaken for 2–3 minutes in a 50-ml separating funnel. The water phase is decanted into a 50-ml beaker, and the organic phase is again treated with ammonia and ammonium carbonate solutions. The aqueous solution which contains U(VI) ions is evaporated to dryness in the beaker on a sand bath; the residue is treated with nitric acid, and the solution is again evaporated to dryness. The residue is dissolved in water, acidified to pH 3–4 with nitric acid and transferred to a 25-ml volumetric flask. The determination of uranium is carried out on an aliquot of this solution.

Gravimetric determination of uranium as uranyl 8-hydroxyquinolate [10]

Uranium is precipitated as uranyl 8-hydroxyquinolate in the presence of Complexone III which masks a large number of accompanying elements.

The method is suitable for the determination of uranium in the presence of thorium, zirconium, REE, and phosphates and also for the separation of uranium from vanadium.

REAGENTS

Complexone III, saturated aqueous solution.
8-*Hydroxyquinoline*, 4% and 0.01% solutions in ethanol.
Ammonium acetate, 20% solution.
Acetic acid, dilute (1:1).
Ammonia, 4 *N* solution.
Methyl red, 0.1% solution in ethanol.

Sulfuric acid, $d = 1.84\ g/cm^3$.
Hydrochloric acid, $d = 1.19\ g/cm^3$.
Nitric acid, $d = 1.4\ g/cm^3$.
Thymol Blue, 0.1 solution in 20% ethanol.

PROCEDURE

Determination of uranium in the presence of thorium, REE, and zirconium

To a solution containing 20–50 mg of uranium and not more than 100 mg each of thorium, REE, and zirconium, are added 10 ml of Complexone III solution and the mixture is neutralized with ammonia solution until yellow to methyl red. To the solution are added 1.1 ml of acetic acid and 25 ml of ammonium acetate solution. The mixture is diluted to 150–175 ml with water and heated on a sand bath to 70°C. Five ml of 4% solution of 8-hydroxyquinoline (or 6–7 ml in the presence of large amounts of zirconium) are added and the mixture is placed on a water bath and held for 5 minutes at 30°C. The precipitate is then filtered through No. 3 sintered glass crucible, washed with 0.01% solution of 8-hydroxyquinoline, dried to constant weight at 110°C and weighed. The conversion factor to uranium is 0.3386.

In the determinations of uranium in the presence of 2- to 5-fold amounts of REE, thorium, or zirconium, the experimental error does not exceed 0.2–0.5%.

Determination of uranium in the presence of thorium, REE, and phosphates

To the solution containing about 40 mg of uranium, not more than 30 mg of REE and not more than 2- to 5-fold amounts of thorium and phosphate ions are added 2 ml of sulfuric acid. The solution is diluted to 125 ml with water; 10 ml of Complexone III solution and 7.5 ml of 4% 8-hydroxyquinoline solution are added. The solution is heated to 70°C, neutralized with ammonia solution until yellow to Thymol Blue, 25 ml of ammonium acetate solution are added, and the determination is continued as above.

If zirconium and phosphate ions are both present, the method is unsuitable.

Determination of uranium in the presence of vanadium ions

To the sample solution, which should contain about 40 mg of uranium, not more than 100 mg of vanadium (as V_2O_5) and a sufficient amount of nitric acid to prevent precipitation of vanadic acid, is added 0.4–0.5 ml of hydrochloric acid. The solution is diluted to 75 ml with water, after which 10 ml of Complexone III solution are added and the solution is boiled for 10–12 minutes. This results in a gradual development of a blue coloration owing to the reduction of vanadate to vanadyl ions. The solution is neutralized with ammonia to methyl red and 1.1 ml acetic acid and 25 ml of ammonium acetate solution are added. The solution is then diluted to 150–175 ml with water, and heated to 70°C. Five ml of 4% hydroxyquinoline solution are added drop by drop and the determination is continued as described above for the determination of uranium in the presence of thorium and REE. In the presence of 2- to 5-fold amounts of vanadium ions (as V_2O_5), the experimental error does not exceed 0.2%.

Vanadatometric determination of uranium [11, 12]

Uranium (IV) is titrated against ammonium vanadate in the presence of N-phenylanthranilic acid as indicator:

$$U(SO_4)_2 + 2NH_4VO_3 + 2H_2SO_4 = UO_2SO_4 + 2VOSO_4 + 2H_2O + (NH_4)_2SO_4.$$

In 6 *N* and more concentrated solutions of sulfuric acid uranium (IV) is rapidly oxidized, so that the titration can be carried out directly with dilute (0.001 *N*) vanadate solution. Phosphoric acid is added to the solution prior to titration in order to accelerate the oxidation of N-phenylanthranilic acid by vanadate.

Uranium (IV) in concentrations above 2.5 μg/ml can be titrated amperometrically against a solution of ammonium vanadate with the aid of a dropping mercury electrode.

Uranium (VI) is reduced to uranium (IV) in a reductor filled with electrolytically reduced cadmium in 3 *N* sulfuric acid; the amount of uranium (III) produced thereby is insignificant. The ions Fe (III), Ti (IV), Mo (VI), V (V), Sn (IV), and Nb (V), which are reduced together with uranium, interfere. Uranium is determined in pure solutions after separation.

REAGENTS

Uranyl sulfate, 0.025 *M* solution.
Sulfuric acid, d = 1.84 g/cm^3, 0.2 *N*, 1, *N* and 6 *N* solutions.
Metallic cadmium, electrolytically reduced.
Ammonium vanadate, 0.02 *N* solution.
Phosphoric acid, d = 1.6 g/cm^3.
N-Phenylanthranilic acid, 0.4% solution in 0.4% sodium carbonate solution.

PROCEDURE

The sample solution of uranyl sulfate is diluted to the mark with water in a 50-ml volumetric flask. A 15 ml aliquot, containing 5–20 mg of uranium, is diluted with an equal volume of 6 *N* sulfuric acid. The cadmium reductor is washed with 1 *N* solution of sulfuric acid and the sample solution is passed through it three or four times at the rate of 1–2 drops per second. When the reduction is complete, the reductor is washed twice with 20 ml of 1 *N* sulfuric acid and twice with 10 ml of 0.2 *N* sulfuric acid.

The cadmium layer should be covered with liquid during the washing of the reductor and during the reduction. The penetration of air will interfere with the complete reduction of uranium.

After the reduction the uranium (IV) solution is collected in a 250-ml conical flask, sulfuric acid ($d = 1.84\ g/cm^3$) is added to a final concentration of 10–12 N (40–45 ml), and then 5 ml of phosphoric acid and 5–6 drops of N-phenylanthranilic acid. The solution is allowed to cool and then slowly titrated against a solution of ammonium vanadate until the last drop turns the greenish solution a permanent cherry-red.

The reduction and the titration of uranium (IV) is repeated until agreeing results are obtained. One ml of 0.0200 N solution of ammonium vanadate is equivalent to 2.3807 mg of uranium.

Complexometric determination of uranium (IV) ions [4, 13]

Complexone III reacts with uranium (IV) ions in the molar ratio of 1:1 to form a complex compound with a stability constant of $4.2 \cdot 10^{25}$. The titration is conducted at pH 1.7 ± 0.1 in the presence of Arsenazo I as indicator. The uranyl ions are reduced by formamidinesulfinic acid:

$$\begin{matrix} H_2N & & & & O \\ & \diagdown & & \diagup\!\!\diagup & \\ & & C-S & & \\ & \diagup\!\!\diagup & & \diagdown & \\ HN & & & & OH \end{matrix}$$

At the titration end point the color of the solution changes from the characteristic blue of the compound of tetravalent uranium with Arsenazo I to the pink of Arsenazo I.

The method is suitable for the determination of uranium in alloys with magnesium, aluminum, iron, and silicon, in technical grade salts (nitrates, sulfates, fluorides, acetates, oxalates, etc.), and also in oxides and industrial liquors.

REAGENTS

Complexone III, 0.01 M solution.
Arsenazo I, 0.05% solution.
Formamidinesulfinic acid, 0.66% solution in 0.25 N sulfuric acid, freshly prepared.
Thymol Blue, 0.1% solution in 5% of ammonia.
Ammonia, 1:5 solution.

PROCEDURE

Fifteen ml of the sample solution, which should contain 40–100 mg of U(VI) in a volume of 50 ml, are withdrawn with a pipet into a 300-ml conical flask; 2–3 drops of Thymol Blue and then ammonia solution are added until the color turns orange-yellow (pH 2.5–2.7). Thirty ml of formamidinesulfinic acid solution are then added. The solution is heated to the boil and boiled for 5 minutes, after which 175 ml of water and 2 ml of Arsenazo I solution are added. Uranium (IV) is titrated against a solution of Complexone III until the last drop of the reagent turns the blue solution pink.

If the pH of the titrated solution has a different value, the color change of the indicator will not be sharp.

Complexometric titration of uranium (VI) ions [14–16]

The end point of the complexometric titration of uranium (VI) is located with the aid of 1-(2-pyridylazo)-2-naphthol (PAN). The reagent forms a red-violet compound with uranyl ions, with an absorption maximum at 560 mμ. Under certain conditions this compound is dissociated to a smaller extent than is uranyl ethylenediaminetetra-acetate. Uranyl ions are titrated against Complexone III at pH 4.4–4.6 in the presence of PAN as internal indicator. The reaction with Complexone III is slow and the solution is accordingly heated to 80°C. Because of the low solubility of the compound of uranyl ion with PAN and of the indicator itself in water, an organic solvent must be added: ethanol (50% by volume) or isopropanol (2 volumes per 1 volume of solution). The method is best employed for the determination of uranium in solutions of pure uranium salts. In the determination of 10–30 mg of uranium, the experimental error is $\pm$0.3 mg.

REAGENTS

Nitric acid, 0.1 *M* solution.
Urotropin, 25% solution.
Isopropanol.
Complexone III, 0.01 *M* solution.
1-(2-*Pyridylazo*)-2-*naphthol*, 0.1% solution in methanol.

PROCEDURE

This solution, which should contain 10–30 mg of uranium (VI) in 50 ml, is acidified to pH 2–3 with nitric acid; 2 ml of urotropin solution and double the volume of isopropanol are added. The mixture is heated to the appearance of water vapor (80–90°C) and 2–3 drops of 1-(2-pyridylazo)-2-naphthol are added. The bright red solution is titrated against Complexone III until the solution turns pure yellow on the addition of one drop of the reagent.

Photometric determination of uranium (VI) by Arsenazo I [17–22]

Arsenazo I reacts with uranyl ions in weakly acid medium to form a soluble blue-violet complex compound. The optical density of the aqueous solutions of the compound remains practically constant between pH 3.5 and pH 8.0. The molar extinction coefficient is 22,800. The reagent is very sensitive: the minimum detectable amount is 0.2 μg of U(VI) [per ml], limiting dilution $1:5 \cdot 10^6$.

Th, V(IV), Al, Fe(III), Zr, Ti, Cu, Cr(III), Ga, Pd and REE, which form violet colored compounds of various tinges with Arsenazo I, interfere. Phosphate, fluoride, arsenate and vanadate ions and also hydrogen peroxide in solution weaken the coloration. Salicylate, sulfosalicylate and ethylenediaminetetraacetate ions have little effect.

REAGENTS

Standard solution of uranyl nitrate, 10 μg of U/ml.
Arsenazo I, 0.05% aqueous solution.
Urotropin, 25% solution.
Borate buffer solution, pH 4.2. Sodium tetraborate (19.07 g) is dissolved in 100 ml of nitric acid and the solution is diluted to one liter with water.
Hydrogen peroxide, 30% solution.

CONSTRUCTION OF CALIBRATION CURVE

Into each of five 50-ml volumetric flasks is introduced one ml of Arsenazo I solution. Water is added to a volume of 10–15 ml, and 2, 4, 6, 8, and 10 ml of uranyl nitrate solution and then 10 ml of borate buffer solution or 2 ml of urotropin solution are added to each flask with stirring. The solutions are made up to the mark with water.

The optical density of the solutions is measured using a photoelectrocolorimeter with an orange filter relative to the reagent solution prepared in the same manner. The results are used to plot the calibration curve.

PROCEDURE

A solution of the sample is prepared exactly as the solutions used to plot the calibration curve and its optical density is determined. The amount of uranium is read off the calibration curve.

Photometric determination of uranium (VI) ions by Arsenazo I [23]

Arsenazo I reacts with uranium (IV) in acid medium to form a blue violet complex compound with an absorption maximum at 555 mμ. The solution of the reagent hardly absorbs in this range. The compound is quite stable; the instability constant is $6 \cdot 10^{-17}$. The determination is best performed at pH 1.5–1.8. At this acidity Fe(II), Be, Al and REE do not react with Arsenazo I. Potassium iodide in acid medium is used to reduce the uranyl ions.

REAGENTS

Standard solution of uranyl nitrate, 10 μg of U/ml.
Potassium iodide, 10% solution.
Hydrochloric acid, $d = 1.17–1.19$ g/cm³ (redistilled) and 0.1 *N* solution.
Arsenazo I, 0.05% aqueous solution.
Sodium thiosulfate, 0.002% solution.

CONSTRUCTION OF CALIBRATION CURVE

Portions of 2, 4, 6, 8, and 10 ml of standard uranyl nitrate solution are introduced into five 50-ml beakers; 2 ml of hydrochloric acid ($d = 1.17–1.19$ g/cm³) and 0.8 ml of potassium iodide solution are added to each beaker and the solutions are evaporated to dryness on a water bath. After cooling, 10 ml of 0.1 *N* hydrochloric acid are added to each beaker and then sodium thiosulfate solution is cautiously introduced drop by drop to the disappearance of the color of iodine (not more than 2–3 drops of thiosulfate solution in excess). Then 2 ml of Arsenazo I solution are added to each beaker, the solution is transferred to 25-ml volumetric flasks, diluted to the mark with water and the optical densities are measured on a photocolorimeter with a red filter, or at 555 mμ if another instrument is used.

PROCEDURE

The sample solution is prepared under conditions similar to those used in the preparation of the calibration curve. The measurements are effected on universal photometer or photocolorimeter relative to the solution of the reagent at 555 mμ. The uranyl content is found from the calibration curve.

Extraction-photometric determination of uranium (VI) by Arsenazo III [24–27]

Arsenazo III reacts with uranyl ions in acid medium to form a soluble green complex compound, with an absorption maximum at 655 mμ. The sensitivity of the determination is 0.01–0.02 μg of U(VI); the molar extinction coefficient is 75,500. The optimum pH range is 1.7–2.5. Th, Zr, Al, Cr(III) and REE interfere with the determination; sulfate, fluoride and phosphate ions do not interfere.

The colored compound is extracted with butanol in the presence of diphenylguanidine hydrochloride and Complexone III. Under these conditions many cations are masked by Complexone III. Chlorides, nitrates and fluorides do not interfere if present in amounts not above 1 mg. Sulfates and phosphates are precipitated by diphenylguanidine hydrochloride and are floated. The precipitate is separated by the filtration of the extracts; the effect of thorium is eliminated by the addition of fluoride (5 mg KF in 5 ml solution).

REAGENTS

Standard uranyl nitrate solution, 5 μg of U/ml.

Hydrochloric acid, 0.05 *N* solution.

Complexone III, 5% solution.

Arsenazo III, 0.05% solution.

Diphenylguanidine hydrochloride, 20% solution. About 105 g of diphenylguanidine base are mixed with 80 ml of 6 *N* hydrochloric acid, and the mixture is diluted with water to 500 ml. The undissolved residue is filtered off and the filtrate is acidified to pH 3–4 with a few drops of concentrated HCl.

Butanol.

CONSTRUCTION OF CALIBRATION CURVE

Into five 25-ml tubes are introduced 1, 2, 3, 4, and 5 ml portions of standard uranyl nitrate solution. The tubes are immersed in a boiling water bath, evaporated to dryness while bubbling through air with the aid of capillaries, and the dry residues are dissolved in 2.0 ml of hydrochloric acid. To each solution is added 2.5 ml of Complexone III solution, 1.0 ml of Arsenazo III solution, 0.5 ml of diphenylguanidine hydrochloride solution, and 5.0 ml of butanol.

The solutions are extracted by vigorous shaking for about 30 seconds, the phases are allowed to separate and the optical density of the extract is measured on an electrophotocolorimeter at 655 mμ relative to the blank solution.

Note. Uranium (IV) may also be preliminarily extracted with 20% solution of tributyl phosphate in chloroform, using ammonium nitrate as the salting-out agent and Complexone III to bind the interfering elements in the aqueous phase. Uranium can then be reextracted with Arsenazo III solution and determined by measuring the optical density of the aqueous phase.

PROCEDURE

The sample solution is prepared under conditions similar to those employed in the construction of the calibration curve. The amount of uranium is found from the calibration curve.

Photometric determination of uranium (VI) ions by 1-(2-pyridylazo)-resorcinol [28, 29]

1-(2-Pyridylazo)-resorcinol (PAR) is one of the most sensitive reagents in the photometric determination of U(VI) ions. It reacts with uranyl ions at pH 3–10 in the molar ratio of 1:1, forming a soluble compound. The absorption maximum of the solutions is at 510 mμ (Figure 5). The sensitivity of the reaction is 0.02 μg of U/ml; the optimum pH value is 7.5 (borate buffer solution). Beer's law is obeyed in the concentration range between 2 and 400 μg of U(VI) in 25 ml of solution.

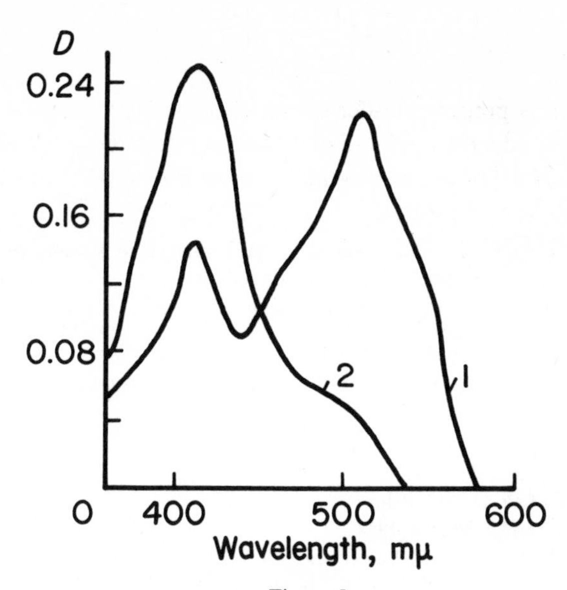

Figure 5

Light absorption by solutions:

1) compound of uranium with 1-(2-pyridylazo)-resorcinol; 2) 1-(2-pyridylazo)-resorcinol.

Alkali metals and alkaline-earth metals, Be, Ti(IV), Sn(IV), Mo, W, Ir, Rh, As, Se, and Te do not interfere; Cu, Cr(III), Pb, Bi, Hg, Sb, Fe, La, Co, Ni, Mn(II), Zr, Zn, and Th interfere. The method is used after separation of uranium from interfering ions.

REAGENTS

Standard uranyl nitrate solution, 5 μg of U/ml.

Borate buffer solution, pH 7.5. Boric acid (10.54 g) and 2.87 g of sodium tetraborate are dissolved in water and the solution made up to one liter with water.

1-(2-*Pyridylazo*)-*resorcinol*, 0.01% solution.

CONSTRUCTION OF CALIBRATION CURVE

To each of five 25-ml volumetric flasks are added 2 ml of 1-(2-pyridylazo)-resorcinol solution, 10 ml of borate buffer solution, 1, 2, 3, 4, and 5 ml of standard uranyl nitrate solution and 5 ml of borate buffer solution. The solutions are made up to the mark with water and are thoroughly mixed.

The optical densities of the solutions are determined relative to a solution of the reagent at 510 mμ, using an electrophotocolorimeter or a universal photometer and the calibration curve is plotted.

PROCEDURE

The sample solution is prepared under conditions just described. To the acid solution urotropin is first added to pH 4–5 and then borate buffer solution. The optical density is measured at 510 mμ. The uranium content is read off the calibration curve.

Note. In the determination of uranium in various materials by the Complexone–phosphate or the diethyldithiocarbamate–Complexone method, the carbonate complexes of uranium are first decomposed by the addition of hydrochloric or nitric acid. The solution is evaporated to a small volume and cautiously neutralized to pH 4–4.5 with urotropin solution, an aliquot is then taken for the determination.

BIBLIOGRAPHY

1. BUDESINSKY, B.—*Coll. Czech. Chem. Comm.*, **27**, 226. 1962.
2. DZHOSHI, M.K.—*ZhAKh*, **11**, 495. 1956.
3. MIKHAILOV, V.A.—*ZhAKh*, **13**, 494. 1958.
4. MARKOV, V.K., E.A. VERNYI, A.V. VINOGRADOV, S.V. ELINSON, A.E. KLYGIN, and I.V. MOISEEV. *Uran metody ego opredeleniya (Methods of Determination of Uranium)*, 2nd ed.—Atomizdat. 1964.
5. *Analiticheskaya khimiya urana (Analytical Chemistry of Uranium)*,—Izd. AN SSSR. 1962. [English translation available as part of the *Analytical Chemistry of the Elements* series, Ann Arbor–Humphrey Sci. Pub. 1970]
6. *Metody opredeleniya radioaktivnykh elementov v mineral'nom syr'e (Determination of Radioactive Elements in Mineral Raw Materials)*,—Gosgeoltekhizdat. 1958.
7. CHERNIKHOV, YU.A. and B.M. DOBKINA.—*Zav. Lab.*, **15**, 1143. 1949.
8. PALEI, P.N.—In: *Proc. Int. Conf. Peaceful Uses of Atomic Energy, Geneva*, 1955.
9. PRZHEVAL'SKII, E.S., E.R. NIKOLAEVA, and N.S. KLIMOVA.—*Vestnik MGU*, math., mech., astron., phys., chem. ser., No. 3, 217. 1958
10. SEN SARMA, R.N. and A.K. MALLIK.—*Z. anal. Chem.*, **148**, 179. 1955.
11. SYROKOMSKII, V.S. and YU.V. KLIMENKO. *Vanadatometriya (Vanadatometry)*.—Metallurgizdat. 1951.
12. MORACHEVSKII, YU.V. and I.A. TSERKOVNITSKAYA.—*ZhAKh*, **13**, 337. 1958.
13. KLYGIN, A.E., I.D. SMIRNOVA, and N.A. NIKOL'SKAYA.—*ZhNKh*, **4**, 2766. 1959.
14. CHENG, K.L.—*Anal. Chem.*, **30**, 1027. 1958.
15. GILL, H.H., R. ROLF, and W. ARMSTRONG.—*Anal. Chem.*, **30**, 1788. 1958.
16. LASSNER, E. and R. SCHARF.—*Z. anal. Chem.*, **164**, 398. 1958.
17. KUZNETSOV, V.I.—*Doklady AN SSSR*, **31**, 895. 1941; *ZhAKh*, **7**, 226. 1952; *ZhAKh*, **13**, 220. 1958.
18. FRITZ, J.S. and M. JOHNSON-RICHARD.—*Anal. chim. acta*, **20**, 164. 1959.
19. KUZNETSOV, V.I. and T.N. KUKISHEVA.—*Zav. Lab.*, **26**, 1344. 1960.
20. KUZNETSOV, V.I., S.B. SAVVIN, and V.A. MIKHAILOV.—*Uspekhi Khimii*, **29**, 525. 1960.
21. LUK'YANOV, V.F., L.M. MOISEEVA, and N.M. KUZNETSOV.—*ZhAKh*, **16**, 448. 1961.
22. MOISEEVA, L.M., N.M. KUZNETSOVA, V.F. LUK'YANOV, and G.L. SEL'MANOVA.—*ZhAKh*, **16**, 585. 1961; TITOV, V.I. and E.P. OSIKO.—*ZhAKh*, **17**, 129. 1962.
23. KUZNETSOV, V.I. and I.V. NIKOL'SKAYA.—*Zav. Lab.*, **26**, 266. 1960.
24. SAVVIN, S.B.—*Doklady AN SSSR*, **127**, 1231. 1959.
25. LUK'YANOV, V.F., S.B. SAVVIN, and I.V. NIKOL'SKAYA.—*ZhAKh*, **15**, 311. 1960.
26. KUZNETSOV, V.I. and S.B. SAVVIN.—*Radiokhimiya*, **2**, 682. 1960.
27. PALEI, P.N., A.A. NEMODRUK, and A.V. DAVYDOV.—*Radiokhimiya*, **3**, 181. 1961.
28. POLLARD, F.H., P. HANSON, and W.J. GEARY.—*Anal. chim. acta*, **20**, 26. 1959.
29. BUSEV, A.I. and V.I. IVANOV.—*Vestnik MGU*, chem. ser. II, No. 3, 52. 1960.

Titanium Ti

Titanium has the valencies +2, +3, and +4. The compounds of titanium (IV) are the most stable. Trivalent titanium compounds are readily oxidized by atmospheric oxygen. Bivalent titanium compounds disproportionate in water with the liberation of metallic titanium and formation of Ti (III) compounds.

The standard electrode potentials at 25°C in aqueous solution, relative to standard hydrogen electrode, have the following values.

$Ti \rightleftharpoons Ti^{2+} + 2e$	−1.630 V
$Ti + H_2O \rightleftharpoons TiO + 2H^+ + 2e$	−1.306 V
$Ti + 6F^- \rightleftharpoons TiF_6^{2-} + 4e$	−1.19 V
$Ti^{2+} \rightleftharpoons Ti^{3+} + e$	−0.368 V
$Ti^{2+} + H_2O \rightleftharpoons TiO^{2+} + 2H^+ + 2e$	−0.135 V
$Ti^{3+} + H_2O \rightleftharpoons TiO^{2+} + 2H^+ + e$	−0.100 V

The solutions contain the violet ions Ti^{3+}, the colorless ions Ti^{4+}, TiO^{2+}, $[TiHal_6]^{2-}$, and different polymeric ions.

In solutions of titanium (IV) sulfate, and in solutions acidified with sulfuric acid, the fairly stable complex ions $[TiO(SO_4)_2]^{2-}$ are formed, with the result that certain analytical reactions for titanium (IV) become negative.

Titanium (IV) ions are masked by oxalic, tartaric, and citric acids which form complexes with a small degree of dissociation.

The reactions of titanium (III) ions are similar to those of iron (III) and aluminum (III) ions.

Ti(III) ions are readily hydrolyzed and are oxidized by atmospheric oxygen to Ti(IV) ions.

In an inert atmosphere solutions of titanium (III) salts are employed as reducing agents (titanometry). They reduce the ions Fe^{3+}, NO_3^-, ClO_3^-, and ClO_4^- in the heat.

Potassium hydroxide solutions precipitate the dark-violet hydroxide $Ti(OH)_3$. The precipitate separates out at pH > 3; $pL_{Ti(OH)3} = 35$. The action of atmospheric oxygen converts $Ti(OH)_3$ into titanic acid. In the presence of sodium potassium tartrate $Ti(OH)_3$ does not precipitate out, but an intense blue color develops, and at small concentrations a green color develops which eventually vanishes.

When sodium acetate is added to a solution of a Ti(III) salt, a dark-green color develops; on heating a grayish-blue precipitate separates out.

When concentrated KSCN or NH_4SCN solutions are added to titanium (III) solutions, an intense violet color develops.

The colorless Ti^{4+} cations are stable only in strongly acid solutions. TiO^{2+} and other ions predominate at lower acidities. The formation of sparingly soluble basic salts begins at pH 1.9; at pH 4 the hydrolysis is almost complete. When dilute acid solutions of Ti(IV) salts are boiled, a white precipitate of titanic acid separates out:

$$Ti^{4+} + 3H_2O \rightleftharpoons H_2TiO_3 + 4H^+.$$

Complexone III reacts with Ti^{4+} ions to form an unstable compound [1], which is hydrolyzed at high pH values. A more stable compound is obtained in the presence of hydrogen peroxide [2, 3]. The peroxytitanyl ions $[OTi(H_2O_2)]^{2+}$ thus formed react with Complexone III in a molar ratio of 1:1, when the compound $Na_2[OTi(H_2O_2)Y]$ is formed.

Titanium (IV) ions are reduced to titanium (III) by metallic zinc, cadmium, or tin in acid solutions. Titanium (IV) ions are not reduced by sulfite or sulfide ions.

Ammonia solutions precipitate the white hydroxide $Ti(OH)_4$ out of titanium (IV) salt solutions; the precipitate is practically insoluble in excess precipitant and is sparingly soluble in alkali hydroxide solutions. $Ti(OH)_4$ also separates out when strongly alkaline solutions of titanates are diluted with water; on being boiled $Ti(OH)_4$ rapidly ages and its solubility decreases.

Potassium fluoride yields a TiF_4 precipitate, which is converted into the rather sparingly soluble complex compound K_2TiF_6 in the presence of excess precipitant.

Hydrofluoric acid reacts with titanium (IV) to form the acid H_2TiF_6. When titanium (IV) salts are evaporated with hydrofluoric acid and concentrated sulfuric acid, the nonvolatile TiF_4 precipitate remains; it can be converted to TiO_2 by ignition and treatment with sulfuric acid.

Ferrocyanide ions form an orange-gray precipitate with titanium (IV) ions.

Acid sodium phosphate and arsenate react with titanium (IV) ions to form white precipitates which are sparingly soluble in acetic acid; aluminum and ferric ions give similar precipitates. Unlike zirconium, the precipitation does not take place in the presence of hydrogen peroxide.

When hydrogen peroxide is added to acid solutions of titanium (IV) salts, a yellow or orange color develops due to the formation of peroxytitanic acid. When the solution is made alkaline, the color disappears owing to the formation of a colorless complex. When crystalline potassium fluoride KF is added to a colored solution of a titanium peroxy compound, the colorless anions TiF_6^{2-} are formed.

Titanium (IV) ions in weakly acid solutions react with aliphatic and aromatic hydroxy compounds with the groupings

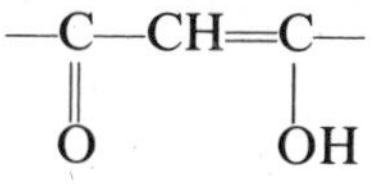

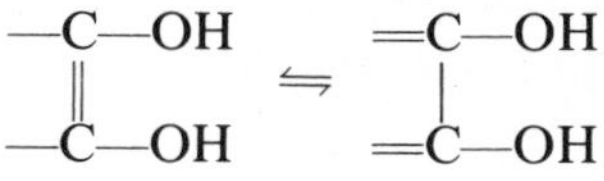

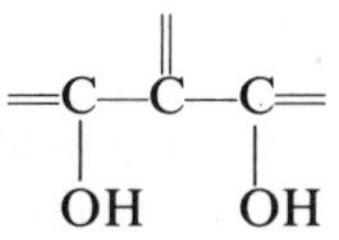

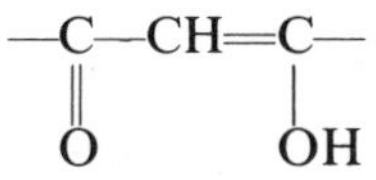

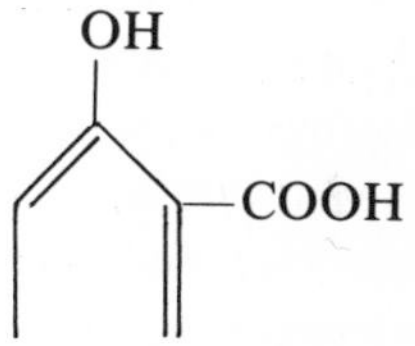

$$-\underset{\substack{\| \\ O}}{C}-CH=\underset{\substack{| \\ OH}}{C}- \qquad \begin{matrix} -C-OH \\ \| \\ -C-OH \end{matrix} \rightleftharpoons \begin{matrix} =C-OH \\ | \\ =C-OH \end{matrix}$$

$$=\underset{\substack{| \\ OH}}{C}-\overset{\substack{\| \\ }}{C}-\underset{\substack{| \\ OH}}{C}= \qquad -\underset{\substack{\| \\ O}}{C}-CH=\underset{\substack{| \\ OH}}{C}- \qquad C_6H_4(OH)COOH$$

to form colored compounds [4–6].

Soluble yellow or yellow-orange compounds in weakly acid solutions are formed by o-diphenols, o-phenolcarboxylic acids, hydroxy-γ-pyrones, flavones, aliphatic 1,3-diketones and enediols. At pH 3.3–4.7 peri-diphenols (1,8-dihydroxynaphthalene, chromotropic acid) form soluble red-violet compounds. In sodium acetate medium (pH 3.8–5.5) the red-violet compounds are converted into stable orange colored compounds.

Chromotropic acid (1,8-dihydroxynaphthalene-3,6-disulfonic acid) reacts with TiO^{2+} in weakly acid solutions to form a soluble reddish brown compound [7]; in sodium acetate medium an orange-red compound is formed, and in concentrated sulfuric acid a red-violet compound is produced [8]. Chromotropic acid is used for the detection of titanium and its photometric determination (p. 123).

2,7-Dichlorochromotropic acid [9] reacts with titanium (IV) ions similarly to chromotropic acid. 2,7-Dichlorochromotropic acid is used for the photometric determination of microgram amounts of titanium in uranium [10], steels [11], vanadium-containing aluminum alloys [12], metallic beryllium [13], and ores and rocks [14].

At pH 2.0 the ions of titanium (IV) react with 2,7-dichlorochromotropic acid in the molar ratio of 1:3. The absorption maximum of the resulting compound is at 490–500 mμ. At pH 4.5 the reaction takes place in the molar ratio of 1:4, and the absorption maximum of the solution is at 470 mμ [15].

Titanium (IV) reacts with pyrocatechol to yield compounds differing according to the pH and the excess of the reagent present [16, 17]. When pyrocatechol is present in a large excess, the complex $[Ti(C_6H_4O_2)_3]^{2-}$ is formed. Diantipyrylmethane increases the stability of the complex of titanium (IV) with pyrocatechol in acid solutions [18]. A sparingly soluble compound is formed in which the molar ratio diantipyrylmethane: titanium (IV): pyrocatechol = 1:1:2. This compound is extracted with chloroform or dichloroethane from 1–2 *N* HCl medium. The extracts have the absorption maximum at 330 mμ. At lower acidities of the aqueous phase, the absorption maximum of the extracts shifts toward longer wavelengths.

When photometrically determining titanium [19] as the compound with pyrocatechol and diantipyrylmethane, the extraction is carried out with dichloroethane from 1 *N* hydrochloric acid. Ferric iron is reduced by ascorbic acid. V(IV) does not interfere, but V(V), Mo(VI), and W(VI) interfere strongly. If the optical density is measured at 320 mμ, the sensitivity of the determination is 6 μg of Ti in 25 ml of extract and if at 420 mμ, 10 μg of Ti in 25 ml of extract.

In concentrated sulfuric acid solution titanium (IV) forms colored compounds with phenols and polyphenols: an orange-yellow color is given by thymol and a red color by salicylic acid.

9-Methyl-2,3,7-trihydroxy-6-fluorone reacts with titanium (IV) ions in the molar ratio of 2:1 with the formation of a pink-colored complex with absorption maximum at 520 mμ [20].

The molar extinction coefficient is 60,400. Beer's law is obeyed between 0.025 and 0.6 μg of Ti/ml. The optimum pH value is at 1.7–2.1.

9-(2′,4′-Disulfophenyl)-2,3,7-trihydroxy-6-fluorone (disulfophenylfluorone) reacts with titanium ions [21] at pH 6 in the molar ratio of 2:1. The colored solutions have an absorption maximum at 570 mμ and the optical density of the solutions is proportional to the concentration of the titanium compound. The sensitivity is 0.01 μg of Ti/ml. The reagent is used in the determination of more than $5 \cdot 10^{-6}\%$ of titanium in germanium and silicon.

Ascorbic acid reacts with titanium (IV) ions in weakly acid solutions (pH 3.5–6) with the formation of yellow or orange-colored soluble compounds. Ascorbic acid as a reagent for titanium is less sensitive than polyphenols, but it is more selective and is used in the photometric determination of this element [22–24].

Cupferron precipitates the yellow salt (I), while 8-hydroxyquinoline precipitates the yellow-orange salt (II) from acid solutions containing titanium ions.

$$\left(\begin{array}{l} C_6H_5\text{—N—O} \\ \quad\;\;\;\; | \quad\;\; \searrow \\ \quad\;\;\;\; \text{N=O} \nearrow \end{array}\right)_4 Ti \qquad (C_9H_6NO)_2TiO \cdot 2H_2O$$

I II

Both reagents are used for the separation of titanium ions from other ions, aluminum ions in particular. Titanium cupferronate is quantitatively extracted with chloroform; titanium is separated from aluminum in this manner.

Diantipyrylmethane reacts with titanium (IV) in hydrochloric acid solutions to form colored compounds and is used in its photometric determination in various alloys (p. 127). It is one of the most selective and highly sensitive reagents for titanium. It is used for the determination of titanium in vanadium and in vanadium oxychloride, in niobium, molybdenum, aluminum and magnesium alloys, steels and heat-resistant alloys based on nickel and iron.

Titanium (IV) reacts with SCN^- ions and with diantipyrylmethane in 2 *N* HCl to form an intensely colored precipitate, which is readily extracted with chloroform and with dichlorethane [25]. The probable formula of the compound is $[TiDiant(SCN)_4]$. The absorption maximum of the extract is at 420 mμ and the molar extinction coefficient reaches 60,000.

The formation of this compound constitutes the basis for the extraction-photometric method for the determination of titanium in steels [26] (p. 130).

N-Benzoyl-N-phenylhydroxylamine reacts with titanium (IV) in the molar ratio of 1:1 in 60% ethanolic solution [27] containing $1.06 \cdot 10^{-2}$ mole/liter of tartaric acid at pH 1.8 ± 0.1. The colored solutions obey Beer's law in the concentration range between 3 and 160 μg of Ti/10 ml of solution at 340, 370, and 400 mμ, with a deviation which does not exceed 2%. Iron (III), vanadium (IV), and molybdenum (VI) interfere. Titanium may be determined [27] in the presence of Nb(V), Ta(V), Zr(IV), and W(VI). N-Benzoyl-N-phenylhydroxylamine is used in the gravimetric determination [28] of titanium (IV) in the presence of aluminum during the separation and determination [29] of Nb(V), Ta(V) and Ti(IV).

3-Hydroxy-1-(p-chlorophenyl)-3-phenyltriazine forms with TiO^{2+} an orange-colored precipitate which is stable on heating. This reagent [30] makes it possible to separate titanium (IV) from Al(III), Be, Mg, Th, etc.

For a survey of the methods used in the analysis of titanium, see [31].

Complexometric determination of titanium in alloys [32–36]

In the presence of hydrogen peroxide Complexone III (Na_2H_2Y) forms with titanium (IV) a compound which is stable below pH 2:

$$[TiO(H_2O_2)]^{2+} + Na_2H_2Y \rightarrow Na_2[TiO(H_2O_2)Y] + 2H^+.$$

The instability constant is $K \approx 10^{-20}$ (p$K \sim 20.4$).

Titanium is determined indirectly, by titrating the excess Complexone III against a solution of ferric salt in the presence of salicylic acid or, which is better, against a solution of a bismuth salt in the presence of Xylenol Orange. In the latter case the optimum acidity is 0.3 *N* HCl. At concentrations above 0.5 *N* HCl the negative error increases; at acidities below 0.2 *N* HCl niobium interferes, as it also forms a stable compound with Complexone and hydrogen peroxide.

The method is suitable for the determination of titanium alloyed with uranium, zirconium, molybdenum, tungsten, and niobium.

The following metals do not interfere: Mg, Mn, Zn, La, Ce(III), Al, and W (up to

50 mg), Ta (up to 70 mg), Mo (up to 100 mg), Cu (up to 5 mg), Nb (up to 60 mg), and U (up to 200 mg). Ferric iron and zirconium, which form acid-stable complexonates, interfere. When determining titanium (IV) in the presence of iron (III) and zirconium (IV), Fe (III) + Zr (IV) are titrated first in the absence of hydrogen peroxide; hydrogen peroxide is then introduced and the sum of Ti (IV), Fe (III), and Zr (IV) is titrated.

The method is suitable for the determination of titanium in different materials.

REAGENTS

Complexone III, 0.01 *M* solution.
Bismuth nitrate, 0.01 *M* solution.
Hydrogen peroxide, 30% solution.
Xylenol Orange, 0.5% aqueous solution.
Sulfuric acid, $d = 1.84$ g/cm^3, and 1 *N* solution.
Ammonium sulfate, crystalline.
Hydrochloric acid, $d = 1.17$–1.19 g/cm^3.
Nitric acid, $d = 1.4$ g/cm^3.

PROCEDURE

Determination of titanium in niobium-based alloys

A 0.1–0.2 g sample of the alloy in a 100-ml heat-resistant beaker is treated with 2–3 ml of sulfuric acid ($d = 1.84$ g/cm^3) in the presence of 1–2 g of ammonium sulfate. The mixture is heated until the metal dissolves and the solution is evaporated on a sand bath to a residual volume of 0.5–1 ml. To the residue 3–5 ml of hydrogen peroxide are added. The residue is dissolved in 1 *N* sulfuric acid. The solution is transferred to a 100-ml volumetric flask and diluted to the mark with 1 *N* sulfuric acid. A 20 ml (1–5 mg of Ti) aliquot is transferred with a pipet into a 300-ml conical flask; 10 ml of 1 *N* sulfuric acid are added. The solution is diluted to 100 ml with water; 3–5 drops of hydrogen peroxide and then 15 ml of 0.01 *M* Complexone III solution are added. The solution is stirred and left to stand for 15 minutes, 3–4 drops of Xylenol Orange solution are added, and the excess Complexone III is titrated against the bismuth nitrate solution until the yellow solution turns orange-red.

One ml of 0.01 *M* Complexone III solution is equivalent to 0.479 mg of titanium.

Note. If the aliquot contains more than 20 mg of Nb, the excess Complexone III is titrated against the bismuth nitrate solution, and there is a sharp color change at the end point.

Determination of titanium in Ti–Nb–Mo, Ti–Nb–W, Ti–U, Ti–U–Al, Ti–U–Mo, Ti–U–W, Ti–Nb–Al *alloys*

The weighed sample of the alloy, which should contain 5–25 mg of Ti, is dissolved in a hot mixture of 8–10 ml of hydrochloric and 1–2 ml of nitric acids in a heat-resistant, 100-ml beaker on a sand bath. Three ml of sulfuric acid ($d = 1.84$ g/cm^3) are added to the resulting solution and the solution is evaporated to an abundant evolution of sulfuric acid fumes. The residue is dissolved in water and the solution is transferred to a 100-ml volumetric flask and diluted to the mark with water. A 20 ml aliquot of the resulting solution is transferred with a pipet to a 300-ml conical flask; 30 ml of 1 *N* sulfuric acid are added; the solution is diluted with water to 100 ml, and the determination is continued as described for the determination of titanium in niobium-based alloys.

Photometric determination of titanium in alloys by chromotropic acid [37, 38]

Chromotropic acid (1,8-dihydroxynaphthalene-3,6-disulfonic acid) forms two compounds with titanium (IV): at pH 1–4 a red-colored compound with a 2:1 molar ratio of the components (absorption maximum at 470 mμ, molar extinction coefficient 12,000), and at pH 3–7 a yellow compound with a 3:1 molar ratio of the components (absorption maximum at 423 mμ, molar extinction coefficient 18,400). The more stable red complex is preferred for the photometric determination of titanium; the optical density is measured at 520 mμ, since the difference between the absorptions of the red and yellow complexes is greater at this wavelength.

The detectable minimum is 3 μg of Ti; the limiting dilution is 1:10,000.

The light absorption is proportional to the concentration of titanium between 5 and 200 μg of Ti/50 ml. Uranium (VI) and iron (III) interfere as they react with chromotropic acid to form gray and green compounds, respectively. Prior to the determination of Ti (IV), the Fe (III) and U (VI) ions are reduced to Fe (II) and U (IV) with sodium sulfite or sodium bisulfite.

Vanadium (V) ions do not interfere. The determination can also be carried out in the presence of small amounts of tartrates, chlorides, sulfates, and molybdates. Nitrite, nitrate, fluoride and oxalate ions interfere and must be removed by evaporation with sulfuric acid.

Using this method it is possible to determine titanium in different objects such as high-alloy steels. Chromotropic acid is rapidly oxidized and for this reason solutions must be freshly prepared and the pH of the solution should be strictly controlled.

REAGENTS

Standard titanyl sulfate solution, in 1 *N* sulfuric acid, 10 μg of Ti/ml.
Ammonia, 1 *N* solution.
Chromotropic acid, 2.5% aqueous solution. Chromotropic acid (2.5 g) is mixed with 0.15 g sodium sulfite and diluted with water to 100 ml. The solution is stable for 2–3 days.
Monochloroacetic acid, 5% solution.

Hydrochloric acid, dilute (1:1) solution.
Cupferron, 0.6% and 6% solutions.
Potassium pyrosulfate, crystalline.
Ammonium thiocyanate, 5% solution.
Sodium sulfite or sodium bisulfite, crystalline.
Oxalic acid, 5% solution.

CONSTRUCTION OF CALIBRATION CURVE

Into five 50-ml volumetric flasks are introduced 1, 2, 3, 4, and 5 ml of standard titanyl sulfate solution and 0–4 ml ammonia solution and the solutions are mixed; two ml of chromotropic acid solution and 10 ml of monochloroacetic acid solutions are added to each flask. The solutions are diluted with water to 50 ml and thoroughly mixed. The optical densities of the solutions are measured with a universal photometer at 520 mμ or an electrophotocolorimeter, relative to a solution of all the above reagents in the same amounts except titanyl sulfate.

PROCEDURE

To 0.5–1.0 g of steel (0.1–1.0% Ti) are added 30 ml of hydrochloric acid and the solution is heated until hydrogen bubbles are no longer evolved. The solution is diluted with water to 200 ml and cooled to 5–10°C; 6% cupferron solution is added to precipitate vanadium, titanium, and part of the iron. The extent of the precipitation is checked and the solution is left at this temperature for $1\frac{1}{2}$–2 hours, with frequent stirring. The precipitate is filtered through White Ribbon filter paper, washed with 50–100 ml of 0.6% Cupferron solution, transferred to a quartz or a platinum crucible, carefully ashed and ignited. The residue is fused with 3–4 g of potassium pyrosulfate while gradually increasing the temperature. The crucible with the homogeneous melt is transferred to a 200-ml beaker; 30 ml of water are added and then 1–2 drops of ammonium thiocyanate solution and a few crystals of sodium sulfite or sodium bisulfite to decolorize the solution. Thirty ml of oxalic acid solution are added and the solution is diluted to the mark with water in a 200-ml volumetric flask and is mixed.

Five ml of the solution are withdrawn with a pipet to a 50-ml volumetric flask and all the operations needed for the construction of the calibration curve are performed. The amount of Ti in the aliquot is found from the calibration curve.

Photometric determination of titanium (IV) in aluminum alloys by 2,7-dichlorochromotropic acid [9–12]

The rate of oxidation in daylight of 2,7-dichlorochromotropic acid solutions is 400 times slower than that of chromotropic acid solutions.

2,7-Dichlorochromotropic acid reacts with titanium (IV) in acid medium above pH 0.2 to form a raspberry-colored complex; the optimum pH range is between

1.5 and 2.5 (Figure 6). At pH 2 the reagent is 14 times more sensitive to Ti(IV) ions than hydrogen peroxide. The absorption maximum of the solution of the compound is at 490 mμ. The molar extinction coefficient is 11,200 at pH 2 and 8000 at pH 1. The optical density is proportional to concentration in the range between 0.1 and 5 μg of Ti/ml.

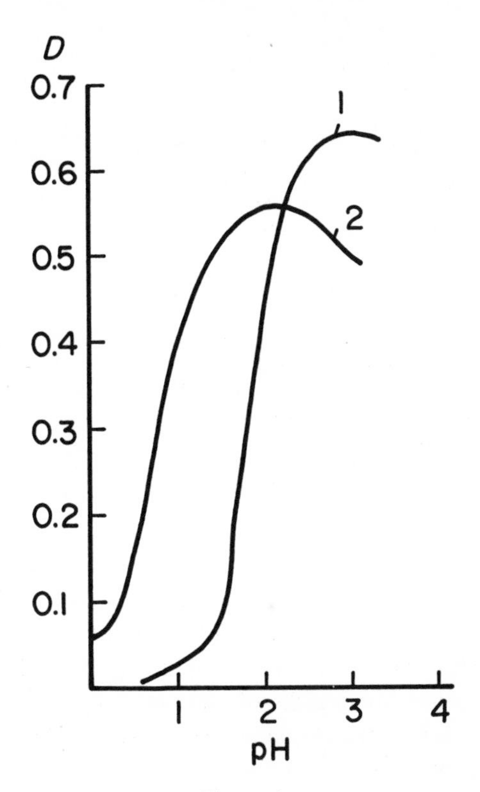

Figure 6

Effect of pH on the optical density of solutions of titanium compounds:
1) with chromotropic acid (at 450 mμ); 2) with 2,7-dichlorochromotropic acid (at 490 mμ). Concentrations of chromotropic and dichlorochromotropic acids $5 \cdot 10^{-4}$ M; concentration of titanium $5 \cdot 10^{-5}$ M; layer thickness 1 cm.

The following elements do not interfere: Ag, Al, Ba, Be, Bi, Ca, Co, Cr(III), Cu(I, II), Fe(II), Ga, Hg(I, II), In, Mg, Mn(II), Ni, Pb, Pt(IV), Sb(III), Se(VI), Sn(II, IV), Sr, Te(IV), Th, Tl(III), rare earths, Zn, and Zr.

The only ions which interfere with the determination at pH 1 are Fe(III), Cr(VI), V(V), Mo, and W. Molybdenum and tungsten ions are masked by phosphate, and the remaining ions are reduced by ascorbic acid, which may be present in 100-fold excess. The interfering anions include ethylenediaminetetraacetate and fluoride ions; 100-fold amounts of tartrate and phosphate ions may be present.

2,7-Dichlorochromotropic acid is acid to determine titanium in steels, uranium, vanadium-containing aluminum alloys, and in water by a photometric method.

REAGENTS

Standard titanyl sulfate solution, 2 μg of Ti/ml. The reagent is dissolved in 0.1 *N* sulfuric acid.
2,7-*Dichlorochromotropic acid,* disodium salt, 0.5% solution.
Sulfuric acid, dilute (1:3) solution.
Hydrochloric acid, 1 *N* and dilute (1:1) solutions.
Reducing agent, for Fe(III) and V(V), prepared by mixing 5% solution of hydroxylamine hydrochloride with an equal volume of 5% solution of ascorbic acid.
Sodium acetate, 50% aqueous solution.

CONSTRUCTION OF CALIBRATION CURVE

Metallic aluminum ("pure") is dissolved in 100-ml volumetric flasks; standard titanium salt solution is added in amounts indicated in the procedure, and the optical densities of the colored solutions are measured at 490 mμ. The calibration curve is plotted from the results.

PROCEDURE

An aluminum sample of 1 g which contains between 0.005 and 0.4% of Ti (and 0.01–0.2% of V) is used. The alloy is dissolved in a conical flask in a mixture of 20 ml of sulfuric acid and 2 ml of hydrochloric acid ($d = 1.12$ g/cm^3). The solution is diluted to 30 ml with water and is filtered into a 100-ml volumetric flask (for titanium contents of a few hundredths of one per cent) or into a 250-ml flask (for titanium contents of a few tenths of one per cent), using a White Ribbon filter paper. The flask used for the precipitation and the filter paper are thoroughly washed with hot water and the solution in the volumetric flask is cooled, diluted to the mark with water and mixed. A 10 ml or 5 ml aliquot is withdrawn (for a few hundredths and a few tenths of one per cent Ti, respectively) into a 100-ml volumetric flask. A small volume of water and then 5 ml of reducing agent solution are added and the mixture is neutralized with sodium acetate solution until violet to Congo Red paper. Ten ml of 1 *N* HCl and 5 ml of 2,7-dichlorochromotropic acid solution are added; the solution is diluted to the mark with water and is mixed. The optical density of the resulting solution is measured in a photometer or a FEK-N-57 photocolorimeter at 490 mμ (blue filters) in a 50-mm cell for a few hundredths of one per cent and in a 30- to 50-mm cell for a few tenths of one per cent of titanium.

The comparison solution is the aluminum solution described above which has been taken through all the analytical operations.

The content of titanium is found from the calibration curve.

Extraction-photometric determination of titanium in steels by 2,7-dichlorochromotropic acid [11]

The extraction-photometric method determines very small amounts of titanium (0.01–1 μg/ml). The added diphenylguanidine forms a virtually undissociated compound with the sulfo groups of 2,7-dichlorochromotropic acid, which is extracted with n-butanol or isoamyl alcohol. Even smaller amounts (0.001–0.01 μg/ml) are determined by coprecipitating the titanium compound with 2,7-dichlorochromotropic acid with diphenylguanidine α-anthracenesulfonate. The precipitate is filtered off and dissolved in n-butanol.

Molybdenum (VI) and tungsten (VI) are extracted together with titanium; Cr(III), Ni, Cu, and V(IV) are not extracted. Compounds of Fe(III), V(V), and Cr(VI) are reduced by ascorbic acid.

This method can be used for the determination of titanium in multicomponent steels.

REAGENTS

Standard titanium salt solution, 5 μg of Ti/ml.
Sulfuric acid, dilute (1:8).
Nitric acid, $d = 1.4\ g/cm^3$.
Sodium hexametaphosphate, 1% solution (stable for 1 week).
Ascorbic acid, 2% solution.
2,7-*Dichlorochromotropic acid*, 1% solution.
Diphenylguanidine hydrochloride, 20% solution.
n-Butanol.

CONSTRUCTION OF CALIBRATION CURVE

The series of solutions should contain 5, 10, 15, 20, and 25 μg of Ti/ml and should be prepared as described in the procedure.

PROCEDURE

A 0.25 g sample of steel, containing 0.02–0.1% Ti, is dissolved in 20 ml of sulfuric acid by heating in a heat-resistant, 100-ml beaker on a sand bath. One ml of nitric acid is added. The solution is boiled to expel nitrous oxides, it is then transferred to a 50-ml volumetric flask, diluted to the mark with water and thoroughly mixed. Five ml portions of the resulting solution are transferred with a pipet into two 50-ml separating funnels and 5 ml of sodium hexametaphosphate solution is added to one of the funnels. To each funnel are added 10 ml of ascorbic acid solution and 2 ml of 2,7-dichlorochromotropic acid solution. The solutions are mixed and 4 ml of diphenylguanidine hydrochloride solution is added drop by drop to each funnel for 5 minutes. The solutions are thoroughly mixed; 10 ml of n-butanol are added to each solution, and the colored compound is extracted by shaking for 2 minutes. The optical density of the extract is determined in an electrophotocolorimeter relative to the blank solution and the titanium content is found from the calibration curve.

Note 1. If hexametaphosphate is introduced into the blank solution, titanium can be determined in the presence of a large number of different cations. Tungsten (VI), which forms a colored compound with 2,7-dichlorochromotropic acid under these conditions, cannot be masked by hexametaphosphate.

Note 2. It is preferable to extract from sulfuric acid and not from hydrochloric acid medium.

Photometric determination of titanium by diantipyrylmethane [39–43]

Diantipyrylmethane reacts with titanium ions in acid medium to form a soluble yellow compound with an absorption maximum at 380 mμ. Variations in acidity between 0.5 and 4 *N* HCl do not affect the optical density of the solutions. The sensitivity is 0.01 μg of Ti/ml, and the molar extinction coefficient is 18,000. The color develops within 45 minutes and is stable for several months. The optical density of the solutions is proportional to the concentration of titanium in a wide concentration range.

The following ions do not interfere: Mg, Al, Zn, Cd, Mn, Cu, Zr, REE, Mo, Nb, Ta, and V(IV). The ions of Fe(III) and V(V) are reduced by hydroxylamine.

The method is suitable for the determination of titanium in light alloys, ferrous alloys, and nonferrous alloys.

DETERMINATION OF TITANIUM IN THE PRESENCE OF VANADIUM IN NICKEL-BASED ALLOYS

REAGENTS

Standard titanium salt solution, 5 μg of Ti/ml.

Diantipyrylmethane, 1% solution. Ten g of diantipyrylmethane are dissolved in 300–400 ml of water, and acidified with 15 ml sulfuric acid (d = 1.84 g/cm^3). The solution is diluted to one liter with water.

Sodium acetate, 50% solution.

Mixture of concentrated nitric and hydrochloric acids, 1:3.

Sulfuric acid, dilute (1:1).

Hydroxylamine, 10% solution.

Hydrochloric acid, dilute (1:1).

CONSTRUCTION OF CALIBRATION CURVE

Into five 25-ml volumetric flasks are introduced 1, 2, 3, 4, and 5 ml of the standard titanyl salt solution; 6 ml of diantipyrylmethane solution and 2.5 ml of hydrochloric acid are added to each flask and the solutions are made up to the mark with water. The optical density of the series of solutions is measured relative to water in an electrophotocolorimeter and the calibration curve is plotted.

PROCEDURE

A 0.25 g sample of the alloy, which should contain 0.01–0.1% Ti, is dissolved in 15 ml of nitric acid–hydrochloric acid mixture; 15 ml sulfuric acid is added to the solution and the mixture is evaporated until thick white fumes appear. The residue is cooled, and 30 ml of water are added. The solution is stirred until solubilization is complete, it is transferred to a 50-ml flask and diluted to mark with water.

An aliquot containing 5–25 μg of Ti is withdrawn with a pipet; sodium acetate solution is added until red to Congo Red paper, and then hydrochloric acid solution is added drop by drop until the solution becomes lilac-colored. Hydroxylamine solution (2.5 ml) is added and after 5 minutes 2.5 ml of hydrochloric acid and 6 ml of diantipyrylmethane solution.

The blank solution is prepared as above from the solution of the alloy, except that diantipyrylmethane is not added.

After 45–50 minutes the optical density of the sample solution is measured relative to the blank solution. The titanium content in the aliquot is read off the calibration curve.

DETERMINATION OF TITANIUM IN THE PRESENCE OF IRON AND ALUMINUM

REAGENTS

Standard titanyl salt solution, 5 μg of Ti/ml.
Diantipyrylmethane, 5% solution.
Hydrochloric acid, $d = 1.17$–1.19 g/cm^3.
Chloroform, pure.
Sulfuric acid, dilute (1:1).
Mixture of concentrated nitric and hydrochloric acids, (1:3).

CONSTRUCTION OF CALIBRATION CURVE

Into five 25-ml volumetric flasks are introduced 1, 2, 3, 4, and 5 ml of the standard titanyl salt solution; 5 ml of diantipyrylmethane solution and 5 ml of hydrochloric acid are added to each flask. The solutions are made up to the mark with water and the optical density is determined in an electrophotocolorimeter relative to water. The calibration curve is then plotted.

PROCEDURE

Ferric ions in the presence of halides or thiocyanates are precipitated by diantipyrylmethane. The resulting compound is extracted with chloroform. In this way iron is separated from titanium. Titanium is determined photometrically in the aqueous phase by diantipyrylmethane. If the titanium content is very small, the determination may be terminated by the extraction-photometric technique, the titanium compound of diantipyrylmethane being extracted with chloroform after the separation of iron.

The sample of the material to be analyzed, which should contain 100–200 μg of Ti, is dissolved with heating in 10 ml of sulfuric acid on a sand bath. The solution is evaporated to a residual volume of 1–2 ml and the residue is dissolved in water. The solution is diluted to 15 ml with water, and 15 ml of hydrochloric acid, 10 ml of diantipyrylmethane solution and 10 ml of chloroform are added. The iron compound is extracted for 5–6 minutes. The organic phase is separated. Chloroform and 10 ml of diantipyrylmethane solution are added to the aqueous phase and the extraction of the iron is repeated.

The aqueous phase is transferred to a 50-ml volumetric flask and is diluted to the mark with water. Ten ml of the resulting solution is withdrawn with a pipet into a 25-ml flask; 5 ml of diantipyrylmethane solution are added and the solution is diluted to the mark with water. The optical density is measured and the amount of titanium is found from the calibration curve.

Extraction-photometric determination of titanium by thiocyanate and diantipyrylmethane [26]

The complex of diantipyrylmethane, thiocyanate, and titanium (IV), which is sparingly soluble in aqueous solutions, is extracted with purified chloroform from 2–4 *M* hydrochloric acid. The optical density of the extract is measured at 420 mμ; the molar extinction coefficient is 60,000 at this wavelength. The effects of ferric iron, cupric copper, and other elements are eliminated by adding sodium thiosulfate to the solution prior to the extraction.

The sensitivity of the determination is 2 μg of Ti/25 ml of the chloroform extract. The optical density of the extract is proportional to the concentration of titanium between 2 and 20 μg/25 ml of chloroform extract.

Ferrous iron, tantalum, chromium, aluminum, manganese, zinc, cadmium, tin, antimony, zirconium, Complexone III, fluorides and phosphates, and small amounts of nickel and vanadium do not interfere. Cobalt, tungsten, molybdenum, and niobium interfere.

In the determination of 0.03–0.6% concentrations of titanium in steels, the experimental error does not exceed ± 5%.

REAGENTS

Hydrochloric acid, $d = 1.19$ g/cm^3 and dilute (1:5).

Nitric acid, $d = 1.40$ g/cm^3.
Ammonium thiocyanate, 20% aqueous solution.
Sodium thiosulfate, 20% solution.
Diantipyrylmethane, 5% solution in 2 *N* HCl.
Purified chloroform. Chloroform is shaken in a separating funnel with concentrated sulfuric acid and is washed first with water and then KOH solution.
Standard titanium salt solution, 10 μg of Ti/ml.

CONSTRUCTION OF CALIBRATION CURVE

A 0.1 g sample of chrome–nickel steel, free from titanium (e.g., standard specimen of steel 125) is dissolved as described below in the procedure. The resulting solution is diluted with water in a 100-ml volumetric flask; a series of 5 ml portions of this solution are withdrawn into separating funnels. Standard titanium salt solutions of 2.4–24 μg of Ti are introduced and the sequence of operations is continued as described below in the procedure. The calibration curve is plotted from the resulting data.

PROCEDURE

A 0.1 g sample of the steel is dissolved with heating in 30 ml of 1:5 hydrochloric acid solution. The carbides are oxidized by heating with nitric acid. When the steel has dissolved, the solution is diluted with water to 40–50 ml; it is boiled to expel nitrous oxides, and is cooled. The solution is transferred to a 100-ml volumetric flask and is made up to the mark with water.

If the content of titanium in the steel is between 0.1 and 0.5%, 5 ml of the above solution are placed in a separating funnel, 5 ml of ammonium thiocyanate solution are added and then the thiosulfate solution is gradually introduced to remove the color of ferric thiocyanate. Five ml of diantipyrylmethane solution and 2 ml of concentrated hydrochloric acid are then added to the solution.

The titanium compound is extracted with two 10 ml portions of chloroform. The extracts are filtered through a dry filter paper into a 25-ml volumetric flask and the solution is made up to the mark with chloroform.

The optical density of the extract is determined in a photocolorimeter with a blue filter or a SF-4 spectrophotometer in a 20-mm cell at 420 mμ relative to the extract of the reagents.

BIBLIOGRAPHY

1. Přibil, R. *Komplexony v chemicke analyse,* 473pp.—Prague, Nakl. Českosl. Akad. Věd. 1957.
2. Wilkins, D. H.—*Anal. chim. acta,* **20**, 113. 1959.
3. Beiber, B. and Z. Večeřa.—*Coll.,* **26**, 2081. 1961.
4. Okač, A. and L. Sommer.—*Anal. chim. acta,* **15**, 345. 1956; *Coll.,* **22**, 464. 1957.
5. Sommer, L.—*Coll.,* **22**, 414, 453. 1957.
6. Sathe, R. M. and Ch. Venkateswarlu.—*Coll.,* **27**, 701. 1962.

7. TANANAEFF, N. A. and G. A. PATSCHENKO.—*Z. anorg. Chem.*, **150**, 163. 1926.
8. OKAČ, A. and L. SOMMER.—*Coll.*, **22**, 433, 464. 1957.
9. KUZNETSOV, V. I. and N. N. BASARGIN.—*ZhAKh*, **16**, 573. 1961.
10. KUZNETSOV, V. I., N. N. BASARGIN, and T. N. KUKISHEVA.—*ZhAKh*, **17**, 457. 1962.
11. BASARGIN, N. N., A. N. TKACHENKO, L. R. STUPA, and L. N. BORODAEVSKAYA.—*Zav. Lab.*, **28**, 1311. 1962.
12. BUDANOVA, L. M. and S. N. PINAEVA.—*Zav. Lab.*, **29**, 149. 1963.
13. BASARGIN, N. N., T. N. KUKISHEVA, and N. V. SOLOV'EVA.—*ZhAKh*, **19**, 553. 1964.
14. KLASSOVA, N. S. and L. L. LEONOVA.—*ZhAKh*, **19**, 131. 1964.
15. BASARGIN, N. N. and T. V. PETROVA.—*ZhAKh*, **19**, 835. 1964.
16. BABKO, A. K. and O. I. POPOVA.—*ZhNKh*, **2**, 147. 1957.
17. SHNAIDERMAN, S. YA. and I. E. KALINICHENKO.—*ZhNKh*, **6**, 1843. 1961.
18. BABKO, A. K. and M. M. TANANAIKO.—*ZhNKh*, **7**, 2549. 1962.
19. TANANAIKO, M. M. and G. N. VINOKUROVA.—*ZhAKh*, **19**, 316. 1964.
20. MAJUMDAR, A. K. and C. P. SAVARIAR.—*Anal. chim. acta*, **21**, 584. 1959.
21. NAZARENKO, V. A. and E. A. BIRYUK.—*ZhAKh*, **15**, 306. 1960.
22. HINES, E. and D. F. BOLTZ.—*Anal. Chem.*, **24**, 947. 1952.
23. KORKISCH, J. and A. FARAG.—*Mikrochim. acta.*, 659. 1958.
24. SOMMER, L.—*Coll.*, **28**, 449. 1963.
25. BABKO, A. K. and M. M. TANANAIKO.—*ZhNKh*, **7**, 562. 1962.
26. TANANAIKO, M. M. and S. L. NEBYLITSKAYA.—*Zav. Lab.*, **28**, 263. 1962.
27. SCHWARBERG, J. E. and R. W. MOSHIER.—*Anal. Chem.*, **34**, 525. 1962.
28. SHOME, S. C.—*Analyst*, **75**, 27. 1950.
29. LANGMYHR, F. J. and T. HONGSLO.—*Anal. chim. acta*, **22**, 301. 1960.
30. SOGANI, N. C. and S. C. BHATTACHARYYA.—*Anal. Chem.*, **28**, 1616. 1956.
31. MELENT'EV and A. I. PONOMAREV.—In: *Metody opredeleniya i analiza redkikh elementov*, p. 238.—Izd. AN SSSR. 1961.
32. CHULKOV, YA. I.—*Zav. Lab.*, **26**, 272. 1960.
33. LIEBER, W.—*Z. anal. Chem.*, **177**, 429. 1960.
34. WILKINS, D.—*Anal. chim. acta*, **20**, 113. 1959.
35. BIEBER, B. and Z. VIČEŘA.—*Coll.*, **26**, 2081. 1961.
36. ELINSON, S. V. and L. I. POBEDINA.—*Zav. Lab.*, **29**, 139. 1963.
37. GENEROZOV, B. A.—*Zav. Lab.*, **18**, 161. 1952.
38. BRANDT, W. and A. E. PREISER.—*Anal. Chem.*, **25**, 567. 1953.
39. MININ, A. A.—*Uchenye Zapiski Permskogo Universiteta*, **9**, No.4, 177. 1955; **15**, No.4, 96. 1958; **19**, No. 1, 97. 1961.
40. ZHIVOPISTSEV, V. P. and A. A. MININ.—*Zav. Lab.*, **26**, 1346. 1960.
41. ZINCHENKO, V. A. and S. I. RUDINA.—*Zav. Lab.*, **27**, 956. 1961.
42. POLYAK, L. YA.—*ZhAKh*, **17**, 206. 1962.
43. POLYAK, L. YA.—*ZhAKh*, **18**, 956. 1963.

Zirconium Zr and Hafnium Hf

The valency of zirconium and hafnium in their compounds is +4. Halide compounds of zirconium and hafnium are known where these elements have valencies of +2 and +3, but they are of no importance in analytical chemistry. There are no methods based on a valency change for determining zirconium and hafnium.

Zirconium and hafnium have very similar chemical properties. No reagent will ensure a reliable separation of zirconium from hafnium ions; the ions of these elements react with inorganic and organic compounds under almost identical conditions and the properties of the reaction products greatly resemble each other. Physical methods—X-ray, spectroscopic, and neutron activation methods—can give a reliable detection and quantitative determination of hafnium in the presence of zirconium. When both elements are present, they can under certain conditions be determined photometrically or by organic reagents (p. 148).

Aqueous solutions of zirconium and hafnium salts are readily hydrolyzed.

Solutions of zirconium salts may contain together the colorless ions Zr^{4+}, zirconyl ions ZrO^{2+}, and different polymerized ions. The degree of polymerization increases with increasing concentration of zirconium salt and varies with the acidity of the solution. If hydrofluoric acid is added, the degree of polymerization of zirconium ions decreases.

The information on the limiting ranges of existence of the various ionic species of zirconium is not fully conclusive. In 2 *N* perchloric acid solutions and at higher concentrations, only the simple Zr^{4+} ions exist. Solutions with pH not above 0.7 contain mainly the ions $Zr(OH)^{3+}$ and solutions with pH higher than 0.7 contain $Zr(OH)_2^{2+}$ ions.

The Zr^{4+} ions react with numerous analytical reagents much more rapidly than do the oxycations ZrO^{2+} and various isopolyoxycations. Zirconium can be obtained in solution predominantly as the Zr^{4+} ions by adding not less than a 10-fold amount of an aluminum salt as coprecipitant; zirconium and aluminum are precipitated by ammonia and the precipitate is dissolved in dilute hydrochloric acid [1]. The sensitivity and the reproducibility of the photometric determination of zirconium by Arsenazo I are improved in this way.

The dioxides ZrO_2 and HfO_2 are insoluble in water. The corresponding hydrates are obtained either by the hydrolysis of aqueous salt solutions, which results in the formation of meta-acids $ZrO(OH)_2$ or H_2ZrO_3, or else by the addition of alkali solutions to solutions of zirconium or hafnium salts, which results in the formation of the ortho-acids $Zr(OH)_4$ or H_4ZrO_4.

The resulting products are invariably contaminated with basic zirconium salts.

The dioxides ZrO_2 and HfO_2 are solubilized by fusion with alkali or potassium bisulfate. When ZrO_2 and HfO_2 are fused with alkalis, zirconates and hafnates are formed which are hydrolyzed in the presence of water. If ZrO_2 and HfO_2 are fused with SiO_2 or TiO_2, zirconium and hafnium silicates and titanates are formed.

Zirconium dioxide ZrO_2 melts at temperatures from 2700°C, whereas the melting point of hafnium dioxide HfO_2 is 2990°C.

The hydrolysis products of zirconium salts may remain in solution as colloids.

Ammonium and alkali carbonates precipitate the white basic salts of zirconium and hafnium. The precipitate is soluble in excess ammonium carbonate and separates out again when the solution is boiled.

If the pH of a solution containing zirconium and hafnium ions is gradually raised by the addition of alkali, the precipitate consists at first mainly of the basic zirconium salt and then of the basic hafnium salt.

At about pH 2 the white basic zirconium salt precipitates out of 0.01 *M* zirconyl chloride solution.

Hydrated zirconium and hafnium oxides are amphoteric; the acidic and basic properties are weak. These compounds are soluble in strong acids, but due to aging the solubility decreases with time.

Hydrogen sulfide does not precipitate zirconium or hafnium from aqueous solutions.

Zirconium and hafnium form a large number of compounds insoluble in dilute acids, such as phosphate, arsenate, selenite, iodate, arylarsonates, etc., and complex compounds. Halide, sulfate, oxalate, tartrate, citrate, ethylenediaminetetra-acetate, and other complexes are known. Information on the compositions and instability constants of many of these compounds is contradictory.

The halide complexes have the formula Me_2ZrHal_6, where Me is any monovalent metal and Hal is F, Cl, Br, I.

The fluoride complex ions of zirconium ZrF_6^{2-} and hafnium HfF_6^{2-} are very stable; the corresponding solids are soluble in water.

When an alkali fluoride is added to solutions of various colored compounds of zirconium and hafnium with organic reagents, the colorless soluble fluorozirconate (fluorohafnate) is formed and the organic reagent is liberated; this is accompanied by a change in the color of the solution. Alkali fluorides are used to mask zirconium.

The sulfate complexes $[ZrO(SO_4)_2]^{2-}$, $[Zr(SO_4)_4]^{4-}$, and others are relatively stable.

Ammonium phosphate $(NH_4)_2HPO_4$ quantitatively precipitates zirconium and hafnium ions from solutions containing 10 vol.% of sulfuric or hydrochloric acid. The composition of the precipitates varies with the method of precipitation. The precipitates are difficult to filter and, for this reason, the separation of zirconium and hafnium from other elements is not always satisfactory. If titanium, niobium, or tantalum is present, zirconium is precipitated in the presence of hydrogen peroxide; stannic ions interfere.

Disubstituted ammonium arsenate quantitatively precipitates zirconium ions from 2.5 *N* HCl solutions and from 3.75 *N* HNO_3 solutions. When thorium, cerium and titanium are present, the zirconium arsenate must be reprecipitated. The precipitate is ignited to ZrO_2.

Selenious acid quantitatively precipitates zirconium ions as the basic selenite from hydrochloric acid solutions. The filtered and washed precipitate is ignited to ZrO_2. Selenious acid will separate zirconium from aluminum, a number of REE, small amounts of iron, and, in the presence of added hydrogen peroxide, from small amounts of titanium. Thorium and phosphorus contaminate the precipitate.

Under certain circumstances it is possible to prepare zirconium selenite with a composition of $Zr(SeO_3)_2$.

KIO_3 precipitates the double salt $2Zr(IO_3)_4 \cdot KIO_3 \cdot 8H_2O$ out of nitric acid solutions. Under certain conditions the composition of the precipitate will be approximately constant and will correspond to the above formula. Thorium and cerium ions are precipitated by potassium iodate under the same conditions as zirconium. Zirconium may be separated from aluminum by potassium iodate. In the titrimetric determination of zirconium the filtered and washed precipitate is dissolved in hydrochloric acid, potassium iodide is added and the liberated iodine is titrated against a solution of thiosulfate.

Potassium ferrocyanide precipitates zirconium ions; the precipitates are of variable composition.

Hydrofluoric acid and alkali fluorides precipitate the white, voluminous ZrF_4 and HfF_4 out of concentrated solutions of zirconium and hafnium salts; the precipitates dissolve in excess precipitant. Hydrofluoric acid is used to separate REE and thorium from zirconium and hafnium, the latter remaining in solution.

Ammonium oxalate precipitates the white zirconium oxalate, which is soluble in excess precipitant and in strong acids, unlike cerium and thorium oxalates. Sulfate ions interfere with the precipitation of zirconium oxalate owing to the formation of $[ZrO(SO_4)_2]^{2-}$ ions.

Arylarsonic acids are highly selective precipitants of Zr^{4+} ions in dilute mineral acid solutions. Of these, phenylarsonic acid $C_6H_5AsO_3H_2$ is the most frequently employed. Other precipitants studied include p-hydroxyphenylarsonic, m-nitrophenylarsonic, p-dimethylaminoazophenylarsonic, and other arsonic acids. The precipitates are white.

Phenylarsonic acid quantitatively precipitates [2] zirconium and hafnium ions dissolved in 1:9 sulfuric acid and it is used to separate these elements from iron, aluminum, REE, thorium, and many other elements (p. 139).

p-Hydroxyphenylarsonic acid [3] will determine zirconium in the presence of Fe(II), Fe(III), Al, Zn, Co, Ni, Be, Cr, Mn, Ca, Mg, Be, Tl, Ce(III), Th, V, U, etc. Zirconium is precipitated from 2.5–3.0 *N* HCl or from 2.5–3.0 *N* H_2SO_4. When titanium ions are present, hydrogen peroxide is added to the solution. Ceric and stannic ions interfere.

Zirconium and hafnium ions are quantitatively precipitated by mandelic, phthalic, fumaric, and other organic acids, and their various derivatives.

Mandelic acid $C_6H_5CH(OH)COOH$ will separate zirconium from Ti, Fe, Al, Cr, V, Mo, REE, and other elements. If zirconium is present in small amounts (about 10 mg), the precipitate of $[C_6H_5CH(OH)COO]_4Zr$ is weighed directly; if more than 25 mg of zirconium is present, the precipitate is first ignited [4, 5] to the dioxide ZrO_2.

Thiodiglycolic acid $HOOCCH_2SCH_2COOH$ quantitatively precipitates [6] zirconium ions from solutions in 0.2–0.4 *N* HCl or HNO_3. The composition of the crystalline precipitate is approximately $ZrO(OOCCH_2SCH_2COO)$. For this reason the precipitate is ignited to ZrO_2. Co, Ni, Zn, Mg, Cd, Mn, Al, Pb, Ti, Be, REE, and small amounts of thorium do not interfere.

Thiomalic acid $HOOCCHSHCH_2COOH$ is a highly selective precipitant [7] for zirconium in dilute solutions of HCl or HNO_3; only bismuth ions interfere. If the solution is boiled, a precipitate with the approximate composition $(OOCCHSHCH_2COO)ZrO$ separates out.

Cupferron quantitatively precipitates zirconium and hafnium ions from sulfuric acid solutions and is used in the separation of zirconium from a number of other elements (p. 141).

N-Benzoylphenylhydroxylamine $C_6H_5CON(OH)C_6H_5$ quantitatively precipitates zirconium ions from acid solutions [8, 9]. The complex obtained from solutions in ≤ 0.5 *N* sulfuric acid has the constant composition [2] of $Zr(C_{13}H_{10}O_2N)_4$, and is suitable as a gravimetric form after being dried at 110°C. The conversion factor to zirconium is 0.0970. The precipitate obtained from 5% hydrochloric acid solutions has a variable composition and is ignited to ZrO_2. The reagent is highly sensitive: 1 μg of Zr/ml can be detected and 0.2 mg of Zr can be quantitatively determined. Thorium and REE do not interfere; the ions Sn(IV), V(V), Ce(IV), and Ti(IV) interfere. Zirconium(IV) ions are not precipitated in the presence of alkali fluorides.

Various hydroxy compounds form colored compounds with zirconium and are used in its detection and photometric determination. Such reagents include hydroxyanthraquinones (alizarin, quinalizarin, purpurin, rufigallic acid, etc.), hydroxy derivatives of flavone (quercetin, morin, etc.), carminic acid, hematoxylin, Xylenol Orange, Pyrocatechol Violet, and many others.

Hydroxy derivatives of flavone are used for the fluorimetric determination of zirconium.

Xylenol Orange is a highly selective reagent for zirconium and hafnium ions in strongly acid solutions (0.8–1.0 *N* $HClO_4$). The reagent is used for the photometric determination of small amounts of zirconium. Under certain conditions zirconium and hafnium may be individually determined in the presence of each other (p. 148). The reagent is also used for the photometric determination of microgram amounts of zirconium in metallic niobium [10]. Colored zirconium compounds are obtained in 0.4 *N* sulfuric acid in the presence of hydrogen peroxide which has been added in the minimum amount required to keep niobium in solution.

Xylenol Orange is also used as complexometric indicator.

Methylthymol Blue, which is an analog of Xylenol Orange, behaves similarly (p. 148).

Pyrocatechol Violet is used [11–13] for the photometric determination of zirconium (p. 146).

Arsenazo I reacts [1] with Zr^{4+} ions in 0.08–0.1 *N* HCl with formation of a violet-colored compound; the reagent itself is pink under these conditions. The optimum pH value is 1.5–1.8. Hafnium (IV) and titanium (IV) ions as well as fluoride and phosphate ions and hydroxyacids, interfere. The addition of a small amount of a gelatine solution prevents the precipitation of the colored zirconium compound. Arsenazo I is used for the photometric determination of zirconium in aluminum and magnesium alloys, and also in ores.

Arsenazo III is used for the photometric determination [14, 15] of zirconium (p. 145).

Inositolhexaphosphoric acid precipitates a white compound out of zirconium salt solutions and is used for the gravimetric determination of zirconium (p. 140).

Datiscin is used for the fluorimetric determination of zirconium (p. 150).

Complexometric methods are very important in the analytical chemistry of zirconium. A number of direct and indirect methods for the determination of zirconium are known (p. 143). The excess Complexone III may be titrated against a solution of bismuth nitrate in the presence of chemical indicators or in the presence of a dropping mercury electrode (reduction current of bismuth ions).

To determine 3–5% of zirconium in niobium-based alloys [16–18], excess standard Complexone III solution is added to the solution of the alloy; the pH is adjusted to 2 and the excess reagent is amperometrically titrated against a solution of bismuth nitrate, using a dropping mercury electrode (at −0.3 V relative to saturated calomel electrode). One mg of Zr can be titrated in the presence of 50 mg of Nb.

Tri-n-octylamine (0.1 *M* solution) quantitatively extracts zirconium ions from solutions in 9–12 *N* hydrochloric acid [19]. Zirconium is determined directly in the extract by adding an alcoholic solution of Xylenol Orange and acetic acid; the optical density of the resulting colored solution is measured at 550 mμ (absorption maximum). Under these conditions zirconium reacts with Xylenol Orange in the molar ratio of 1:1. The molar extinction coefficient at 550 mμ is 53,000. Zirconium can be determined in the presence of 10,000-fold amounts of alkali metals, alkaline-earth metals and REE, which cannot be extracted from aqueous solution, and also in the presence of aluminum, and 100-fold amounts of numerous other extractable ions, e. g., Mn(II), Sn(II), Bi(III), Cd, Pb(II), Cr(III), Hg(II), Tl(III), Au(III), Pd(II), Se(IV), Zn, Th(IV), La, In.

Hafnium and zirconium are determined by indirect chemical methods. For example, zirconium and hafnium can be precipitated by p-bromomandelic acid $BrC_6H_4CH(OH)COOH$ from concentrated hydrochloric acid solution in the presence of sulfuric acid at 85–95°C; the precipitate of $Zr(C_8H_6O_3Br)_4$ and $Hf(C_8H_6O_3Br)_4$ is dried at 120–130°C and weighed. The precipitate is then ignited to ZrO_2 and HfO_2 and is again weighed. The zirconium and hafnium content in the mixture can be calculated from the data thus obtained [20]. If the mixture of hafnium with zirconium contains more than 10% Hf, the absolute error is $\pm 0.5\%$. A determination takes 5–6 hours.

Zirconium and hafnium are separated by ion-exchange chromatography and by extraction methods.

Hafnium is determined in zirconium by spectroscopic and neutron activation methods.

For a survey of quantitative methods for zirconium and hafnium, see [21, 22].

Chromatographic separation of aluminum and zirconium [23]

The maximum concentration of hydrochloric acid at which aluminum and zirconium ions are fully absorbed by the different cation exchangers is shown in Table 4.

The largest difference in the HCl concentrations is noted for sulfonated coal and for KU-1 cation exchanger. The use of sulfonated coal is preferable, because the solution volume required for complete elution of aluminum ions is less (175–200 ml) than when KU-1 cation exchanger is used (300–350 ml).

Satisfactory results were obtained in the separation of 40–80 mg of zirconium from 10–60 mg of aluminum.

REAGENTS

Hydrochloric acid, dilute (1:4) and 1 *N* solution.
Silver nitrate, 0.05 *M* solution.
Nitric acid, 2 *N* solution.

Table 4

CONDITIONS FOR COMPLETE ABSORPTION OF ALUMINUM AND ZIRCONIUM IONS BY CATION EXCHANGERS

Cation exchangers	Maximum concentration of HCl, *N*	
	absorption of Zr(IV)	absorption of Al(III)
Sulfonated coal	1.0	0.25
Wofatit	0.75	0.25
KU-1	1.5	0.75
SBSR	1.25	0.75
MSF-1	1.0	0.25

PROCEDURE

The solution, which should contain 40–80 mg of zirconium and not more than 60 mg of aluminum, is diluted to 50 ml with hydrochloric acid to obtain a final acid concentration of 1 *N*.

Sulfonated coal is treated with a solution of hydrochloric acid (1:4), then is heated to 70–80°C, and washed with distilled water until neutral to methyl orange and until the reaction for chlorides is negative (test with silver nitrate in nitric acid). Sulfonated coal is placed in a 100-ml buret with a stopcock, 25 ml of 1 *N* hydrochloric acid are run through, and then a mixture of aluminum and zirconium salts dissolved in 1 *N* HCl is passed through at 5 ml/min. The cation exchanger is washed with 100–120 ml of 1 *N* hydrochloric acid. The solution issuing from the column, which contains the aluminum ions, is discarded.

Zirconium (IV) ions are eluted from the sulfonated coal column with 20–30 ml of hydrochloric acid (1:4). Zirconium (IV) is determined in the resulting solution by any suitable method.

Gravimetric determination of zirconium by phenylarsonic acid [2]

When phenylarsonic acid is added to strongly acid solutions of zirconium and hafnium salts, a white flocculent precipitate separates out; after drying in the air it has the approximate composition $(C_6H_5OAsO_2)_2Zr \cdot xH_2O$. To complete the determination, the precipitate is ignited to ZrO_2.

The ions Sn(IV), Ti(IV), Nb, Ta, and Ce(IV) which form similar precipitates with phenylarsonic acid, interfere. Tin is preliminarily precipitated by hydrogen sulfide. The thorium phenylarsonate precipitate is soluble in mineral acids and thus thorium does not interfere; the effect of other tetravalent elements is eliminated by adding hydrogen peroxide.

Fluoride ions, which bind zirconium as a stable soluble complex, interfere.

Reprecipitation is necessary when iron and aluminum are present in major amounts. The ignited zirconium dioxide is fused with potassium pyrosulfate, the melt is dissolved in 10% hydrochloric acid and the precipitation by phenylarsonic acid is repeated.

The method is not suitable for the analysis of materials containing major amounts of niobium and tantalum.

The precipitation of zirconium by phenylarsonic acid is one of the most accurate methods for its determination. The method is suitable for the analysis of minerals, rocks, and special brands of steel containing more than 0.1% zirconium.

REAGENTS

Hydrochloric acid, $d = 1.19$ g/cm^3 and 1% solution.
Hydrogen peroxide, 30% solution.
Phenylarsonic acid $C_6H_5AsO(OH)_2$, 10% solution in 1:1 hydrochloric acid.

PROCEDURE

The solution, which should contain 10–15 mg of zirconium, is acidified with 30 ml of hydrochloric acid ($d = 1.19$ g/cm^3), and is diluted with water to 100 ml in a 250- to 300-ml beaker; 3 ml of hydrogen peroxide are added, and the solution is heated to the boil. The phenylarsonic acid solution (6–10 ml) is added to the hot solution. The mixture is heated another few minutes to coagulate the precipitate, cooled, and a little macerated filter paper is added to the beaker. The precipitate is filtered through White Ribbon filter paper, washed with 1% solution of hydrochloric acid, and the filter paper is partly dried. The filter paper with the precipitate is then placed in a tared crucible; the paper is carefully ashed over a gentle flame in a fume hood; the flame is increased and the precipitate is vigorously heated. The crucible is then placed in a muffle furnace at 1000°C and ignited to constant weight. The conversion factor to zirconium is 0.7407.

Gravimetric determination of zirconium in steels by calcium magnesium inositolhexaphosphate [24]

Calcium magnesium inositolhexaphosphate (phytin) reacts with zirconium ions to form a white flocculent precipitate, which is soluble in oxalic acid and sodium fluoride and practically insoluble in mineral acids and in alkalis. The limiting dilution is 1:200,000. The solubility of the zirconium compound in water is less than

10^{-5} mole/liter at room temperature. The precipitate is easily filtered and washed. The last stage is the ignition of the zirconium compound at 1000–1050°C to $2ZrO_2 \cdot 3P_2O_5$. The empirical conversion factor is $2ZrO_2/2ZrO_2 \cdot 3P_2O_5 = 0.3932$.

Ferric ions interfere if present in major amounts and for this reason it is recommended that the precipitate be dissolved in oxalic acid and reprecipitated.

The method is suitable for the determination of zirconium in steels, alloys, and rocks.

REAGENTS

Calcium magnesium inositolhexaphosphate, 2% solution in 0.5 *N* nitric acid.
Hydrochloric acid, $d = 1.19$ g/cm³ and dilute (1:4, 1:1) solutions.
Oxalic acid, crystalline.

PROCEDURE

The alloy sample, which should contain not less than 10 mg of zirconium, is dissolved in 80 ml of 1:1 hydrochloric acid by heating on a sand bath. The solution is evaporated to a residual volume of 10 ml and an equal volume of hydrochloric acid ($d = 1.19$ g/cm^3) is added to the residue. The mixture is heated to 70–80°C and the solution of the reagent is added drop by drop with stirring. The solution with the precipitate is held for 10 minutes on the water bath and the extent of the precipitation of zirconium ions is tested.

The precipitate is washed by decantation with 30 ml of 1:1 hydrochloric acid, transferred onto Blue Ribbon filter paper and washed with 50 ml of 1:4 hydrochloric acid and then with water.

The washed precipitate is flushed from the filter paper into the beaker which was used for the precipitation; 2 g of crystalline oxalic acid and an equal volume of HCl ($d = 1.19$ g/cm^3) are added to the solution and the precipitation is repeated.

The precipitate is dried, the filter paper ashed, and the precipitate ignited at 1000–1050°C in a muffle furnace until a constant weight is obtained. The zirconium content in the steel is then calculated.

Determination of zirconium by cupferron in the presence of large amounts of molybdenum and tungsten [25–27]

The determination of zirconium in the presence of major amounts of molybdenum and tungsten ions, without their previous separation, is based on the precipitation of zirconium cupferronate at pH 6.8 in the presence of tartaric and oxalic acids, which are added to prevent molybdenum and tungsten cupferronates from precipitating. The precipitate is filtered off and ignited to ZrO_2.

The reagent will separate zirconium ions from Al, Cr, U(VI), Zn, Be, Mg ions and also from boric acid and small amounts of phosphates. The ions Ti, Ce, Fe(III), V, Nb, Ta, U(VI), W, which form sparingly soluble compounds under the experimental conditions employed, and silicic acid interfere.

REAGENTS

Hydrofluoric acid.
Nitric acid, $d = 1.4\ g/cm^3$.
Sulfuric acid, $d = 1.84\ g/cm^3$.
Oxalic acid, 4% solution.
Cupferron, freshly prepared 6% solution in water.
Tartaric acid, 10% solution.
Ammonia, dilute (1 : 1) solution.
Bromocresol Purple, indicator solution.

PROCEDURE

Determination of zirconium alloyed with molybdenum

The alloy sample, which should contain 4–40 mg of zirconium, is treated in a platinum dish with 3–5 ml of hydrofluoric acid. Nitric acid is cautiously introduced, drop by drop, until the alloy is dissolved. Sulfuric acid (5–6 ml) is added to the solution, the mixture is heated on a sand bath until sulfuric acid fumes appear and held for 7–10 minutes at this temperature. It is then cooled to room temperature, 25 ml of oxalic acid solution are added, the mixture is stirred until the salts are dissolved, transferred to a 300-ml beaker and the residue on the walls of the dish is rinsed with cold water.

The solution is diluted to 100 ml with water, 3–4 drops of Bromocresol Purple solution are added, the yellow solution is neutralized with NH_3 until it turns purple, and 10 ml of cupferron solution are added dropwise with constant stirring.

The precipitate is filtered through White Ribbon filter paper with a little macerated filter paper inside the filter cone, washed 5–6 times with cold water and transferred to a tared crucible. The filter paper is ashed and the precipitate ignited at 1000°C to constant weight.

The conversion factor is $Zr/ZrO_2 = 0.7403$.

The method will determine 4–40 mg Zr in the presence of 27–54 mg Mo with an error of ±0.2 mg.

Determination of zirconium in tungsten-based alloys

A sample of the alloy, which should contain 4–40 mg of zirconium, is treated in a platinum dish with 10–15 ml of hydrofluoric acid, then nitric acid is introduced drop by drop until the alloy is dissolved; 5–10 ml of sulfuric acid are added to the solution and the mixture is heated on a sand bath for 7–10 minutes after the incipient evolution of sulfuric acid fumes. The contents of the dish are cooled, 10 ml of tartaric acid solution and 5 ml of oxalic acid solution are added and the mixture is transferred to a 300-ml beaker.

The residue of tungstic acid in the dish is treated with 20 ml of water and enough ammonia solution to fully dissolve it, and the solution is transferred to the original beaker. It is made alkaline with an ammonia solution, 3–4 drops of Bromocresol Purple are added, and then sulfuric acid is added drop by drop until the solution becomes yellow; ammonia solution is then added drop by drop with stirring, until the color becomes purple. Subsequent operations are the same as described for the determination of zirconium alloyed with molybdenum.

The method will determine 4–40 mg Zr in the presence of 550 mg W with an error of ± 0.2 mg.

Complexometric determination of zirconium [17, 28]

Complexone III reacts with zirconium (IV) ions in strongly acid medium (2 *N* HCl) to form a stable complex compound with a molar component ratio of 1:1. The titration is conducted in the presence of p-sulfobenzeneazopyrocatechol, Eriochrome Black T or other indicators.

The solution of Complexone III should be standardized against a solution of zirconium salt in 2 *N* HCl.

The method is highly selective (large amounts of ferrous ions, aluminum, bismuth, REE, titanium, uranium, etc., do not interfere) and is applicable to the analysis of a great variety of products: ores, concentrates, half-products of ore dressing, alloys, etc., for zirconium contents exceeding 0.2%.

REAGENTS

Hydrochloric acid, $d = 1.19$ g/cm^3 and 2 *N* solution.
Complexone III, 0.01 *M* and 0.05 *M* solutions.
p-Nitrobenzeneazopyrocatechol, 0.02% solution in ethanol.
p-Sulfobenzeneazopyrocatechol, 0.02% solution.
Eriochrome Black T, in mixture with NaCl; ratio Eriochrome Black T:NaCl = 1:50.
Hydroxylamine hydrochloride, 25% solution.

PROCEDURE

Determination of zirconium by p-sulfobenzeneazopyrocatechol

The determination may be conducted in 1.5–2.2 *N* HCl. At the end point of the titration there is a sharp color change from raspberry pink to yellow. The method will determine 0.5–40 mg of Zr/10 ml. If zirconium is present in larger amounts, the solution should be suitably diluted.

The ions Fe(II), Th, Ti, Sn(IV), Mo, Nb, Al, Bi, Sb, Ni, Ge, present in amounts equal to that of zirconium, and also alkali metal, alkaline earth metals and REE ions do not interfere.

Oxidizing agents such as ferric ions, which decompose the indicator, are previously reduced with hydroxylamine. Tantalum and tungsten, which precipitate out as acids and adsorb the compound of zirconium with the indicator, interfere. Up to 10–20 mg of sulfates and up to 10 mg of tartrates may be present. Oxalate, fluoride and phosphate ions must be absent.

The solution, containing 40–80 mg Zr as chloride or perchlorate, is diluted to 50 ml with hydrochloric acid and is thoroughly stirred. A 15-ml aliquot is withdrawn with a pipet into a 100–150 ml conical flask, the mixture is heated to boiling, 5 drops of indicator solution are added and the mixture is titrated against 0.05 *M* solution of Complexone III until the raspberry-pink solution turns yellow on the addition of the last drop of the reagent. If the solution has already cooled by the time the titration is finished it should be brought again to the boil and the titration continued.

If iron is present, a 15 ml aliquot of the solution is evaporated to dryness, the solution treated with 2 ml of concentrated HCl, again evaporated to dryness and 2 ml of concentrated HCl and then 2 ml of hydroxylamine hydrochloride solution are added to the residue. The mixture is diluted with 6 ml of water, boiled for 1 minute, 5 drops of indicator solution are added and the solution is titrated against 0.05 *M* solution of Complexone III.

One ml of 0.0500 *M* solution of Complexone III is equivalent to 4.56 mg of zirconium.

Determination of zirconium with the use of Eriochrome Black T

Zirconium ions in 1.5–2 *N* HCl react with Eriochrome Black T to form a blue-violet compound which is less stable than zirconium complexonate. The titration is conducted with 0.01 *M* solution of Complexone III, which increases the sensitivity of the determination. However, if the titration is performed in 2 *N* HCl, there is no exact stoichiometric ratio between the amount of zirconium taken for the determination and the amount of Complexone III consumed in the titration. For this reason the solution of Complexone III must be standardized against a solution of a zirconium salt.

The advantage of using Eriochrome Black T rather than other complexometric indicators is that with this indicator zirconium can be titrated in the presence of large (about 200 mg) amounts of sulfate. This is very important, since zirconium-containing materials are most frequently solubilized with sulfuric acid. Stannous tin does not interfere; in its presence the experimental error in the determination of zirconium in the presence of 100 mg of copper, vanadium, molybdenum or iron decreases, as a result of the reduction of these elements. Up to 500 mg of Mg, Ca, Ba, Zn, Sn(II), up to 200 mg of Mn, UO_2^{2+}, Ti(IV), Ni, up to 100 mg of Al(III), up to 75 mg of Cr(III), up to 50 mg of Bi and La, up to 30 mg of Nb and up to 4 mg of Th may be present.

Hafnium, which is titrated together with zirconium, and fluoride, phosphate, oxalate and tartrate ions, which form undissociated compounds with zirconium, interfere.

The method will determine 1–100 mg of Zr in its salts and in different zirconium-containing materials in a volume not exceeding 100 ml. The experimental error is about 1%.

The solution containing 15–30 mg of Zr as chloride, nitrate or sulfate is diluted to mark with 2 *N* HCl in a 50-ml volumetric flask and is thoroughly mixed. Fifteen ml of the solution are placed in a 100-ml conical flask, heated to boiling, and as much indicator as will cover the tip of a spatula (1:50 mixture of Eriochrome Black T with NaCl) is added. The mixture is titrated against 0.01 *M* solution of Complexone III until the blue-violet solution becomes pink.

One ml of 0.0100 *M* Complexone III solution is equivalent to 0.9122 mg of zirconium.

Photometric determination of zirconium by Arsenazo III [29, 30]

Arsenazo III is a highly sensitive and selective reagent for the photometric determination of zirconium. The solution of the reagent, which is red in acid medium, rapidly turns blue or blue-violet, depending on the ratio between the reagent and zirconium ions. The color intensity is constant between 1 *N* and 3 *N* in HCl, but varies greatly with time. The color is stabilized by the introduction of a solution of gelatin. The light absorption by the solutions obeys Beer's law between 5 and 30 μg of Zr/50 ml solution.

To obtain reproducible results, zirconium is previously converted into a single ionic species (zirconyl ion) by boiling with 2 *N* hydrochloric acid.

The determination of microgram amounts of zirconium is possible in the presence of more than 100 mg of aluminum, up to 20 mg of stannous or stannic tin, up to 10 mg of beryllium, nickel and titanium, up to 5 mg of chromium, REE and niobium, and up to 50 mg of tartrates and up to 100 mg of sulfates. If stannous chloride is added, the determination may be performed in the presence of 15 mg of iron and 5 mg of copper.

The following ions interfere: uranium and thorium, which form colored compounds with the reagent in acid medium; fluoride, phosphate and oxalate ions, which form stable complexes with zirconium ions and decompose the colored compound of zirconium with Arsenazo III.

The method is suitable for the determination of zirconium in ores and side products of ore dressing (tails, slurries, and fractions with high contents of titanium, iron, aluminum and tin).

REAGENTS

Standard solution of zirconium oxychloride $ZrOCl_2$ in 2 *N* HCl, 5 μg of Zr/ml.
Hydrochloric acid, 2 *N* solution.
Arsenazo III, 0.05% aqueous solution.
Gelatin, 1% solution.
Mixture of sodium carbonate and sodium perborate, 3:2.

CONSTRUCTION OF CALIBRATION CURVE

Into five 50-ml volumetric flasks are introduced 1, 2, 3, 4 and 5 ml of standard solution of zirconium oxychloride and the solutions are made up to 10 ml with hydrochloric acid. They are heated to boiling on a sand bath, cooled, 3 ml gelatin solution and 2 ml Arsenazo III solution are added to each flask, and the solutions are made up to mark with hydrochloric acid. The comparison solution should contain the same amounts of reagents except for the zirconium salt. The optical densities of the solutions of zirconium compound with Arsenazo III are measured in an electrophotocolorimeter with a red filter relative to a comparison solution containing all the reagents, and the calibration curve is plotted.

PROCEDURE

A 0.1–0.2 g sample of the material to be analyzed, which should contain between 0.1 and 0.5% Zr, is placed in a platinum dish containing 5 g of sodium carbonate–sodium tetraborate mixture, then mixed and fused at 900°C in a muffle furnace for 7–10 min. The melt is distributed in a thin layer on the walls by rotating the dish, water is poured into the dish and the mixture is heated on a sand bath. The solution with the precipitate is transferred to a 300-ml beaker, then diluted with water to 100–150 ml and heated to coagulate the precipitate. The precipitate is filtered off, washed with hot water and dissolved on the filter in hydrochloric acid. The filtrate is collected into a 50-ml volumetric flask (for 0.2-g sample size and 0.1% zirconium content in the sample) and 100-ml flask (for 0.1 g sample size and 0.5% zirconium content in the sample) and diluted to mark with hydrochloric acid. An aliquot of 5 ml or one corresponding to 10–15 μg of Zr is withdrawn for the determination of zirconium; transferred to a 50-ml volumetric flask, and the determination is continued as described for the construction of the calibration curve. The amount of zirconium is read off the calibration curve.

Photometric determination of zirconium in phosphorites by Pyrocatechol Violet [11, 12]

Pyrocatechol Violet reacts with zirconium ions to form a blue compound with an absorption maximum at 625 mμ. The yellow aqueous solution of the reagent has an absorption maximum at 450 mμ (Figure 7). The colored compound is formed in

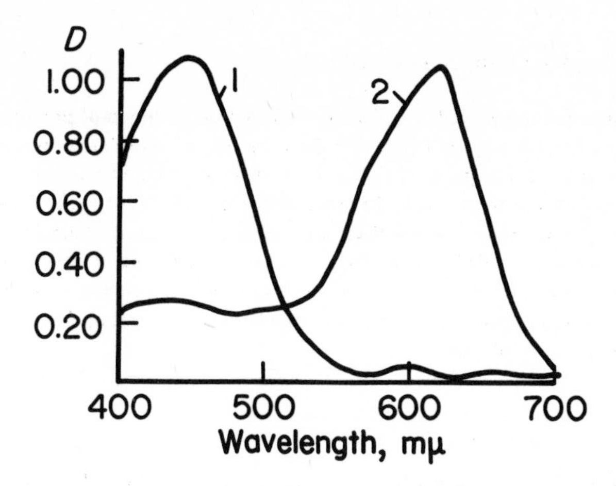

Figure 7

Light absorption by solutions of Pyrocatechol Violet and its zirconium compound at pH 5.4: 1) $1 \cdot 10^{-5}$ *M* solution of Pyrocatechol Violet; 2) solution of complex 30 minutes after mixing $1 \cdot 10^{-4}$ *M* zirconium oxychloride with $1 \cdot 10^{-4}$ *M* Pyrocatechol Violet.

hydrochloric, nitric and sulfuric acid media. The optimum pH value for the determination is 5–5.5. The color develops within 30 minutes and is stable for 2–3 hours. The absorption by the solutions obeys Beer's law between 0 and 100 μg of Zr/50 ml.

The compound between zirconium ions and the reagent is stable in the presence of Complexone III. Thus, Complexone III can be used to mask the ions of other elements.

Microgram amounts of zirconium may be determined in the presence of 5 mg of Al, 15 mg of Fe(II), up to 2 mg of Fe(III), 3 mg of Ti, 1 mg of Th, Be, U, Bi, V, Mo, W and Co (not more than 5 mg of each ion and not more than 0.5 mg of Cu and Ni).

The method will determine a few hundredths of one per cent of zirconium in metallic uranium and its alloys, in ores (phosphorites, monazites) and other objects with an error of ± 10%.

REAGENTS

Standard solution of zirconium salt, 10 μg of Zr/ml.
Nitric acid, 1:3 solution.
Potassium bifluoride, crystalline.
Sulfuric acid, $d = 1.84$ g/cm³.
Hydrochloric acid, $d = 1.19$ g/cm³ and 1:2 solution.
Complexone III, 0.05 *M* solution.
Ammonia, 1:1 solution.
Methyl Red, 0.001% alcoholic solution.
Pyrocatechol Violet, 0.04% alcoholic solution.
Acetate buffer solution, pH 5.2–5.4; 32 ml acetic acid ($d = 1.05$ g/cm³) are mixed with 34 ml of 25% ammonia solution and diluted to one liter with water.

CONSTRUCTION OF CALIBRATION CURVE

Into five 50-ml beakers are introduced 1, 2, 3, 4 and 5 ml standard solution of zirconium salt, 5 ml of hydrochloric acid ($d = 1.19\ g/cm^3$) are added to each flask and the solutions are evaporated on a sand bath to a residual volume of 1–1.5 ml. One ml of Complexone III solution is added to each beaker, the solutions are mixed and neutralized with ammonia in the presence of 1–2 drops of Methyl Red and then with 1:2 hydrochloric acid until faintly pink. The solutions are transferred to 50-ml volumetric flasks, 4 ml of Pyrocatechol Violet solution is added to each flask and the solutions are made up to mark with acetate buffer solution and mixed. The comparison solution is prepared in a similar manner.

After 30 minutes the optical density of the series of standards is measured on an electrophotocolorimeter and the calibration curve is plotted.

PROCEDURE

A finely ground 0.05–0.2-g sample of the material in a 100-ml beaker is treated with 10 ml of nitric acid, the beaker is covered with a watch glass and heated for 15–20 minutes on a sand bath. The insoluble residue, which contains all the zirconium as $ZrSiO_4$, is separated on Blue Ribbon filter paper, and carefully washed with hot water; the paper with the residue is placed in a platinum dish and dried. The filter paper is ashed and ignited for 15–20 minutes.

The residue is cooled, 0.5–1.5 g of potassium bifluoride is added and the mixture is fused at 900°C in a muffle furnace until a clear melt is obtained. To the cooled melt are added 10 ml of sulfuric acid, the solution is evaporated on a sand bath to abundant evolution of sulfuric acid fumes, the residue is cooled and dissolved in a small amount of water and the solution is quantitatively transferred to a 50-ml flask and diluted to the mark with water.

An aliquot of the solution, which contains between 15 and 45 μg of Zr, is transferred to a 50-ml beaker and all the operations described for the construction of the calibration curve are performed. The amount of zirconium in the aliquot is read off the calibration curve. The data serve to calculate the percentage content of zirconium in the sample.

Photometric determination of hafnium in the presence of zirconium by Xylenol Orange or Methylthymol Blue [31]

Xylenol Orange and its analog Methylthymol Blue react with zirconium and hafnium ions to form colored compounds (Table 5) which are used in the photometric determination of these elements.

Hafnium ions in the presence of sulfate ions, are unlike zirconium ions which can be masked by hydrogen peroxide under certain conditions. Thus, hafnium can be photometrically determined in the presence of zirconium.

Compounds of hafnium with both these reagents are formed practically instantly

Table 5

DESCRIPTION OF COMPOUNDS OF ZIRCONIUM AND HAFNIUM WITH XYLENOL ORANGE AND METHYLTHYMOL BLUE

Dissociation constants

Xylenol Orange		Methylthymol Blue	
$pK_1 = 2.61$	$pK_4 = 10.5$	$pK_{1-3} = 4.5$	$pK_5 = 11.2$
$pK_2 = 3.2$	$pK_5 = 12.3$	$pK_4 = 7.2$	$pK_6 = 13.4$
$pK_3 = 6.4$			

	Xylenol Orange		Methylthymol Blue	
	Zr	Hf	Zr	Hf
Absorption maximum, mμ	535	530	580 (600)	570 (590)
Optimum acidity	0.5–1.0 *N* $HClO_4$	0.2–0.4 *N* $HClO_4$	0.3–1.2 *N* $HClO_4$	pH 3
Formation constant of the 1:1 complex	$4.0 \cdot 10^7$ (0.8 *N* $HClO_4$)	$3.2 \cdot 10^6$ (0.3 *N* $HClO_4$)	$1.0 \cdot 10^5$ (1.0 *N* $HClO_4$)	—
Molar extinction coefficient	24.200 (0.8 *N* $HClO_4$)	15.700 (0.2 *N* $HClO_4$)	21.700 (1.0 *N* $HClO_4$)	18,700 (pH 3)
Sensitivity in μg/cm³ log $(I_0/I) = 0.001$)	0.004	0.011	0.004	0.010
Color of compound	Red	Red	Red (1:1) Violet (2:1)	Red (1:1) Violet (2:1)
Masking substances	EDTA*, F^-, citrate ions H_2O_2, PO_4^{3-}	EDTA, F^-, citrate ions PO_4^{3-}	EDTA, F^-, citrate ions H_2O_2, PO_4^{3-}	EDTA, F^-, citrate ions PO_4^{3-}

* EDTA = ethylenediaminetetraacetic acid

when the solutions of the components are combined; a threefold molar excess is recommended for Xylenol Orange, while a twofold molar excess is recommended for Methylthymol Blue. The color intensity is constant for several days. Beer's law is obeyed from 0–0.65 μmole of Hf/25 ml for Xylenol Orange and 0.1–0.5 μmole of Hf/25 ml for Methylthymol Blue.

The ions Bi, Fe(III), Sn(IV), Th, Nb, Mo(VI), V(IV) and Ti(IV) react with these reagents in a similar manner; Ce(IV) oxidizes the reagents; Se(IV), Te(IV), Pt(IV), Ge, Ta, W and certain other elements do not form colored compounds with the reagents; all these elements except niobium are masked with Complexone III.

Ethylenediaminetetraacetate ions, phosphates, citrates, fluorides and more than twofold molar amounts of sulfates interfere; perchlorates, chlorides and nitrates and also hydrogen peroxide (5 ml of 30% H_2O_2/0.5 μmole of Hf) do not interfere.

The method is best employed after the separation of zirconium and hafnium from most accompanying elements.

REAGENTS

Standard solution of zirconium perchlorate in 0.2 N $HClO_4$, 10 μg Zr/ml.
Standard solution of hafnium perchlorate in 0.2 N $HClO_4$, 10 μg Hf/ml.
Sodium sulfate, 10^{-3} M solution.
Hydrogen peroxide, 30% solution.
Xylenol Orange, 10^{-3} M solution.
Perchloric acid, 5 N solution.

CONSTRUCTION OF CALIBRATION CURVE

Into five 25-ml volumetric flasks are placed 1, 2, 3, 4 and 5 ml of the standard solution of hafnium (zirconium) perchlorate and 1.5 ml perchloric acid and 2 ml Xylenol Orange are placed in each flask; the solutions are made up to mark with water, mixed and their optical densities are determined after 30 minutes in an electrophotocolorimeter relative to a comparison solution prepared in a similar manner. Calibration curves for hafnium and zirconium are then plotted.

PROCEDURE

The sample solution containing 5–40 μg of hafnium and not more than 25 μg of zirconium, and all the other reagents in the amounts and in the sequence used in plotting the calibration curve are placed in a 25-ml volumetric flask; the optical density D of the solution is then measured.

An equal volume of the solution is introduced into another 25-ml volumetric flask, with 1.5 ml of perchloric acid, 5 ml of hydrogen peroxide, 1 ml of sodium sulfate solution and 2 ml of Xylenol Orange solution, and the mixture is made up to the mark with water. The optical density D_1 is measured and the amount of hafnium is read off the calibration curve.

The difference between the optical densities $D_2 = D - D_1$ corresponds to the light absorption by the compound of zirconium with the reagent. The zirconium content is found from the calibration curve obtained with pure solutions of zirconium perchlorate.

Fluorimetric determination of zirconium by datiscin [32, 33]

Datiscin—the glucoside of 3,5,7,2′-tetrahydroxyflavone—reacts with zirconium ions in solutions in mineral acids (0.5–8 N HCl) to form a compound which fluoresces when irradiated with UV light. The maximum intensity of the fluorescence lies within the visible range (520 mμ).

The intensity of the fluorescence attains the maximum value 15–20 minutes after the solutions have been combined and is stable for 12 hours. Alcohol (20 vol. %) is practically without effect on the intensity. The fluorescence intensity is proportional to the concentration between 0.005 and 3 μg of Zr/ml.

Zirconium in 6 *N* HCl can be determined in the presence of any amounts of Mg and Zn ions, 100,000-fold amounts of Al and 100-fold amounts of Ag, Cd, Mn, Cu, Pb, Hg(II), Be, Co, In, Cr(III), Y, Fe(II), Ta, V(V), Ni, Nb, W, U(VI), Ce(III), La, and also in the presence of 10-fold amounts of Fe(III), Mo(VI), Ti(IV), Sb(V), Th and Ga.

REAGENTS

Standard solution of zirconium salt, 100 μg Zr/ml, in 6 *N* HCl.
Hydrochloric acid, 6 *N* solution.
Datiscin, 0.03 % solution in acetone or alcohol.

PROCEDURE

Into five 15-ml glass or quartz tubes are placed portions of standard solution corresponding to 5, 10, 20, 40 and 70 μg of Zr; 1 ml of datiscin solution is placed in each tube and the solutions are made up to mark with hydrochloric acid.

The solution containing the unknown amount of zirconium is prepared in a similar manner.

After 15–20 minutes the fluorescence intensity of the sample solution is visually compared with that of the series of standards in UV light (PRK-2 or PRK-4 mercury lamp, UFS-3 filter).

BIBLIOGRAPHY

1. Kuznetsov, V. I., L. P. Budanova, and T. V. Matrosova.—*Zav. Lab.*, **22**, 406. 1956.
2. Alimarin, I. P. and O. A. Medvedeva.—*Zav. Lab.*, **11**, 254. 1945.
3. Simpson, C. T. and G. C. Chandlee.—*Ind. Eng. Chem., Anal. Ed.*, **10**, 642. 1938.
4. Kumins, C. A.—*Anal. Chem.*, **19**, 376. 1947.
5. Astanina, A. A. and E. A. Ostroumov.—*ZhAKh*, **6**, 27. 1951.
6. Sant, S. B. and B. R. Sant.—*Anal. chim. acta*, **21**, 221. 1959.
7. Sant, S. B. and B. R. Sant.—*Anal. chim. acta*, **32**, 379. 1960.
8. Ryan, D. E.—*Canad. J. Chem.*, **38**, 2488. 1960.
9. Alimarin, I. P. and Chien Yün-hsiang.—*ZhAKh*, **14**, 574. 1959.
10. Elison, S. V. and T. I. Nezhnova.—*Zav. Lab.*, **30**, 396. 1964.
11. Flaschka, H. and M. Farah.—*Z. anal. Chem.*, **152**, 401. 1956.
12. Chernikhov, Yu. A., V. F. Luk'yanov, and T. M. Knyazeva.—*ZhAKh*, **14**, 207. 1959.
13. Goryushina, V. G. and T. A. Archakova.—*Trudy GIREDMET*, **3**, 13. Metallurgizdat. 1961.
14. Goryushina, V. G. and E. V. Romanova.—*Ibid.*, **3**, 5. 1961.

15. Goryushina, V. G., E. V. Romanova, and T. A. Archakova.—*Zav. Lab.*, **27**, 795. 1961.
16. Vladimirova, V. M.—*Zav. Lab.*, **22**, 529. 1956.
17. Goryushina, V. G. and E. V. Romanova.—*Zav. Lab.*, **23**, 781. 1957.
18. Chernikhov, Yu. A. and V. M. Vladimirova.—*Trudy GIRDMET*, **3**, 69. Metallurgizdat. 1961.
19. Cerrai, E. and C. Testa.—*Anal. chim. acta*, **26**, 204. 1962.
20. Hahn, R. B.—*Anal. Chem.*, **23**, 1259. 1951.
21. Elison, S. V. and K. I. Petrov. *Analiticheskaya khimiya tsirkoniya i gafniya (Analytical Chemistry of Zirconium and Hafnium)*.—Izd. "Nauka". 1965. [English translation available as part of the *Analytical Chemistry of the Elements* series, Ann Arbor–Humphrey Sci. Pub. 1969.]
22. Glaassen, A.—In: *Handbuch der analytischen Chemie*, Vol. IVb, part 3, p. 170. Berlin, Göttingen, Springer Verlag. 1950.
23. Usatenko, Yu. I. and L. I. Gureeva.—*Zav. Lab.*, **22**, 781. 1956.
24. Alimarin, I. P. and L. Z. Kozel'.—In: *Khimiya redkikh elementov*, No. 3, 114. Izd. AN SSSR. 1957.
25. Sheskol'skaya, A. Ya.—*ZhAKh*, **17**, 949. 1962.
26. Popova, O. I. and V. I. Kornilova.—*ZhAKh*, **16**, 651. 1961.
27. Elinson, S. V. and T. I. Nezhnova.—*ZhAKh*, **15**, 73. 1960.
28. Poluektov, N. S., L. I. Kononenko, and T. A. Surichan.—*Zav. Lab.*, **23**, 660. 1957.
29. Goryushina, V. G. and E. V. Romanova.—*Zav. Lab.*, **26**, 415. 1960.
30. Goryushina, V. G., E.V. Romanova, and T. A. Arkiakova.—*Zav. Lab.*, **27**, 795. 1961.
31. Cheng, K. L.—*Talanta*, **2**, 61. 1959; **3**, 81. 186, 267. 1959: *Anal. chim. acta*, **28**, 41. 1963.
32. Golovina, A. P., I. P. Alimarin, E. A. Boshevol'nov, and L. B. Agasyan.—*ZhAKh*, **17**, 591. 1962.
33. Golovina, A. P. and A. E. Martirosov.—*Vestnik MGU*, chem. ser. II, No. 6, 64. 1962.

Vanadium V

In its compounds vanadium displays valencies of +2, +3, +4 and +5. The cations V^{2+}, V^{3+} and VO^{2+} are oxidized to vanadate ions in solutions. The most stable compounds in solution are those of V(V). In acid solutions, the compounds of V(IV) are also stable.

The standard electrode potentials in aqueous solutions at 25°C, relative to the potential of the standard hydrogen electrode, have the following values:

$$V \rightleftharpoons V^{2+} + 2e \qquad -1.175\ \text{V}$$

$$V^{2+} \rightleftharpoons V^{3+} + e \qquad -0.255\ \text{V}$$

$$VO^{+} \rightleftharpoons VO^{2+} + e \qquad -0.044\ \text{V}$$

$$V^{3+} + H_2O \rightleftharpoons VO^{2+} + 2H^{+} + e \qquad +0.337\ \text{V}$$

$$VO^{2+} + H_2O \rightleftharpoons VO_2^{+} + 2H^{+} + e \qquad +1.004\ \text{V}$$

$$VO^{+} + 3H_2O \rightleftharpoons VO_4^{3-} + 6H^{+} + 2e \qquad +1.256\ \text{V}$$

$$VO^{2+} + 3H_2O \rightleftharpoons H_2VO_4^{-} + 4H^{+} + e \qquad +1.314\ \text{V}$$

The most important forms of vanadium in the different valency states and their corresponding formal redox potentials are shown on the following page in Figure 8.

Numerous versions of the titrimetric determination of vanadium are based on reactions in which V(V) is reduced or lower valency vanadium is oxidized.

The redox potential of the system

$$VO_2^+ + 2H^+ + e \rightleftharpoons VO^{2+} + H_2O,$$

$$E = 1.01 + 0.06 \log \frac{[VO_2^+][H^+]^2}{[VO^{2+}]} \quad \text{volt (at 20°C)},$$

varies with the variation of the acidity of the solution over a wide range. In 9 *N* perchloric acid or 8 *N* sulfuric acid $E_0' = 1.30$ volt. In alkaline medium the potential becomes very low ($E_0' = -0.74$ volt at pH 14).

In strongly acid media, such as 6 *N* sulfuric acid, vanadium (V) displays strong oxidizing properties. Vanadate solutions are employed in the titrimetric determination of many reducing agents.

At lower acid concentrations the oxidizing properties of pentavalent vanadium become weaker. In weakly acid solutions many reducing agents are no longer oxidized by vanadates.

Vanadyl salts (VO^{2+}) are oxidized in moderately acid solutions by strong oxidizing agents. They are quantitatively oxidized by permanganate in ~5% sulfuric acid and also by solutions of potassium bromate in dilute hydrochloric acid in the presence of ammonium salts.

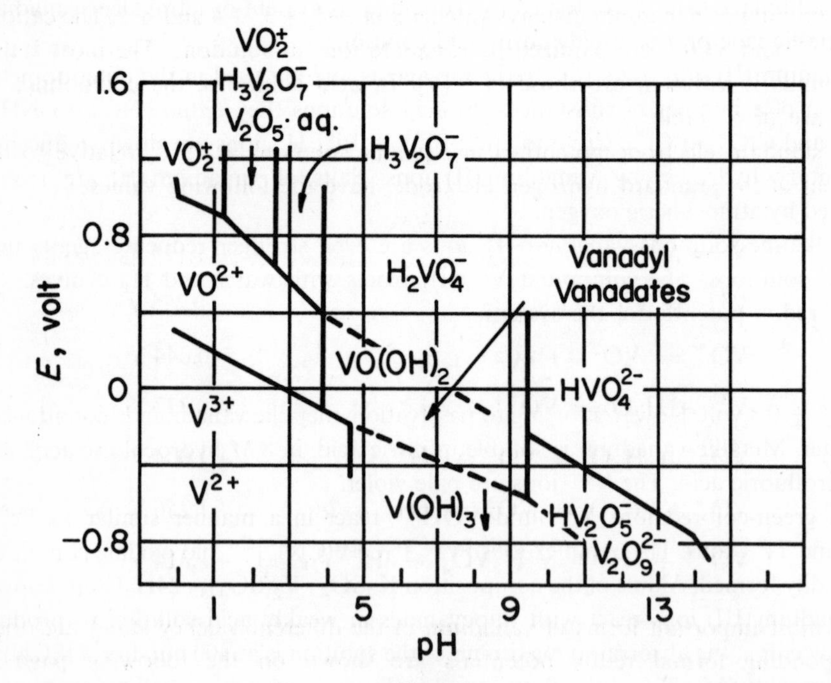

Figure 8

Formal redox potentials and vanadium species existing at different pH. Concentration of vanadium in the initial solution 10^{-2} *M*.

With increasing pH the tendency of V(IV) to become oxidized increases. In alkaline medium it is oxidized even by atmospheric oxygen.

As an example we may consider the reaction between vanadium and iron:

$$V(V) + Fe(II) \rightleftharpoons V(IV) + Fe(III).$$

In acid medium the reaction proceeds from left to right practically quantitatively. In alkaline medium the reaction proceeds from right to left [1], i.e., V(IV) is oxidized by Fe(III) to V(V); the Fe(II) ions formed in the reaction are readily detectable by dimethylglyoxime and o-phenanthroline.

The formal redox potential of the system

$$VO^{2+} + 2H^+ + e \rightleftharpoons V^{3+} + H_2O$$

in 1 N sulfuric acid is 0.34 volt.

The ions V^{3+} (VOH^{2+}) are oxidized by atmospheric oxygen to VO^{2+}. In neutral and alkaline media V(III) is rapidly oxidized to V(V).

The formal potential of the system

$$V^{3+} + e \rightleftharpoons V^{2+}$$

varies between -0.21 and -0.25 volt.

Vanadium (II) salts are obtained by reducing pentavalent or tetravalent vanadium by metallic zinc or zinc amalgam in acid solution.

Vanadium (II) salts are strong reducing agents and are used in the potentiometric titration of a number of substances. In acid solutions they reduce NO_3^- to NH_4^+; ClO_3^- and ClO_4^- to Cl^-; Cu^{2+} to Cu^0; Bi^{3+} to Bi^0; H^+ ions are slowly reduced to elementary hydrogen by vanadium (II) ions. Salts of vanadium (II) are readily oxidized by atmospheric oxygen.

In alkaline solutions vanadium (II) ions are even stronger reducing agents than in acid solutions. The compound $V(OH)_2$ reacts with water and H_2 evolves.

The redox potential of the system

$$V^{2+} + 2e \rightleftharpoons V$$

is -1.5 ± 0.3 volt. However, owing to passivation, metallic vanadium is not attacked by water. Metallic vanadium is soluble in nitric acid, in 8 M hydrochloric acid, and in hydrofluoric acid. The V^{2+} ions are pale violet.

The green-colored ions V^{3+} and VOH^{2+} react in a manner similar to Fe^{3+}, Al^{3+} and Ti^{3+} ions. The complex ions $[VF_5]^{2-}$, $[V(CN)_6]^{3-}$ and oxalate complexes are readily formed. Alums of the composition $K_2SO_4 \cdot V_2(SO_4)_3 \cdot 24H_2O$ are known.

Vanadium (III) ions react with thiocyanates in weakly acid solution to produce a yellow color; the absorption maximum of the solution is at 400 mμ. For $V(SCN)^{2+}$ ions the value of pK = 1.7–2.0 at $\mu = 2.6$. There is reason to believe that other thiocyanate complexes of vanadium (III) are formed. Vanadium (III) may be spectrophotometrically determined as the thiocyanate complex in 60% acetone [3].

At pH 4–5 alkali hydroxides precipitate the green $V(OH)_3$, which is rapidly oxidized when its color changes to gray. The hydroxide $V(OH)_3$ is basic.

Vanadium (III) salts reduce Ag^+, Cu^{2+}, Hg^{2+} ions.

Vanadium (IV) ions, VO^{2+}, are blue in aqueous solutions; their color is noticeable even at concentrations as low as 0.1 mg V/ml. In acid solutions V(IV) compounds are not oxidized by atmospheric oxygen and are perfectly stable, but in alkaline solutions they are readily oxidized to vanadates.

Vanadium (IV) salts are obtained by reducing vanadates with hydroxylamine, ascorbic acid, etc., while boiling. Vanadium (IV) chloride is formed when vanadate solutions are evaporated with hydrochloric acid.

Ammonium sulfide precipitates vanadyl ions VO^{2+} as the oxysulfide VOS.

At about pH 4 the grayish brown $VO(OH)_2$ precipitates out of 0.01 *M* solutions of vanadium (IV) salts; $pL_{VO(OH)_2} = 22.1$ (Figure 9). The precipitate is appreciably soluble in caustic alkalis with the formation of poly-ions of tetravalent vanadium, $HV_2O_5^-$ or $V_4O_9^{2-}$, which is reddish gray and then becomes red. For the reaction

$$2VO(OH)_2 \rightleftharpoons HV_2O_5^- + H_2O + H^+,$$

$pK = 10.5$.

The vanadyl hydroxide $VO(OH)_2$ is thus amphoteric.

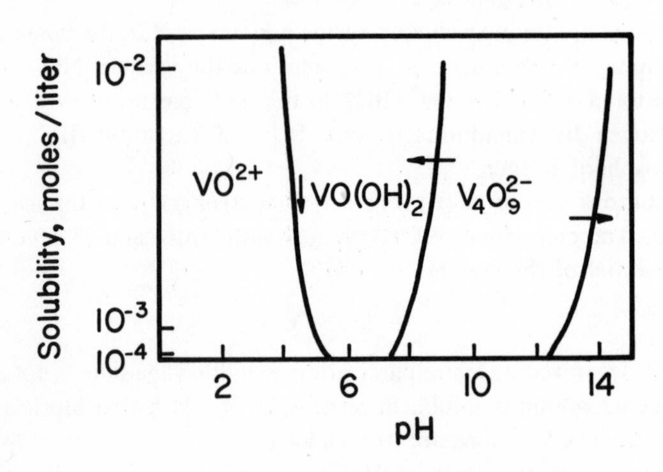

Figure 9

Solubility of $VO(OH)_2$ as a function of pH

Sodium phosphate reacts to give a blue-green precipitate; potassium cyanide gives a light green precipitate. Both precipitates are soluble in excess precipitant with the formation of green-colored solutions.

Vanadium (IV) forms strongly dissociated fluoride complexes and very strongly dissociated thiocyanate complexes. The oxalate and tartrate complexes are fairly stable:

$$VOF^+ \rightleftharpoons VO^{2+} + F^- \qquad pK = 3.2$$

$$VO(SCN)^+ \rightleftharpoons VO^{2+} + SCN^- \qquad pK = 0.9\ (\mu = 2.6)$$

$$VO(C_2O_4)_2^{2-} \rightleftharpoons VO^{2+} + 2C_2O_4^{2-} \qquad pK = 12.0$$

$$VO(OH)C_2O_4^- \rightleftharpoons VOOH^+ + C_2O_4^{2-} \qquad pK = 6.5$$

$$VO\ Tart \rightleftharpoons VO\ Tart^- + H^+ \qquad pK = 2.8$$

Vanadium (IV) ions form blue thiocyanate complexes ($VOSCN^+$, etc.) [2, 4] and the color of the blue solution is stronger by a factor of 4–5 than the color of a solution of vanadyl salt of equal molar concentration. In the photometric determination of vanadium a thiocyanate complex of vanadium (IV) in water–acetone medium is obtained; the optical density is measured at 750 mμ [4].

Vanadium (IV) ions typically react with hydroxylated and some other organic reagents [5], but little is known about these reactions.

Pyrocatechol reacts with vanadium (IV) ions to form a blue-colored complex compound which is used in the photometric determination of vanadium (p. 168).

The hydroxylated azo compound, Acid Chrome Blue K, reacts with vanadium (IV) ions to form a violet-colored compound. The reagent is employed in the photometric determination of vanadium (p. 169).

β-Hydroxynaphthylethylamine and β-hydroxynaphthylaldoxime quantitatively precipitate vanadium (IV) as the crystalline compounds $VO[OC_{10}H_6CHNC_2H_5]$ and $VO[OC_{10}H_6CHNOH]_2$ which are yellow-green and gray-green, respectively [6]. The precipitates are applied to determine vanadium in ferrovanadium by the gravimetric method. Iron ions do not interfere; copper and tetravalent titanium ions do interfere.

Complexone III reacts with vanadium (IV) ions in weakly acid solutions and is used in its titrimetric determination (p. 163).

Formaldoxime is a highly sensitive reagent for V(IV) ions and is used in the photometric determination of vanadium [7]. It forms a soluble yellow compound with V(IV) ions.

In alkaline solutions the pentavalent vanadium ion, VO_4^{3-}, is the predominant species. Metavanadate ions exist in acid solutions:

$$VO_4^{3-} + H^+ \rightleftharpoons VO_3^- + OH^-.$$

Polymetavanadate ions, such as $V_2O_7^{4-}$, $V_3O_9^{3-}$, $V_4O_{11}^{2-}$, $V_{10}O_{28}^{6-}$, etc., are also formed. Below pH 3 the yellow cations VO_2^+ and some VO^{3+} are also present (Figure 10).

The reactions given by VO_4^{3-} ions resemble certain reactions of PO_4^{3-} and AsO_4^{3-} and also of hexavalent molybdenum, tungsten and chromium ions.

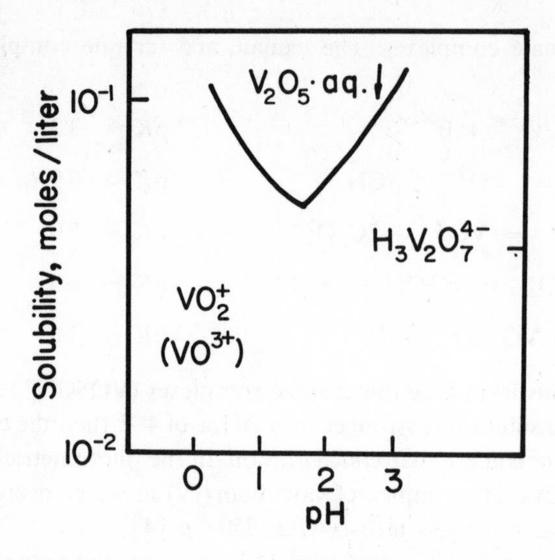

Figure 10
Solubility of vanadium pentoxide hydrate $V_2O_5 \cdot$ aq. as a function of pH

The salts of orthovanadic acid H_3VO_4 and metavanadic acid HVO_3 are usually colorless or faintly colored; polyvanadates are yellow or orange-colored.

Vanadium pentoxide V_2O_5 is a nonvolatile dark red solid, m.p.658°C. The pentoxide is formed when vanadium compounds are ignited with cupferron; it is suitable for use in gravimetric determination. The pentoxide is readily soluble in solutions of alkalis, less so in acids, and is sparingly soluble in solutions of ammonia.

Vanadium pentoxide is distinctly acid, but VO^{3+} and VO_2^+ cations exist in acid solutions.

The vanadates of most metals are sparingly soluble in water, except for sodium and potassium vanadates. The solubility of ammonium metavanadate NH_4VO_3 decreases on the addition of major amounts of ammonium chloride, when the white crystalline NH_4VO_3 precipitates out.

Lead acetate gives a yellow precipitate of a vanadate of variable composition, which is soluble in nitric acid.

The colorless solution of the metavanadate becomes orange-colored when acidified with acetic acid. When silver nitrate is added to such a solution after heating, the orange-colored silver pyrovanadate $Ag_4V_2O_7$ precipitates out.

Vanadium (V) ions are not precipitated by ammonia. However, if sufficient amounts of aluminum, iron or titanium salts are present in solution, ammonia quantitatively precipitates vanadium (V) ions at pH 6–7 together with the hydroxides of these elements, probably as vanadates [8]. Advantage is taken of this fact in chemical analysis to effect separation of small amounts of vanadium.

If ammonium sulfide is added to the solution of a vanadate, a brick-red coloration appears owing to the formation of VS_4^{3-} ions. When the resulting solution is acidified with sulfuric acid, the brown-colored sulfide V_2S_5 precipitates out, but the precipitation is not quantitative. The V_2S_5 precipitate is sparingly soluble in very dilute acids, but dissolves in solutions of alkalis and sulfides, forming orange-red sulfosalts.

Hydrogen sulfide does not precipitate vanadium sulfide out of acid solutions, but reduces vanadium (V) to vanadium (IV). Vanadium is separated from metals of the hydrogen sulfide group by precipitating the latter from acid solution in the presence of tartaric acid (to prevent coprecipitation of vanadium ions with sulfides of other metals).

Vanadium (V) ions in acid solutions are also readily reduced to vanadium (IV) by sulfur dioxide and by Mohr's salt.

Phosphovanadates and phosphotungstovanadates are yellow and are soluble in organic solvents. Different variants of photometric determination of vanadium (p. 165) are based on the formation of the yellow phosphotungstovanadic acid when phosphoric acid and sodium tungstate are added to a vanadate solution in acid medium.

Hydrogen peroxide in acid solutions reacts with vanadium (V) ions to form red-brown peroxy compounds. In the presence of a large excess of hydrogen peroxide the color intensity is attenuated and the solutions become yellow. In concentrated sulfuric acid medium hydrogen peroxide reduces pentavalent to tetravalent vanadium. The peroxy compounds of vanadium are not extracted with ether, unlike the peroxy compounds of hexavalent chromium. Hydrogen peroxide is used in the photometric determination of vanadium in sulfuric or in nitric acid medium (p. 167).

A number of titrimetric methods for the determination of vanadium in steels, cast irons and other materials are based on the oxidation of ferrous ions in acid solution by pentavalent vanadium ions. The equivalence point is established with the aid of N-phenylanthranilic acid. In analytical practice the equivalence point is often found by potentiometric or amperometric techniques (appearance of oxidation current of ferrous ions on a rotating platinum anode).

Vanadium (V) ions in acid solution oxidize a number of organic compounds with the formation of intensely colored products. Thus, benzidine is oxidized to benzidine blue and aniline is oxidized to aniline black. Diphenylamine in concentrated sulfuric acid is oxidized to a blue compound. p-Phenylene diamine reacts with vanadium (V) ions in boiling dilute hydrochloric acid solutions to give a red coloration, while a yellow coloration is obtained in neutral solutions.

3,3′-Diaminobenzidine is oxidized by vanadate ions in phosphoric acid solutions with formation of reddish brown products. The reagent is used to determine small amounts of vanadium [9]. The optical density is measured at 470 mμ. The method is highly selective and very rapid; Fe(III), Cr(III), Ti(IV), Nb(V), Ta(V), Sn(IV), Mo(VI), Zr(IV), Cu(II), W(VI), Mn(II) ions do not interfere.

Diphenylcarbazide reacts with vanadate solutions in dilute sulfuric acid to give an orange-gray coloration and is a highly sensitive reagent for vanadium. The ions CrO_4^{2-}, MoO_4^{2-}, Fe^{3+}, Hg^{2+} interfere. Ferric and molybdate ions are masked by the addition of NaF, whereas mercuric ions are masked by chlorides.

Cupferron quantitatively precipitates vanadium (V) ions from solutions in 1% hydrochloric acid and is used to separate V(V) from Cr(III), Al, phosphates and other ions [10]. The red-colored precipitate is ignited to V_2O_5 and weighed.

Alcoholic solutions of N-benzoylphenylhydroxylamine give a gray-colored precipitate with dilute vanadate solutions in sulfuric acid [11]. The precipitate is readily extracted with chloroform; the orange-yellow extract has an absorption maximum at 450 mμ and its molar extinction coefficient is 3700. In hydrochloric acid solutions containing pentavalent vanadium ions the reagent gives a purple coloration or forms a precipitate; the purple-colored chloroform extract has an absorption maximum at 530 mμ and the molar extinction coefficient is 4500. If a chloroform solution of the reagent is used, the extracts will be stable, provided that the concentration of HCl in aqueous phase is 5–9 moles per liter. Vanadium (IV) ions are not precipitated and do not form colored compounds under these experimental conditions.

N-Benzoylphenylhydroxylamine is a highly selective reagent for V(V) ions and is used in the photometric determination of vanadium in steels and ores.

Hydroxamic acids (e.g., benzohydroxamic acid) react with vanadium (V) ions with the formation of colored extractable compounds which are used in the photometric determinations of vanadium (p. 170).

Xylenol Orange reacts with vanadium (V) ions in weakly acid solutions to form colored compounds and is used in the photometric determination of vanadium [12]. The reagent is highly selective in the presence of 1,2-diaminocyclohexanetetraacetic acid and will determine trace amounts of vanadium.

8-Hydroxyquinoline reacts with vanadate ions to form two differently colored compounds, the identity of which is not yet known. Chloroform extraction at pH 4 of the red compound of vanadium with 8-hydroxyquinoline is used to separate vanadium from Cr(VI), but not from Fe(III), Mo(VI), W(VI).

Tannin quantitatively precipitates vanadium (V) ions from almost neutral solutions in the presence of ammonium chloride and ammonium acetate. The very bulky precipitate is bluish-black. Tannin is used in the separation of small amounts of vanadium.

Vanadium compounds catalyze the oxidation reactions of a number of organic compounds, e.g., the oxidation of p-phenetidine by potassium chlorate, with phenol as activator. A number of quantitative catalytic (kinetic) methods of determination of small amounts of vanadium have been developed [14]. The catalytic properties of vanadium are probably connected with its capacity to pass from one valency state to another.

Relatively large amounts of vanadium are determined mainly by titrimetric methods based on redox reactions. Gravimetric methods are used less frequently. Small amounts of vanadium are determined photometrically, as peroxy compounds or phosphotungstovanadates. Polarographic methods are of no great practical importance. Vanadium in alloys is often determined by the spectroscopic method.

Gravimetric determination of vanadium [15]

The method is based on the precipitation of vanadium (V) ions by lead acetate from acetic acid solution. The precipitate of lead vanadate is dissolved in nitric acid, lead is removed as $PbSO_4$, while the filtrate containing the vanadium ions is evaporated and the residue ignited to V_2O_5.

The method is applicable to the analysis of ores and other materials. Chromium (VI), molybdenum (VI), tungsten (VI) and phosphorus (V) are precipitated together with vanadium.

REAGENTS

Lead acetate, 10% solution.
Acetic acid, dilute (1:5) solution.
Nitric acid, $d = 1.4$ g/cm^3.
Sulfuric acid, $d = 1.84$ g/cm^3.

PROCEDURE

A small excess of lead acetate solution is added with stirring to the sample solution in acetic acid. The solution with the precipitate is placed on a water bath, when the initial orange color of the precipitate disappears. Lead vanadate is filtered off and washed with acetic acid solution to remove Pb^{2+} ions. The precipitate is rinsed with nitric acid into a porcelain dish, heated until dissolved, a little sulfuric acid is added and the solution is heated to the appearance of sulfuric acid fumes. Water (100 ml) is then added, the precipitate of $PbSO_4$ is filtered off and washed with water until the wash waters no longer contain vanadium ions. The filtrate is evaporated to a small volume in the porcelain dish, transferred to a tared platinum crucible and heated in a Bunsen flame until the residue begins to melt. The resulting V_2O_5 is weighed.

Titrimetric determination of vanadium in ores by Mohr's salt [16]

Vanadium (V) ions are titrated in 5 *N* sulfuric acid medium against a standard solution of Mohr's salt in the presence of excess phosphoric acid which binds ferric ions into a colorless complex. The end point of the titration is established with the aid of an internal indicator such as phenylanthranilic acid or by the potentiometric technique using a platinum leaf or spiral as the indicator electrode.

Vanadium (IV) is oxidized to vanadium (V) in acid solution in the cold by dilute $KMnO_4$ solution, which is added drop by drop, with constant stirring, until a permanent pink coloration appears. The excess of the oxidant is removed a) by heating with HCl or NaCl, b) by sodium nitrite at room temperature, the excess nitrite being decomposed by urea, c) by oxalic acid.

The ore is solubilized in a mixture of sulfuric and phosphoric acids in the heat.

REAGENTS

Mixture of acids. To 300 ml of water 150 ml of sulfuric acid (d = 1.84 g/cm^3) are cautiously added; the solution is allowed to cool and 150 ml of H_3PO_4 (d = 1.70 g/cm^3) are added.
Potassium permanganate, 3% solution.
Hydrochloric acid, dilute (1:2) solution.
Sulfuric acid, dilute (1:1) solution.
N-Phenylanthranilic acid, 0.2 g of reagent and 0.2 g Na_2CO_3 are dissolved in 100 ml water with heating.
Mohr's salt, 1.08 g is dissolved in 5% sulfuric acid and the solution is made up to one liter with the acid. The solution is standardized against a solution of 0.135 g $K_2Cr_2O_7$ in one liter.

PROCEDURE

One gram of the ore is placed in a 250–300-ml flask and it is dissolved in the heat in 20–30 ml acid mixture, the solution being thoroughly stirred from time to time; when completely dissolved, the solution is cooled to room temperature. To the solution, which must be completely cold, potassium permanganate solution is added drop by drop to the appearance of a pink color, stable for 5 minutes; 3–5 drops hydrochloric acid are added and the solution is boiled until quite colorless. The solution is cooled, and 15 ml of sulfuric acid and 5–6 drops of N-phenylanthranilic acid are added; the solution is diluted to 100 ml with water and thoroughly mixed. It is then titrated against a solution of Mohr's salt until the color changes from cherry red to pale green.

Titrimetric determination of vanadium in steel by methyl orange [17]

Methyl orange in 60 vol % sulfuric acid is oxidized by vanadate with the evolution of nitrogen:

$$(H_3C)_2N\text{—}C_6H_4\text{—}N{=}N\text{—}C_6H_4\text{—}SO_3H + 4H_2SO_4 + 4HVO_3 \longrightarrow$$

$$\longrightarrow HO\text{—}C_6H_4\text{—}N(CH_3)_2 + HO_3S\text{—}C_6H_3(OH)\text{—}OH + N_2 + 4VOSO_4 + 5H_2O.$$

The following ions do not interfere: Co(II), Ni(II), Cu(II), Cr(III), Al(III), Mn(II), Fe(III), ferric iron in the presence of phosphoric acid, Ti(IV), Mo(VI), and W(VI). The ions Fe(II), Bi(III), Pb(II), Ca, Sr, Ba, NO_2^-, I^- and oxidizing agents such as chlorine, bromine, permanganates, chromates, etc., interfere.

The method is suitable for the determination of vanadium in steel.

REAGENTS

Sulfuric acid, dilute (1:3) solution.
Nitric acid, dilute (1:1) solution.
Phosphoric acid, $d = 1.7$ g/cm^3.
Potassium permanganate, 2% solution.
Manganese sulfate, 4% solution.
Oxalic acid, 0.1 N solution.
Methyl orange, 10^{-4} M solution in 60% sulfuric acid, standardized against a solution of vanadate of known concentration, as described below.

PROCEDURE

A 0.1–0.5-g steel sample is dissolved with heating in a mixture of 20 ml sulfuric acid with 5 ml of nitric acid. The solution is evaporated to the appearance of sulfuric acid fumes. The residue is cooled, 100 ml of water and 3–5 ml of phosphoric acid are added and the vanadium ions are oxidized by adding potassium permanganate solution drop by drop, until the pink coloration becomes stable for 1–2 minutes. After the oxidation of vanadium ions 1 ml of manganese sulfate solution is added and then oxalic acid solution drop by drop, until the color disappears. If the solution is turbid, it must be filtered. The filtrate is diluted with water to 200–250 ml in a volumetric flask. Two ml of the resulting solution are titrated against methyl orange solution to the appearance of the color.

Complexometric determination of vanadium [18]

The method is based on the formation of the blue complex compound of vanadium (IV) with Complexone III which is stable in weakly acid medium (pK = 18.77). N-Benzoylphenylhydroxylamine, which forms a red complex soluble in 50% ethanol with vanadyl ions at pH 3, is used as indicator. At the end point the red solution turns blue. Vanadium (V) ions are reduced to vanadium (IV) by sodium sulfite in sulfuric acid medium. Less than 10 mg of titanium, manganese and molybdenum do not interfere. Iron must be separated by precipitation with ammonia in the presence of hydrogen peroxide.

The method will determine 3–40 mg of V.

REAGENTS

Sodium sulfite, crystalline.
Acetate buffer solution, pH 3.
N-Benzoylphenylhydroxylamine, 2% solution in 96% ethanol.
Complexone III, 0.01 *M* solution.
Ethanol, 96%.

PROCEDURE

Solid sodium sulfite is added to the acid sample solution of vanadate and the mixture is boiled to expel SO_2. The solution is neutralized to pH 3 and acetate buffer solution is added, after which ethanol is added to a final concentration of 50%. The mixture is titrated against Complexone III solution until the color changes from red to blue.

One ml of 0.01 *M* Complexone III solution is equivalent to 0.5095 mg of vanadium.

Amperometric titration of vanadium (IV) ions [19]

The method is based on the titration of vanadium (IV) ions against a solution of cerium (IV) sulfate. The end point is indicated by the appearance of the reduction current of the excess Ce^{4+} ions (platinum rotating electrode) at +0.5 volt with respect to saturated calomel electrode.

The method is applicable to the determination of vanadium in ferrochrome.

REAGENTS

Sulfuric acid, dilute (1:4) solution.
Nitric acid, $d = 1.4$ g/cm^3.
Cerium (IV) *sulfate*, 0.004 *N* solution.
Mohr's salt, approximately 0.02 *N* solution.

PROCEDURE

The sample of ferrochrome (0.5–1.0 g) is dissolved by heating in 30–40 ml of sulfuric acid, a little nitric acid is added and the mixture is boiled to expel nitrous oxides. The solution is transferred to a 100-ml volumetric flask and diluted to the mark with water. A 25-ml aliquot of the solution is placed in an electrolyzer, the electrodes are immersed and the voltage adjusted to +0.9 volt. The absence of a diffusion current indicates that the solution does not contain ferrous ions. In this case a little of the solution of Mohr's salt is added drop by drop. The solution is then titrated against ceric sulfate solution. When all ferrous ions have reacted with the titrant, no current will pass through the galvanometer. The potential is then adjusted to

+ 0.5 volt and the solution is titrated against ceric sulfate until the diffusion current appears.

Photometric determination of vanadium as phosphotungstovanadic acid [20–22]

In the presence of a sufficient excess of tungstate ($V_2O_5:WO_3 = 1:100$) and phosphate ions vanadium forms phosphotungstovanadic acid which is reduced to the blue form at an acidity of 0.2–0.75 *N*; phosphotungstic acid is not reduced under these conditions. The absorption maximum of the reduced heteropoly acid is at 510–550 mμ. Phosphotungstovanadic acid is most abundantly formed at the boiling point, but its reduction should be conducted at room temperature.

The method is applicable to the analysis of steels and alloys. Vanadium ions are separated from nickel, cobalt and cupric ions, which are themselves colored, by coprecipitation with ferric hydroxide; the interfering chromium (III) is oxidized to the dichromate by ammonium persulfate in the presence of silver nitrate as catalyst. Mg, Ca, Sr, Ba, Zn, Cd, Hg(II), Al, Pb, As, Th, do not interfere.

The method is suitable for the determination of a fraction of one microgram of vanadium in 1 ml of solution.

DETERMINATION OF VANADIUM IN SOLUTION

REAGENTS

Sodium tungstate, 5% solution.
Phosphoric acid, $d = 1.7$ g/cm^3.
Stannous chloride, 0.5% solution. Fifty grams of $SnCl_2 \cdot 2H_2O$ are dissolved in 250 ml of concentrated hydrochloric acid and the solution is diluted to 500 ml with water; the resulting solution is diluted twenty times with water.
Standard solution of ammonium vanadate, 0.2 mg of V/ml.

CONSTRUCTION OF CALIBRATION CURVE

Into five 50-ml volumetric flasks are introduced 1, 2, 3, 4 and 5 ml of standard vanadate solution. These samples are heated to boiling, 4 ml of sodium tungstate, 3 ml of phosphoric acid and 5 ml of water are added to each flask, and the flasks are again heated to 90–95° (temperature checked with a thermometer); the solutions are left to stand for 2 minutes and cooled to room temperature, after which 2 ml of stannous chloride solution are added to each flask and the solutions are diluted to mark with water. The optical density of the solutions is measured in an electrophotocolorimeter or a photometer with a green filter. The calibration curve is then plotted.

PROCEDURE

To the sample solution are added the reagents employed in the construction of the calibration curve, the optical density is determined and the amount of vanadium read off the calibration curve.

DETERMINATION OF VANADIUM IN ORES

REAGENTS

Sodium peroxide, solid.
Acids, H_2SO_4 (1:1), HCl (1:2), HNO_3 (1:1).

CONSTRUCTION OF CALIBRATION CURVE

The calibration curve is plotted from different sample sizes of standard specimens as in the procedure. If a vanadium-free ore with a composition similar to that of the sample is available, the calibration curve is plotted after a known amount of standard solution of NH_4VO_3 has been added to the leached solution.

PROCEDURE

The ore (0.4 g) in an iron crucible is mixed with 4 g Na_2O_2 with the aid of a glass rod. The crucible with the mixture is gradually heated to dull red heat (500–600°C) in a muffle furnace or over a Bunsen flame until fused. The crucible is then grasped with a pair of tongs, its contents rapidly mixed with a swirling motion and the heating is continued for 5 more minutes. Overheating should be avoided or the melt will be difficult to dissolve. The melt is then cooled to 30–40°C. After cooling, it is washed with a jet of warm water from a washbottle into a 100–150-ml beaker. To the resulting solution are added 12 ml of 1:1 sulfuric acid, 4 ml of 1:2 hydrochloric acid and 4 ml of 1:1 nitric acid. The solution is heated and evaporated to a volume of 50 ml. The residue is filtered off and the filter paper is washed several times with hot water. The filtrate is cooled to room temperature and 5 ml of phosphoric acid are added. As a result, the color of the solution should disappear. If phosphoric acid is added to the warm solution, the solution assumes a permanent yellow coloration and the determination is not possible.

To the decolorized solution are added 20 ml of sodium tungstate solution and the solution is heated to 90–95°C; the heating is interrupted, the solution is left to stand for 5 minutes, cooled to room temperature, 2 ml of stannous chloride solution added, the solution transferred to a 100-ml volumetric flask and diluted to the mark with water. The optical density is measured using a green filter.

Photometric determination of vanadium by hydrogen peroxide [23, 24]

The method is based on the formation of the orange-red complex of vanadium (V) with hydrogen peroxide in sulfuric acid medium. The constant of the formation reaction $K = 1.2 \cdot 10^{-5}$.

The absorption maximum of the complex solution is at 450 mμ. The optimum conditions are 0.03% hydrogen peroxide and 0.6–6 N sulfuric acid. If the concentration of the peroxide is higher than 0.09%, the red-brown solution becomes yellow. Titanium (IV) and major amounts of ferric and molybdenum (VI) ions interfere. The effect of titanium (IV) and ferric ions is eliminated by adding fluoride and phosphate ions. Reducing agents, in particular iodides and bromides, interfere.

REAGENTS

Hydrogen peroxide, 3% solution.
Potassium permanganate, 0.01 N solution.
Oxalic acid, 2.5% solution.
Phosphoric acid, $d = 1.7$ g/cm^3.
Hydrochloric acid, $d = 1.19$ g/cm^3.
Nitric acid, $d = 1.4$ g/cm^3.
Sulfuric acid, dilute (1:4) solution.
Standard solution of ammonium vanadate, 0.01 mg of V/ml. 0.1286 g of NH_4VO_3 is dissolved in one liter of water and the resulting solution is diluted ten times.

CONSTRUCTION OF CALIBRATION CURVE

Into five 100-ml volumetric flasks are introduced 1, 2, 3, 4 and 5 ml of standard vanadate solution. These samples are diluted to 50 ml with water, and 3 ml of phosphoric acid, enough potassium permanganate solution to turn the solution pale pink and, after 1–2 minutes, 3–4 drops of oxalic acid solution are added to each flask. The solutions are allowed to stand for 3 minutes, after which 2 ml of hydrogen peroxide solution are added to each flask, the flasks made up to mark with water and the optical density is measured using an electrophotocolorimeter with a green filter.

PROCEDURE

A 0.5 g sample of steel is dissolved in a mixture of 30 ml of hydrochloric acid with 10 ml of nitric acid or in 50 ml of sulfuric acid, while adding nitric acid. If the steel contains tungsten, 3 ml of phosphoric acid are added. The resulting solution is evaporated with 10 ml sulfuric acid to the evolution of sulfuric acid fumes. The salts are dissolved in 50 ml of water, silicic acid is filtered off, the filtrate collected in a 250-ml volumetric flask, cooled, made up to the mark with water and stirred. Fifty ml of the resulting solution are withdrawn into each of two 100-ml volumetric flasks, 3 ml of phosphoric acid are added to each flask (unless this has already been done during the dissolution of the steel) and the determination is continued as described for the construction of the calibration curve, except that hydrogen peroxide is not placed in one of the flasks. The optical density of the sample solution is measured with respect to this solution using a green filter.

Photometric determination of vanadium by pyrocatechol [25]

Vanadium ions react with pyrocatechol in alkaline medium to form a blue-colored complex in which vanadium is tetravalent. The interference of Fe, Co, Ni, Cr, Mo, W, and Cu ions is eliminated by the introduction of ascorbic acid, citrate and cyanide ions and Complexone III. The following elements do not interfere: Zn, Cd, Zr, Th, Sn, Pb, As, La, Ce, U($<20\%$), Sb($<10\%$), Ga($<10\%$), etc.

Niobium, bismuth, titanium and other ions which form yellow complexes do not interfere at all, since the optical density is measured at 800 mμ.

The method is highly selective; up to 1% vanadium can be determined in a great variety of materials.

REAGENTS

Perchloric acid, concentrated.

Pyrocatechol. 35 g of reagent are dissolved in 300 ml of water, 50 ml of citric acid solution (500 g/liter), 2 g of ascorbic acid, 60 ml of ammonia solution ($d = 0.92$ g/cm^3) are added and the solution is diluted with water to 500 ml; the solution is kept under a layer of toluene, in order to prevent oxidation by atmospheric oxygen.

Ascorbic acid, crystalline.

Citric acid, 500 g of the reagent dissolved in one liter.

Solution of potassium cyanide and Complexone III. Twenty-five g of KCN and 12.5 g of Complexone III are dissolved in 250 ml of concentrated ammonia solution.

Standard solution of sodium vanadate, 0.5 mg of V/ml. In a platinum crucible 0.8925 g of CP vanadium pentoxide is fused with 6 g of sodium carbonate, the melt is dissolved in 100 ml of water, the solution is quantitatively transferred to a one-liter volumetric flask and made up to mark with water.

CONSTRUCTION OF CALIBRATION CURVE

Into five 100-ml volumetric flasks are introduced 0.2, 0.4, 0.6, 0.8 and 1.0 ml of standard vanadate solution and all the reagents employed in the procedure are added. A 5-ml aliquot of the solution is placed in a 25-ml volumetric flask, pyrocatechol solution is added to the mark and the optical density is measured at 800 mμ.

The calibration curve is plotted.

PROCEDURE

A few ml of perchloric acid are added to the solution obtained by the decomposition of the sample (0.1–0.5 mg of V) and the solution is evaporated. The residue is rinsed into a 100-ml flask using 35 ml of water. Five ml of citric acid solution, 0.5 g of ascorbic acid and 20 ml of potassium cyanide–Complexone III solution are added, the mixture is boiled for 1–2 minutes, cooled and diluted to 100 ml with water. Potassium perchlorate is filtered off, 5 ml of the clear filtrate are withdrawn into a 25-ml volumetric flask and the mixture is diluted to the mark with pyrocatechol solution. The solution is left to stand for 20 minutes and its optical density relative to the blank solution is measured at 800 mμ (l = 3 cm).

Photometric determination of vanadium in silicates by Acid Chrome Blue K [26]

Vanadium (IV) ions react with Acid Chrome Blue K at pH 3.6–6 to form a violet-colored compound with an absorption maximum at 590 mμ. The optical density of the colored solutions is proportional to the concentration between 0.01 and 0.3 μg V/ml. Vanadium (V) is reduced to vanadium (IV) by ascorbic acid. If tartaric acid is present, 2000-fold amounts of aluminum, 125-fold amounts of molybdenum and 10-fold amounts of tungsten ions do not interfere. Fluorides, chlorides, nitrates, sulfates and phosphates also do not interfere. The interference of chromium (III) ions is eliminated by introducing an equal amount of chromium ions into the comparison solution.

The method is suitable for the determination of vanadium in silicate rocks.

REAGENTS

Standard solution of ammonium vanadate, 5 μg of V/ml.
Acid Chrome Blue K, 0.1% solution.
Acetate buffer solution, pH 4.7.
Ascorbic acid, 3% solution.
Formic acid, concentrated.

CONSTRUCTION OF CALIBRATION CURVE

Into five 50-ml volumetric flasks are introduced 1, 2, 3, 4 and 5 ml of standard vanadate solution, 1 ml of ascorbic acid solution, a few drops of formic acid, 6 ml of acetate buffer solution, and 3 ml of reagent solution are added to each flask and the solutions are diluted to 50 ml with water. The comparison solution should contain all the reagents in the amounts taken for the sample solution. The optical density of the solutions is measured in an electrophotocolorimeter with a green filter

PROCEDURE

A 0.1–0.2 g sample of silicate rock is fused with sodium carbonate and potassium nitrate and the melt is leached with water. The precipitate is filtered off and silicic acid is isolated from the filtrate. The filtrate is diluted to the mark with water in a 25- or 50-ml volumetric flask and vanadium is determined on an aliquot of the solution. The solution is prepared using the procedure described for the construction of the calibration curve.

Extraction-photometric determination of vanadium [27]

Vanadium (V) ions react with benzohydroxamic acid to form a complex which is blue in strongly acid medium and red in weakly acid medium. The complex which is formed at pH 1.2–5.5 is extracted with Hexanol-1. The optical density of the solutions is proportional to the concentration between 0.4 and 8 μg of V/ml. Ta, Cr, Cu, Zn, Cd, F^-, SCN^-, citrates and tartrates do not interfere. Oxidizing and reducing agents should be absent. Ti, Zr, Mn, Fe(II), Th, Al, Sn, Mo(VI) and W(VI), etc., interfere. The effect of elements which form colorless complexes with the reagent is eliminated by introducing the reagent in excess.

REAGENTS

Potassium benzhydroxamate, 0.3% aqueous solution, pH 5.
Sulfuric acid, dilute (1:5) solution.
Hexanol-1.
Standard solution of ammonium vanadate, 0.05 μg V/ml.

CONSTRUCTION OF CALIBRATION CURVE

In five separating funnels are placed 1, 2, 3, 4 and 5 ml of standard vanadate solution and 10 ml of potassium benzhydroxamate are added to each funnel; the solutions are diluted to 30 ml with water, the pH is adjusted to 2 with sulfuric acid and the vanadium complex is extracted with 20 ml of Hexanol-1. The optical densities of the extracts are measured at 450 mμ relative to Hexanol-1 and the calibration curve is plotted.

PROCEDURE

The procedure for the determination of vanadium is the same as that used in preparing standard solutions. The content of vanadium is found from the calibration curve.

Catalytic (kinetic) determination of vanadium [28]

In the presence of vanadium compounds the oxidation of aniline by potassium chlorate at pH 6.7–7.5 is accelerated. The catalytic effect of vanadium ions is intensified in the presence of 8-hydroxyquinoline. Under otherwise identical conditions, the intensity of the resulting grayish color depends on the duration of heating of the solution. For example, the sensitivity is 0.02 μg of V after 5 minutes heating and 0.005 μg of V after 20 minutes heating. The method is quite selective and will determine vanadium in the presence of 10 μg each of Mo, W, Re, Ti, As, Sb, Bi, Zr, Al. Iron is masked by pyrophosphate.

REAGENTS

Standard solution of ammonium vanadate, 0.1 μg of V/ml.
Aniline sulfate, 5% solution.
Potassium chlorate, saturated aqueous solution.
8-*Hydroxyquinoline*, 1% solution in 1 *N* acetic acid.

CONSTRUCTION OF CALIBRATION CURVE

Into five 10-ml graduated tubes are placed 1, 2, 3, 4 and 5 ml of standard solution of ammonium vanadate, and 2 ml of aniline sulfate solution, 1 ml of potassium chlorate solution and 0.1 ml 8-hydroxyquinoline solution are added to each tube. The solutions are diluted to 10 ml with water and heated for 5 minutes on a water bath. The optical density of the solutions is measured in an electrophotocolorimeter with blue filter relative to a comparison solution containing all the reagents except the vanadium salt.

PROCEDURE

All the reagents are introduced into the solution as described for the construction of the calibration curve; the optical density is measured and the content of vanadium in the sample is read off the calibration curve.

BIBLIOGRAPHY

1. EPHRAIM, F.—*Helv. chim. acta,* **14,** 1266. 1931.
2. FURMAN, S. C. and C. S. GARDNER.—*J. Am. Chem. Soc.,* **73,** 4528. 1951.
3. CROUTHAMEL, C. E., B. E. HJELTE, and C. E. JOHNSON.—*Anal. Chem.,* **27,** 507. 1955.
4. FEINSTEIN, H. I.—*Anal. chim. acta,* **15,** 141. 1956.
5. KUZNETSOV, V. I. and L. S. KOZYREVA.—*ZhAKh,* **8,** 90. 1953.
6. GUSEV, S. I., V. I. KUMOV, and E. V. SOKOLOVA.—*ZhAKh,* **15,** 180. 1960.
7. TANAKA, M.—*Mikrochim. acta,* 701. 1954.
8. SUGAWARA, K., M. TANAKA, and H. NAITO.—*Bull. Chem. Soc. Japan,* **26,** 417. 1953.
9. CHENG, K. L.—*Talanta,* **8,** 658. 1961.
10. JONES, G. B.—*Anal. chim. acta,* **17**, 254. 1957.
11. RYAN, D. E.—*Analyst,* **85,** 569.1960.
12. BUDEVSKY, O. and R. PŘIBIL.—*Talanta,* **11,** 1313. 1964.
13. BOCK, R. and S. GORBACH.—*Mikrochim. acta,* 593. 1958.
14. BONTSCHEV, P. R.—*Ibid.,* 577. 1962.
15. SCOTT, W. W. *Standard Methods of Chemical Analysis,* p.1210.—New York. 1962.

16. LYALIKOV, YU. S., V. I. SAKUNOV, and N. S. TKACHENKO. *Analiz zheleznykh i margantsevykh rud (Analysis of Iron and Manganese Ores)*, p. 174.—Metallurgizdat. 1954.
17. ZHIGALKINA, T. S. and A. I. CHERKESOV.—*ZhAKh*, **16**, 505. 1961.
18. KAIMAL, V. R. M. and S. C. SHOME.—*Anal. chim. acta*, **27**, 594. 1962.
19. ALIMARIN, I. P. and S. I. TERIN.—*Zav. Lab.*, **21**, 777. 1955.
20. DAVYDOV, A. L. and Z. M. VAISBERG.—*Zav. Lab.*, **6**, 715. 1940.
21. TIKHONOVA, A. A.—*Zav. Lab.*, **16**, 1168. 1950.
22. LYALIKOV, YU. S., V. I. SAKUNOV, and N. S. TKACHENKO. *Analiz zheleznykh i margantsevykh rud (Analysis of Iron and Manganese Ores)*, p. 177.—Metallurgizdat. 1954.
23. PEN'KOVA, E. F., A. V. GLADKOVA, and T. V. NOVIKOVA.—*Zav. Lab.*, **22**, 918. 1956.
24. SANDELL, E. *Colorimetric Determination of Traces of Metals.*—Interscience, N.Y., 1944.
25. BLANQUET, P.—*Chim. anal.*, **41**, 359. 1959.
26. MORACHEVSKII, YU. V. and I. A. TSERKOVNITSKAYA.—*AhAKh*, **16**, 106. 1961.
27. WISE, W. M. and W. W. BRANDT.—*Anal. Chem.*, **27**, 1392. 1955.
28. NAZARENKO, V. A. and E. A. BIRYUK.—*ZhAKh*, **10**, 28. 1955.

Niobium Nb and Tantalum Ta

Compounds in which niobium and tantalum display a valency of + 5 and also certain compounds of trivalent niobium are important in analytical chemistry.

Compounds of niobium(V), niobates, are not strong oxidizing agents and are reduced only by strong reducing agents, e.g., by metallic zinc in acid solution. In this they differ from vanadates, which are strong oxidizing agents in acid solutions.

In a number of titrimetric methods for the determination of niobium, niobium in acid solution is reduced to the trivalent state by metallic zinc, cadmium or aluminum and is then titrated against a standard solution of an oxidizing agent (p. 184).

Most difficulties are involved in the quantitative reduction of niobium(V) in the presence of tantalum and other elements, and for this reason titrimetric methods are not yet extensively employed in analytical practice.

Compounds of tantalum(V) are not reduced in aqueous solution.

The properties of niobium and tantalum greatly resemble each other and also the properties of titanium, zirconium and tungsten. All these elements are frequently encountered in natural materials.

The analysis of chemically complex minerals of tantalum and niobium which contain titanium, zirconium and tungsten is very laborious, requires a high analytical skill and gives results which are not very reliable. There are no sufficiently reliable and simple methods for the isolation of small amounts of niobium and tantalum in the analysis of rocks, pure metals and alloys, or for the determination of niobium and tantalum present in concentrations of the order of $10^{-5}\%$ in metallic titanium, zirconium, tungsten and other metals. The most satisfactory results are obtained by using extraction methods and chromatographic methods of separation.

The difficulties encountered in analytical chemistry of niobium and tantalum are due to the strong tendency of the compounds of niobium and tantalum, and also of titanium, zirconium and tungsten to form aqueous colloidal solutions, to undergo hydrolysis and to form polynuclear compounds.

Complex formation by niobium and tantalum has not been thoroughly studied. In a number of cases, only the molar ratios in which niobium and tantalum ions reacted with organic reagents were determined, but no attention was paid to the possible formation of the so-called triple complexes. There are no reliable values for the instability constants of the complexes of niobium and tantalum or for the complexing reaction constants.

The usual analytical practice is to work with solutions of oxalate, tartrate, fluoride and peroxide compounds of niobium and tantalum, and also with solutions of niobates and tantalates. Methods for the preparation of a number of such solutions are given below (pp. 186, 188, 190, 194).

Oxalic acid and ammonium oxalate are employed to keep niobium and tantalum in solution. Water-soluble oxalate complexes of niobium and tantalum are stable in the presence of excess oxalate ions. The composition of the complex ions appears to depend on the conditions of their formation. Alkali oxalatoniobates $Me_3[NbO(C_2O_4)_3]$ (where Me is Na^+, K^+, or NH_4^+) have been isolated in the solid state.

Certain α-hydroxycarboxylic acids (tartaric, citric, lactic, trihydroxyglutaric) form complexes with niobium(V) and tantalum(V) ions.

Niobates and tantalates are stable in solution in the presence of excess KOH; they tend to polymerize with the formation of various polyanions.

When potassium niobate or potassium tantalate solution is acidified by dilute sulfuric acid, white amorphous precipitates of niobic acid or tantalic acid separate out. The precipitates are soluble in warm concentrated sulfuric acid; when cooled and diluted with water, niobic acid remains in solution, whereas tantalic acid precipitates out. However, separations of niobium from tantalum conducted in this manner are not quantitative.

Ignition of niobic and tantalic acids yields the white refractory pentoxides Nb_2O_5 and Ta_2O_5. The former oxide becomes yellow on being heated. Both pentoxides are suitable in many respects as gravimetric forms, but they are usually contaminated by oxides of other metals. For this reason the amount of impurities in niobium or tantalum pentoxide is determined after weighing and a correction is introduced.

The pentoxides Nb_2O_5 and Ta_2O_5 are also obtained on igniting niobium and tantalum compounds with pyrogallol, tannin, etc.

In concentrated hydrochloric acid niobium(V) probably exists [1] as $H[Nb(OH)_2Cl_4]$. The absorption maximum [2] of the solutions is at 281 mμ; the apparent molar extinction coefficient is 9000.

Beer's law is obeyed in the concentration range between 0.9 and 10 μg of Nb_2O_5/ml. Large amounts of iron and also molybdenum, uranium, titanium and, to a smaller extent, tantalum interfere. Zirconium, tungsten, REE, tartaric acid and small amounts of sulfates are practically without effect.

Large amounts of niobium in niobium-based alloys containing 3–30% of tantalum and also small amounts of titanium and iron [3] can be determined by differential spectrophotometry. This technique is just as accurate as the gravimetric method (error $\pm 0.3\%$), while being 4–5 times as rapid.

Niobium(V) reacts with concentrated thiocyanic acid to form the yellow complex $H[NbO(SCN)_4]$ [4]. Alkaloids precipitate a typically colored precipitate out of aqueous solutions of the complex compounds. The compound can be extracted with alcohols, ethers, aldehydes and ketones. The amount of niobium is found by determining the optical density of the ether extract. Methods based on the measurement of the optical density of water–acetone solutions of the thiocyanate complex of niobium have also been proposed (p. 186).

Tantalum forms the colorless complex $H_2[TaO(SCN)_5]$. This complex is extracted with organic solvents and is precipitated by alkaloids; unlike niobium, tantalum gives a colorless precipitate.

Numerous variant methods for the detection and photometric determination of niobium in the presence of tantalum and certain other elements [5–8] are based on the formation of the thiocyanate complex. The thiocyanate method is used in the determination of small amounts of niobium in ores which contain titanium, tungsten, molybdenum and chromium (p. 186).

Potassium fluorotantalate K_2TaF_7 and potassium fluorooxyniobate K_2NbOF_5 display different solubilities in aqueous hydrofluoric acid. Thus, niobium and tantalum can be separated by fractional crystallization (Marignac's method). The theoretical justification of this separation has been given [9]. At present other, better methods for the separation of niobium from tantalum are used in analytical chemistry.

Selenious acid quantitatively precipitates niobium and tantalum ions from boiling tartaric acid solutions containing hydrochloric acid. It is also used to separate niobium and tantalum from titanium and also to separate niobium from zirconium [10,11].

Hypophosphorous acid H_3PO_2 quantitatively precipitates tantalum from oxalic acid solutions as the hypophosphite. Under these conditions niobium is not precipitated and the two elements can be separated in this way [12]. Hypophosphorous acid is used in the determination of tantalum in steels and minerals.

Photometric determination of niobium (p. 188) is based on the formation of a phosphomolybdic heteropoly acid by niobium(V) ions.

Hydrogen peroxide reacts with niobium(V) ions in sulfuric and phosphoric acid solutions to form the yellow peracids. A number of variants of the photometric determination of niobium [13–15] are based on their formation.

Phenylarsonic acid [16] (p. 181) and its derivatives are valuable reagents in the isolation of small and large amounts of niobium together with tantalum and their separation from a number of other elements.

Phenylarsonic acid quantitatively precipitates tantalum ions from oxalate solutions at pH $\leqq$ 5.8 (acidity is adjusted [17] by the addition of sulfuric acid). Niobium ions are precipitated above pH 4.8. Tantalum ions are usually precipitated at pH below 3.0, or from 5% sulfuric acid. Niobium ions are subsequently precipitated from the filtrate at pH 5.0, by adding ammonium acetate and phenylarsonic acid. If the filtrate contains large amounts of oxalates, niobium ions are precipitated by cupferron. If the ratio between niobium oxide and tantalum oxide is larger than two, the tantalum compound must be reprecipitated and the results are not always satisfactory.

Phenylarsonic acid will separate tantalum from Fe(III), Cr(III), Al(III), REE, Ce(IV), Be(II), U(VI), Th(IV), Ca, Mg, Sn(II), Cu(II), Cd, Bi(III), Hg(II), Zn, Mn, Ni(II), Co(II), As(III), Sb(III), W(VI) and Mo(VI) by precipitation at pH 2–3 from solutions containing excess Complexone III.

Cupferron quantitatively precipitates Nb, Ta, and also Ti, Zr, Fe(III), V and Sn from solutions in 10% sulfuric acid. The ions Al(III), Cr(VI), U(VI), Be(II), Mn(II), Ni(II), Co(II), Zn(II), P, B remain behind in solution.

The reagent is employed [18] in the separation of niobium from tantalum and in the separation of niobium from a number of other elements in tartrate solutions containing ammonium acetate (pH 4.5–5.0) in the presence of Complexone III. Niobium ions are precipitated by cupferron with tin as coprecipitant. If ferric ions and other trivalent metals are present in 100-fold amounts, reprecipitation is necessary.

To separate niobium from tungsten, niobium ions are precipitated by cupferron from weakly acid solutions in the presence of oxalic or tartaric acids (p. 182).

N-Benzoyl-N-phenylhydroxylamine [19, 21] precipitates niobium out of tartaric acid solutions at pH 3.5–6.5 and is used in the separation of niobium from tantalum (p. 184). Tantalum is precipitated from the filtrate below pH 1.5.

The compound of niobium with N-benzoyl-N-phenylhydroxylamine is extracted with chloroform from tartaric acid solutions at pH 4–6. One portion of chloroform will extract 98–100% of the niobium compound [22]. Under these conditions tantalum ions remain behind in the aqueous phase.

Niobium and tantalum typically react with various hydroxylated organic reagents: pyrogallol, tannin, pyrocatechol, derivatives of 2,3,7-trihydroxy-6-fluorone, 1-(2-pyridylazo)-resorcinol Arsenazo I and other reagents.

Pyrogallol precipitates niobium and tantalum from hydrochloric acid solutions, while titanium and zirconium ions remain behind in solution [23]. In the presence of tartaric or citric acid the precipitation does not take place.

In oxalic acid solutions which contain a mineral acid tantalum ions react with pyrogallol to give a yellow-orange water-soluble compound; niobium ions give no coloration under these conditions [24]. On the contrary in weakly acid medium, niobium ions, but not tantalum ions, form an orange-colored compound [24]. These properties of niobium and tantalum compounds are used to advantage in the different photometric determinations of niobium and tantalum in the presence of each other [24–29].

Tannin precipitates [30,31] tantalum and then niobium from oxalic acid solutions at somewhat different pH values. The precipitate of tantalum is yellow, while that of niobium is red. Complete precipitation of tantalum is accompanied by a partial coprecipitation of niobium. A practically quantitative separation of tantalum from niobium may be attained by repeating the fractional precipitation several times. Titanium ions, if present in concentrations higher than 2% of the concentration of tantalum, interfere with the separation of niobium from tantalum.

Tannin is also employed in the simultaneous isolation of niobium and tantalum from hydrochloric acid solutions and their separation from a number of elements.

In the precipitation of niobium and tantalum ions different masking substances are employed. Thus, niobium (V) ions are quantitatively precipitated by tannin from solution in 5% hydrochloric acid with ascorbic acid present; titanium ions remain in solution [32].

Tannin quantitatively precipitates [33] niobium, tantalum and titanium ions from oxalate–tartrate solutions at pH 4.5 in the presence of Complexone III. Tungsten ions (if less than 2% of the sample) and many other elements remain behind in solution. The reagent has been employed in the determination of the sum total of niobium, tantalum and titanium in different minerals.

In order to separate niobium and tantalum from titanium, the fluorides of these elements are dissolved in hydrochloric acid solution of tannin, when a fairly stable, clear solution is obtained. When boric acid is added to this solution, compounds of niobium and tantalum with tannin precipitate out, whereas titanium ions remain behind in solution [34]. The method is employed in the analysis of different minerals.

Triple complexes of niobium (red) and tantalum (yellow) are formed in solutions containing Complexone III and pyrocatechol at pH 2.5 [35]. Niobium and tantalum ions react with Complexone III in the molar ratio of 1:1; the number of pyrocatechol molecules participating in the reaction is not known, since the complex compound is formed in the presence of a very large excess of the reagent. Under certain conditions the colored compound is formed by niobium only or by tantalum only.

Niobium and tantalum may be individually determined when present together in a mixture of oxides in ratios varying from 1:5 to 5:1.

To determine niobium in the presence of tantalum, the bisulfate melt of the oxides is solubilized in tartaric acid solution; Complexone III and excess pyrocatechol, in that order, are added to the solution, the pH is adjusted to 2.5, the solution brought to the boil and its optical density determined at 470 mμ. The amount of niobium is found from the calibration curve. Tantalum in tartaric acid solutions does not form a colored compound.

To determine tantalum in the presence of niobium the bisulfate melt of the oxides is dissolved in a concentrated solution of ammonium oxalate, Complexone III, a large excess of pyrocatechol and crystalline ammonium oxalate are added, the pH is adjusted to 2.5 and the optical density of the yellow solution is measured. Niobium does not form a colored compound under these conditions.

Pyrocatechol is also employed in extraction separation and subsequent photometric determination of niobium, tantalum and titanium [36].

Complexes of titanium and tantalum with pyrocatechol are extracted with n-butanol from oxalate solutions at pH 3, when niobium ions remain behind in the water phase. Titanium and tantalum can be quantitatively separated from niobium in this way.

9-Substituted derivatives of 2,3,7-trihydroxy-6-fluorone, in particular dimethylfluorone (p. 194), react with niobium, titanium and tantalum ions. Experimental conditions (acidity of medium, complex formers, etc.) can be adjusted so as to render the reaction sufficiently specific for each element. Thus, niobium ions alone react with trihydroxyfluorone derivatives in mineral acid solutions (sulfuric acid, hydrochloric acid, phosphoric acid); under these conditions tantalum ions either do not react at all or react very slowly [37,38].

Arsenazo I reacts with tantalum ions in strongly acid solutions (mainly hydrochloric acid solutions) which contain tartrates or citrates to form a violet-colored compound which is soluble in water [39]. The compound is stable on being heated to 100°C; the colored complex of titanium with Arsenazo I is then fully decomposed and for this reason there is no interference by titanium with the photometric determination of tantalum. Under these experimental conditions, however, niobium also forms a colored compound and interferes. Fluorides and oxalates suppress the formation of colored compounds.

Photometric techniques will determine tantalum present in titanium alloys at concentrations between 1 and 15% without previous separation.

The peroxy compound of niobium reacts with Methylthymol Blue [40,41], 1-(2-pyridylazo)-resorcinol [40], nitrilotriacetic acid, diethylenetriamine-N,N,N′,N″,N″-pentaacetic acid, and N-hydroxyethylethylenediamine-N,N′,N′-triacetic acid [42,43]. This results in the formation of mixed complexes which contain hydrogen peroxide and the organic reagent.

The peroxy compound of niobium reacts [42] with nitrilotriacetic acid and also with N-hydroxyethylethylenediamine-N,N′,N′-triacetic acid in the molar ratio of of 1:1. Niobium may be complexometrically determined [42].

The peroxy compounds of both tantalum and niobium react with different complexones [43] with the formation of mixed complexes.

1-(2-Pyridylazo)-resorcinol reacts with tantalum ions at pH 4–5.5 in the presence of oxalates, fluorides or hydrogen peroxide to form colored complex compounds [44,45] in which the molar ratio is 1:1. The complex with 1-(2-pyridylazo)-resorcinol and oxalate ions is suitable for the photometric determination of tantalum in alloys based on zirconium, molybdenum, tungsten and uranium. The optical density of the solution is independent of the oxalate concentration within a wide range. The absorption maximum is at 535 mμ (for the reagent at 410 mμ). The molar extinction coefficient is 17,000. The sensitivity of the photometric method is 0.2 μg Ta/ml. The method is applicable for the determination of tantalum present in the alloy at concentrations greater than 0.1%.

Niobium in the form of complexes with hydrogen peroxide, oxalates, tartrates or fluorides [46,47] reacts with Xylenol Orange at pH 2–3 to form colored compounds. The absorption maximum of the solutions is at 570 mμ (for the reagent at 445 mμ). The molar extinction coefficient is approximately 14,000. A photometric method is available for the determination of niobium by Xylenol Orange in the presence of ammonium oxalate in uranium-, molybdenum- and tungsten-based alloys. The sensitivity of the method is 0.1 μg Nb/ml. For niobium concentrations in the alloy of 10% the mean square error is $\pm$ 0.2%.

Lumogallion reacts with the tartrate complex of niobium (V) in 0.5–2 *N* sulfuric acid to form a compound with an absorption maximum at 510 mμ [48]. Under these conditions tantalum does not form a colored complex with lumogallion and does not interfere. The effect of zirconium, iron and titanium ions is eliminated by adding small amounts of Complexone III. The reagent has been employed in the determination of niobium in loparite concentrate [49].

8-Hydroxyquinoline quantitatively precipitates niobium ions from tartaric acid solutions at pH 6 [50] and is used in the separation of niobium from tantalum and other elements.

The fluoride complex of tantalum is extracted with benzene from acid solutions in the presence of Methyl Violet [51], Rhodamine 6G [52], Butylrhodamine B [52] and other organic compounds capable of forming cations in acid solutions. The extraction is performed from hydrofluoric acid solutions. The amount of tantalum is found from the optical density of the colored extracts. Small amounts of niobium, titanium, zirconium and hafnium ions do not interfere with the photometric determination of tantalum.

Methyl Violet is used in the extraction-photometric determination of tantalum in ores [51], and in metallic zirconium, hafnium and niobium [53, 54]. The extraction methods and the chromatographic methods of separation of niobium, tantalum and titanium are very important, since sufficiently selective reagents for their determination are not available. Extraction methods will effect a more complete and less laborious separation of niobium and tantalum than will methods based on precipitation reactions. However, such extractions must be performed in special vessels made of plastic material resistant to the action of hydrofluoric acid.

Ketones [55,56], amines [57,58], mixtures of n-butylphosphoric acids (solution in di-n-butyl ether) [59], tributyl phosphate [60,61], and diisobutylcarbinol [62,63] are used in the extraction separation of niobium and tantalum as chloride and fluoride complexes, and also in the separation of niobium from tantalum.

Ketones extract tantalum complexes at a much lower concentration of hydrofluoric acid than niobium complexes.

The extractability of fluoride complexes of niobium and tantalum by diisopropyl ketone from solutions in sulfuric acid and in nitric, hydrochloric and perchloric acids is different, the tantalum complex being extracted much more readily than the niobium complex [55]. Owing to this difference tantalum may be quantitatively separated from niobium.

Niobium ions are readily extracted from solutions in 11 *M* HCl with an 8% solution of tribenzylamine in chloroform or methylene chloride, while the extraction of tantalum ions is insignificant [58]. In this way a rapid quantitative separation of niobium from tantalum is possible. Tribenzylamine may be replaced by methyldioctylamine and other aliphatic or aromatic amines with a long carbon chain dissolved in organic solvents.

Cyclohexanone extracts tantalum from solution in hydrofluoric and sulfuric acids as the fluoride complex, but niobium and titanium are not extracted [64]. Tantalum ions are then reextracted with ammonium oxalate solution containing boric acid and are then photometrically determined by pyrogallol in oxalic acid–sulfuric acid solution. A few hundredths or even a few thousandths of one percent of tantalum can be determined in this way in ores, loparite concentrates and other materials.

Cyclohexanone is used in the separation of tantalum from zirconium by triple extraction of the fluoride complex of tantalum from solutions in 2–4 *M* sulfuric acid [65]. This is one of the best methods for the separation of tantalum from zirconium.

Niobium, tantalum and titanium can also be quantitatively separated by different chromatographic methods [66–72].

Niobium and tantalum are separated by paper chromatography [66] using methyl isobutyl ketone and hydrofluoric acid as solvent.

Niobium and tantalum are detected by 8-hydroxyquinoline. The detectable minimum is 20 μg of each element. The method has been employed in the detection of niobium and tantalum in steels.

Niobium (V) ions, unlike tantalum ions, are reduced on a mercury electrode in solutions of inorganic acids and also in solutions containing organic ligands, giving catalytic waves.

One promising method is the polarographic determination of niobium and titanium in the presence of each other (Nb : Ti = 1 : 20) with pyrophosphoric acid as supporting electrolyte [73] (p. 197).

The determination of niobium and tantalum by the method of isotopic dilution does not necessitate a complete separation of these elements; it is sufficient to isolate only part of the niobium or tantalum in the pure form. The neutron activation method is employed in the determination of tantalum in alloys containing niobium, titanium and zirconium and in ores, and also to determine niobium in alloys.

X-ray spectroscopy is used for the rapid determination of more than 0.1% niobium and tantalum in ores, minerals and alloys.

Emission-spectroscopic techniques for determination of niobium and tantalum are employed in a number of cases but they are not very sensitive. They are used for the quantitative determinations of niobium and tantalum in industrial raw materials and in other industrial products [74].

For surveys of the methods of determination of tantalum and niobium the reader is referred to [75,76].

Gravimetric determination of niobium and tantalum in steels by phenylarsonic acid (in the absence of tungsten) [77]

Niobium and tantalum are quantitatively isolated from dilute hydrochloric acid solution by hydrolysis, with subsequent supplementary precipitation by phenylarsonic acid. Niobium and tantalum pentoxides serve as the gravimetric form.

REAGENTS

Hydrochloric acid, d = 1.19 g/cm^3, dilute (1:1) and 1% solutions.
Nitric acid, d = 1.4 g/cm^3.
Phenylarsonic acid, 3% solution.
Ammonium nitrate, 2% solution.
Sulfuric acid, d = 1.84 g/cm^3 and dilute (1:1) solution.
Hydrofluoric acid, concentrated.
Potassium pyrosulfate, crystalline.

PROCEDURE

A 1–2-g sample of steel is dissolved in 30 ml of concentrated hydrochloric acid and 10 ml of nitric acid in a 300-ml beaker. The solution is evaporated to dryness and the residue is gently ignited. Ten ml of concentrated hydrochloric acid are added to the residue and the mixture is heated. The solution with the precipitate is diluted to 200 ml with hot water and heated to 90°C after which 40 ml of phenylarsonic acid are introduced to precipitate niobium and tantalum, and the mixture is left to stand overnight. The precipitate, which contains niobic, tantalic and silicic acids, is separated on Blue Ribbon filter paper, washed with hot 1% hydrochloric acid and then with ammonium nitrate solution.

The washed precipitate together with the filter paper is placed in a platinum dish, the filter paper is cautiously carbonized and then ignited at 1000°C under a fume hood. In order to remove silicic acid the ignited precipitate is treated with 6 ml of concentrated sulfuric acid and with a few drops of hydrofluoric acid.

The analyst should make sure that sulfuric acid is present in excess during the elimination of silicic acid in order to avoid losses of tantalum. The contents of the dish are evaporated until sulfuric acid fumes are expelled. The residue is gently ignited in a muffle furnace at 800°C and is fused with potassium pyrosulfate. The melt is leached with water and transferred to a 300-ml beaker. The contents of the beaker are diluted with hot water to 200 ml, 8 ml of concentrated hydrochloric acid are added and the mixture is heated for one hour at 90°C. The bulk of niobium and tantalum separates out as white flakes owing to hydrolysis. In order to ensure complete precipitation, 40 ml of phenylarsonic acid solution are added, the mixture is heated for another 30 minutes and left to stand overnight. The precipitate is filtered through a White Ribbon filter paper, washed once with hot 1% hydrochloric acid, then twice with water and ignited in a porcelain dish at 1000°C under a fume hood; the product is weighed as Nb_2O_5 and Ta_2O_5.

Note 1. If tungsten is present in the steel, it separates out together with niobium and tantalum.

Note 2. If titanium is present in the steel, the precipitate of niobium and tantalum pentoxides will contain titanium oxide as impurity. The weighed precipitate is fused with potassium pyrosulfate, leached with 20 ml of 1:4 sulfuric acid, the solution transferred to a 50–100-ml volumetric flask and titanium is determined photometrically. The amount of titanium dioxide found (in grams) is deducted from the weight of niobium and tantalum pentoxides.

Gravimetric determination of niobium (in the presence of tungsten) [32, 78]

Niobium ions are precipitated by cupferron in weakly acid medium in the presence of oxalic (or tartaric) acid, which masks the tungsten ions. If niobium is to be determined in specimens containing tungsten, iron, titanium, vanadium and other elements which are precipitated by cupferron, niobium must first be precipitated by tannin in the presence of ascorbic acid (marked coprecipitation of tungsten ions), and then separated from tungsten in oxalic acid solution.

Satisfactory results have been obtained in determinations of 12–50% Nb in alloys containing 36–70% W.

REAGENTS

Tannin, freshly prepared 1% solution in water.
Ascorbic acid.
Hydrochloric acid, $d = 1.19$ g/cm^3 and 1:2, 1:1 and 4% solutions.
Nitric acid, $d = 1.4$ g/cm^3.
Sulfuric acid, dilute (1:1) solution.
Hydrofluoric acid, 40% solution.
Ammonium chloride, crystalline.
Potassium pyrosulfate, crystalline.
Oxalic acid, 4% solution.
Cupferron, 3% solution.
Washing solution, containing 20 ml of a 3% solution of cupferron and 1 ml of hydrochloric acid ($d = 1.19$ g/cm^3) in one liter.

PROCEDURE

A 0.5-g steel sample is dissolved in 100 ml of hydrochloric acid (1:1) in a 300-ml beaker. Five ml of nitric acid are added and the solution is boiled and evaporated almost to dryness. To the residue are added 30 ml of hydrochloric acid (1:2) and the mixture is boiled to expel nitrous oxides (when the volume of the solution has decreased by one-third, 10–20 ml of water are added). The solution is diluted with water to 180–190 ml (any black precipitate which may be formed is ignored), 1–2 g of ammonium chloride and 0.1–0.2 g of ascorbic acid are added and the mixture is heated to boiling. To the boiling solution 10 ml of tannin solution are added drop by drop, with constant stirring, and the mixture is kept at the boil for 2–3 hours to ensure complete coagulation of the precipitate.

Some macerated filter paper is introduced into the solution, the mixture cooled or left to stand overnight, the precipitate filtered through a White Ribbon filter paper 7–9 cm in diameter and washed 6–8 times with cold 4% hydrochloric acid. The filter paper with the precipitate is placed in a platinum crucible, ashed and the precipitate ignited in a muffle furnace at 500–600°C to effect complete combustion of the organic matter. The residue is wetted with water, 2–3 ml of sulfuric acid (1:1) added, then 2–3 ml of hydrofluoric acid and the crucible is placed on a sand bath and heated until the precipitate has dissolved and the sulfuric acid fumes appear. After sulfuric acid fumes have been evolving for about 10 minutes, the crucible is cooled, 1–2 ml of water are added to the residue and the mixture is again evaporated, leaving behind 2–3 drops (not to dryness!). After cooling, 10 ml of 1:1 hydrochloric acid are added and the mixture is heated until the salts are dissolved. The solution is transferred to a 100-ml beaker, rinsing the crucible several times with cold water, 0.05–0.1 g of ascorbic acid are added, the solution diluted to 80–90 ml with water, and heated to boiling, after which niobium is reprecipitated by tannin as above.

The precipitate is ashed in a platinum dish, the residue fused with 1–2 g of potassium pyrosulfate, 10–20 ml of oxalic acid solution are added and the mixture is heated until the melt has dissolved. The solution is transferred to a 250–300-ml beaker, the walls of the beaker are rinsed with cold water, the volume of the solution is made up to 150–200 ml, 5 drops of HCl ($d = 1.19$ g/cm^3) are added and niobium is precipitated by adding 20–25 ml of cupferron drop by drop, with stirring. A little macerated filter paper is added and the mixture is carefully stirred until the precipitate is fully coagulated. The precipitate is filtered through a White Ribbon paper 7–9 cm in diameter and washed 6–8 times with the washing solution. The filter with the precipitate is placed in a platinum dish, ashed and the residue ignited at 500–600°C. The residue is fused with 1–2 g of potassium pyrosulfate over a small burner flame. To the dish are added 20–30 ml of oxalic acid solution, the mixture is heated, 20–30 ml of water are added and the heating is continued until the melt has dissolved completely. The solution is transferred to the beaker, 5 drops hydrochloric acid ($d = 1.19$ g/cm^3) added and a reprecipitation is conducted as above.

The filter with the precipitate is placed in a tared platinum or porcelain crucible, ashed, the residue ignited at 1000°C in a muffle furnace for 15–20 minutes, cooled in a desiccator and weighed as Nb_2O_5.

The conversion factor to Nb is 0.6990.

Separation of tantalum from niobium by N-benzoyl-N-phenylhydroxylamine [19]

Niobium (V) is precipitated from tartaric acid solution at pH 3.5–6.5 by N-benzoyl-N-phenylhydroxylamine as $NbO(C_{13}H_{10}NO_2)_3$, the niobium precipitate is separated, and tantalum is precipitated from the filtrate by acidifying the solution to pH 1 with sulfuric acid. Satisfactory separation is achieved in the range of $Nb_2O_5:Ta_2O_5$ ratios between 1:16 and 100:1. If the compound of niobium with N-benzoyl-N-phenylhydroxylamine is reprecipitated, niobium may be isolated from mixtures containing 100 times more Ta_2O_5 than Nb_2O_5. In the presence of Fe, U, Th, Cr, Al, REE, Cu, Cd, Bi, Pb, Hg, Zn, Mn, Ni, Co, As, Sn, Sb, Ba, Sr, Ca, Mg and P, the precipitation is effected after adding 2–3 times the theoretical amount of Complexone III.

REAGENTS

N-Benzoyl-phenylhydroxylamine, 10% solution in ethanol.
Tartaric acid, crystalline.
Ammonium acetate, 20% solution.
Sulfuric acid, $d = 1.84$ g/cm^3.
Washing solution, 0.1 g of reagent is dissolved in 100 ml of boiling water.

PROCEDURE

To the sample solution containing 5% tartaric acid, ammonium acetate solution is added to pH 3.5–6.5 (if tantalum is present in large amounts, the pH is adjusted to 6), the solution is diluted with water to 350 ml, heated to boiling and niobium is precipitated by adding the solution of the reagent drop by drop. The precipitate is filtered off, washed with dilute reagent solution, ignited and weighed as Nb_2O_5. The filtrate is acidified to pH 1.0 with sulfuric acid, heated for 45 minutes on a boiling water bath, the precipitate filtered through filter paper, washed with a dilute solution of the reagent, ignited and weighed as Ta_2O_5.

Titrimetric determination of niobium in niobium carbide [79]

Niobium (V) is reduced to niobium (III) in sulfuric and hydrochloric acid medium (15–20 ml of concentrated sulfuric acid and 30–40 ml concentrated hydrochloric acid are added to each 100 ml of solution) by metallic cadmium or aluminum, after which additional reduction is performed in a cadmium reductor.

The olive-green solution of the reduced niobium is added to a solution of ferric ammonium alum; the ferrous iron which is formed in an equivalent amount is titrated against a solution of potassium dichromate in the presence of phenylanthranilic acid. The method is suitable for the determination of niobium in niobium carbide.

REAGENTS

Nitric acid, $d = 1.4\ g/cm^3$.
Hydrochloric acid, $d = 1.19\ g/cm^3$.
Sulfuric acid, $d = 1.84\ g/cm^3$ and dilute (1:20) solution.
Hydrofluoric acid, 40% solution.
Potassium dichromate, 0.1 *N* solution, standardized against a solution of a niobium compound.
Phenylanthranilic acid, 0.1% solution.
Ferric ammonium alum, 0.1 *N* solution.
Cadmium, metallic.

PROCEDURE

A 0.1-g sample of carbide in a platinum dish is dissolved in a mixture of 15–20 ml of nitric acid and 5–10 ml of hydrofluoric acid. When the dissolution has ended, 10 ml of sulfuric acid are added and the solution is evaporated to the appearance of sulfuric acid fumes. The residue is transferred to a conical flask, rinsing the dish with 15–20 ml of water. To the solution are added 15–20 ml of sulfuric acid, and 30–40 ml of hydrochloric acid and the volume is made up to 100 ml. Metallic cadmium (2–2.5 g) is introduced in small portions and the mixture is gently heated to dissolve the metal, when the solution turns green. The solution is cooled and passed through a cadmium reductor with a layer height of not less than 30 cm. The reductor is previously washed 2–3 times with 1:20 sulfuric acid. The sample solution is run through slowly, drop by drop, and is collected in a vessel with 20–25 ml ferric ammonium alum solution. A stream of carbon dioxide is passed through the vessel. The reductor is then washed 2–3 times with 25 ml of sulfuric acid (1:20) and 2–3 times with water saturated with carbon dioxide.

The ferrous ions which have been formed are titrated against the solution of potassium dichromate in the presence of a few drops of phenylanthranilic acid solution.

Photometric determination of niobium in tungsten-containing steels [77]

Niobium is hydrolytically separated in the presence of tannin and stannous chloride which is added to reduce ferric ions and to stabilize the thiocyanic acid in hydrochloric acid solutions. The presence of stannous chloride is without effect on the formation of the thiocyanate compound of niobium. The precipitate, which also contains coprecipitated tungsten, is ignited and the residue is fused with potassium pyrosulfate. The melt is dissolved in tartaric acid and niobium is determined in the tartrate solution by the thiocyanate reaction in water-acetone medium.

Oxalates and fluorides, and also large amounts of sulfates, interfere.

The method is suitable for the determination of niobium in steels and other materials.

REAGENTS

Hydrochloric acid, $d = 1.19\ g/cm^3$ and dilute (1.9) solution.
Nitric acid, $d = 1.4\ g/cm^3$.
Sulfuric acid, dilute (1:1) solution.
Aluminum, shavings.
Tannin, 0.5 and 10% solutions in 1:100 sulfuric acid.
Stannous chloride, 20% solution.
Potassium pyrosulfate, crystalline.
Potassium thiocyanate, 30% solution.
Acetone, C.P.
Tartaric acid, 20% solution.
Standard solution of niobium, 0.1 mg Nb/ml. Spectroscopically pure Nb_2O_5 (0.05 g) is fused with 2 g of $K_2S_2O_7$, the melt is dissolved in 30 ml of 20% solution of tartaric acid and the solution is made up to mark with water in a 500-ml volumetric flask.

CONSTRUCTION OF CALIBRATION CURVE

Into five colorimetric cells are placed 0.1, 0.2, 0.3, 0.4 and 0.5 ml of standard pentovalent niobium compound; 5 ml of HCl ($d = 1.19\ g/cm^3$), 1 ml of stannous chloride solution, 2 ml of water and 5 ml of acetone are placed in each cell, the solutions are well stirred and cooled to 15°C, after which 5 ml of potassium thiocyanate solution are added to each cell.

PROCEDURE

A 0.5-g sample of steel is dissolved in a 100-ml beaker in 30 ml hydrochloric acid ($d = 1.19\ g/cm^3$) and 10 ml of nitric acid, and 10 ml sulfuric acid are added and the solution is evaporated to the appearance of sulfuric acid fumes. The residue is cooled, water is cautiously added and the solution is evaporated to a residual volume of 2–3 ml. The contents of the beaker are rinsed into a 250-ml beaker with hot 1:9 hydrochloric acid. The solution is heated to boiling, 0.1–0.15 g of aluminum shavings are added and the solution is boiled until the shavings are fully dissolved; iron (III) and titanium (IV) are reduced and the solution assumes a violet coloration which is typical of titanium (III) ions.

Ten ml of 10% solution of tannin and 2 ml of stannous chloride solution are added, the mixture is boiled for 10 minutes and left to stand for 10 hours. The niobium-containing precipitate is separated by filtration and washed on the filter paper with cold 0.5% tannin solution. The filter paper with the precipitate is ashed in a porcelain (quartz) crucible and ignited at 900°C. The residue is fused with 2 g of $K_2S_2O_7$; the melt is leached by heating with 50 ml of tartaric acid solution, the solution transferred to a 250-ml volumetric flask, diluted to the mark with water and stirred.

A 1–2 ml aliquot of the solution is placed in the colorimetric cell and 5 ml of HCl (d = 1.19 g/cm^3), 1 ml of stannous chloride solution, 2 ml of water and 5 ml of acetone are added. The solution is mixed, cooled to 15°C and 5 ml of potassium thiocyanate solution are added. After 5 minutes the color of the solution is visually matched against the scale of standards.

Photometric determination of small amounts of niobium in ores containing titanium, tungsten, molybdenum and chromium [80]

Niobium is isolated and at the same time separated from chromium, tungsten, molybdenum, tin, titanium, zirconium and iron by coprecipitation with $MnO_2 \cdot nH_2O$. Niobium is solubilized by fusion with sodium pyrosulfate and the melt is leached with tartaric acid. The niobium in solution is determined photometrically by the thiocyanate method in acetone medium.

REAGENTS

Sodium hydroxide, crystalline and 2% solution.
Sulfuric acid, dilute (2:250 and 1:2).
Hydrogen peroxide, 30% solution.
Manganese sulfate, 1 *N* solution.
Ammonium persulfate or potassium persulfate, crystalline.
Sodium bisulfate, crystalline.
Tartaric acid, 15% solution.

PROCEDURE

The ore sample (0.1–2 g) is fused with 5–10 g of sodium hydroxide in a nickel or an iron crucible. The melt is leached with water, the solution with the precipitate is boiled for 2–3 minutes, the precipitate separated on filter paper and washed 3–5 times with the solution of sodium hydroxide. The filtrate contains Cr, W, Mo, Sn, Si, Al, while the precipitate contains Nb, Ta, Ti, Fe. The filter paper with the precipitate is unfolded on the edge of the beaker and the precipitate is washed into the beaker with a small amount of water. The filter paper is then washed several times with 2:250 sulfuric acid solution containing a few drops of hydrogen peroxide and is then discarded. To the beaker with the filtrate 6–8 ml of sulfuric acid (1:2) are added and the solution is boiled to decompose hydrogen peroxide. To the cooled solution

1 ml of manganese sulfate solution and 2 g of potassium or ammonium persulfate are added. The resulting solution should be about 200 ml in volume. It is then heated on an electric plate. The solution with the separated manganese dioxide is boiled for 7–10 minutes, left to stand for a brief period of time and the precipitate is separated on a small, dense filter paper and washed with 2:250 sulfuric acid. The filtrate contains Mo, Ti, Cr, Zr, Fe. Niobium is quantitatively separated together with manganese dioxide.

For complete separation of interfering elements the manganese dioxide is reprecipitated. To do this the precipitate is washed from the filter into the beaker which was used for the precipitation. The filter is washed with 2:250 sulfuric acid containing a few drops of hydrogen peroxide, and then with pure water until the paper has been washed free from hydrogen peroxide. The filtrate (200 ml) should contain 6–8 ml of 1:2 sulfuric acid. The solution is boiled to effect full decomposition of the hydrogen peroxide, cooled and manganese dioxide is again precipitated as above. The precipitate is filtered through the same filter paper and washed with dilute sulfuric acid.

The precipitate of manganese dioxide together with the coprecipitated niobium is fused with 1.5–3 g $NaHSO_4$, the melt is dissolved in tartaric acid solution and the solution diluted to 25 or 50 ml. The final concentration of tartaric acid should be 7.5%.

Niobium is determined photometrically by the thiocyanate method (p. 187).

Photometric determination of niobium in tantalum pentoxide [81]

The method is based on the determination of the optical density of the blue solution of reduced phosphomolybdoniobic heteropoly acid. The yellow phosphomolybdoniobic acid is most fully formed in 0.6 *N* sulfuric acid; after 15–20 minutes it is reduced by stannous chloride in 2.5–4 *N* sulfuric acid. Under these conditions molybdates and phosphomolybdic heteropoly acid are not reduced. Tantalum, titanium (up to 80 μg of TiO_2/50 ml in the presence of fluorides), iron (up to 50 mg), zirconium (up to 0.15 mg) and silicon (0.10 mg) do not interfere; vanadium ions interfere. In the presence of tungsten ions the amount of the added phosphate must be increased.

The method will determine 5–150 μg of Nb in 50 ml of solution.

REAGENTS

Standard solution of niobium, 20 μg of Nb/ml. Metallic niobium (0.2000 g) is dissolved with heating in a mixture of 1.5 g ammonium sulfate and 3 ml sulfuric acid (d = 1.84 g/cm^3). Another 30 ml sulfuric acid (d = 1.84 g/cm^3) are added, the solution transferred to a 100-ml volumetric flask and diluted to mark with water. One ml of solution is withdrawn into a 100-ml volumetric flask, 16 ml of tartaric acid–ammonium oxalate mixture are added and the solution is made up to the mark.

Disubstituted sodium phosphate, 0.06% solution.

Ammonium molybdate, 5% solution.
Stannous chloride, 0.5% solution: 25 g $SnCl_2 \cdot 2H_2O$ are dissolved in 25 ml of hydrochloric acid (d = 1.19 g/cm^3) and the solution is diluted to 50 ml with boiling water. Ten ml of 1:1 HCl are added to 5 ml of 25% solution and the result is diluted to 250 ml with water. The reagent is stable for 8–10 hours.
Sulfuric acid, d = 1.84 g/cm^3, and 1 N, 3 N and 12 N solutions.
Mixture of tartaric acid with ammonium oxalates, 250 ml of 10% solution of tartaric acid are mixed with 150 ml of 5% ammonium oxalate solution.
Ammonium fluoride, 0.4% solution.
Ammonium sulfate, crystalline.
Diluting solution, 1% in tartaric acid, 0.3% in ammonium oxalate and 1 N in sulfuric acid.

CONSTRUCTION OF CALIBRATION CURVE

In five 50-ml volumetric flasks are placed 1,2,3,4, and 5 ml of standard solution of niobium; 5 ml of 1 N sulfuric acid, 4 ml of ammonium fluoride solution (in the absence of titanium 4 ml of water are added), 2 ml of disubstituted sodium phosphate solution and 5 ml of ammonium molybdate solution are added to each flask; the solutions are thoroughly mixed and left to stand for 20–30 minutes. Eight ml of 12 N sulfuric acid are then added to each solution, the solutions are shaken and 1 ml of stannous chloride solution is added to each flask after one minute and the solutions are made up to mark with 3 N sulfuric acid. The optical densities of the solutions are measured after not more than 10 minutes, using an electrophotocolorimeter with a red filter. From the data thus obtained the calibration curve is plotted.

PROCEDURE

A 50-mg sample of tantalum pentoxide, which should contain 0.1–4% niobium, is dissolved in a mixture of 1.5 g of ammonium sulfate and 3 ml of sulfuric acid (d = 1.84 g/cm^3) by heating in a glass flask covered with a watch glass. The clear solution is cooled, 16 ml of hot tartaric acid–ammonium oxalate mixture is added, the solution cooled, transferred to a 100-ml volumetric flask and diluted to the mark with water.

A 5–10-ml aliquot of the solution is transferred to a 50-ml volumetric flask, 5 ml of diluting solution (for a 5-ml aliquot) are added and the determination is continued as described for the construction of the calibration curve.

Photometric determination of niobium in alloys with zirconium and titanium by 1-(2-pyridylazo)-resorcinol [44]

1-(2-Pyridylazo)-resorcinol reacts with niobium (V) ions in the presence of hydrogen peroxide at pH 5 to form a soluble red-colored complex compound (molar ratio 1:1). The absorption maximum of the solution is at 590 mμ (Figure 11). The molar extinction coefficient is 32,000 at 540 mμ. Beer's law is obeyed at concentrations between 2 and 60 μg of Nb/50 ml solution.

If Complexone III is used as masking agent, the method will determine niobium in

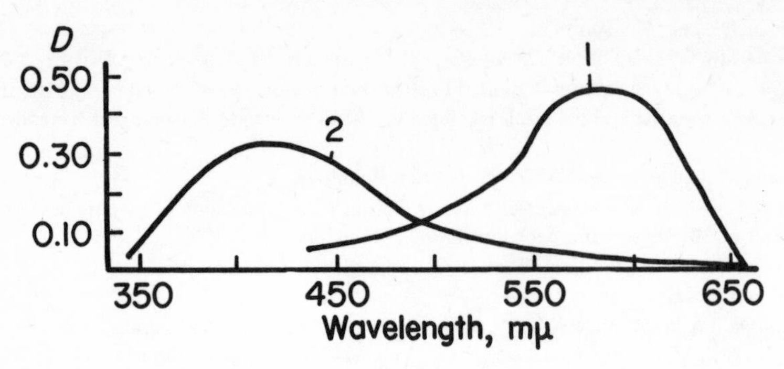

Figure 11

Light absorption by solutions of the compound of niobium (V) with H_2O_2 and 1-(2-pyridylazo)-resorcinol: 1) the complex [$1 \cdot 10^{-5}$ *M* solution of niobium (V), $1 \cdot 10^{-5}$ *M* solution of 1-(2-pyridylazo)-resorcinol, $9 \cdot 10^{-2}$ *M* solution of H_2O_2]; 2) $1 \cdot 10^{-5}$ *M* solution of 1-(2-pyridylazo)-resorcinol, pH 5; $l = 2$ cm.

the presence of up to 5 mg of titanium and up to 10 mg of zirconium. However, the introduction of the Complexone reduces the optical density of the colored solution of niobium. For this reason an appropriate amount of Complexone III must be introduced into the series of standard solutions in the presence of zirconium and titanium ions.

REAGENTS

Standard solution of niobium compound, 5 μg Nb/ml. Spectroscopically pure Nb_2O_5 is fused with potassium pyrosulfate and the melt is dissolved in a volumetric flask.
Sulfuric acid, $d = 1.84$ g/cm^3 and 1 *N* solution.
Potassium sulfate, crystalline.
Hydrogen peroxide, 30% solution.
Potassium pyrosulfate, 2% solution.
Complexone III, 0.025 *M* solution.
Potassium pyrosulfate, crystalline.
Acetate buffer solution, pH 5.
1-(2-*Pyridylazo*)-*resorcinol*, 0.1% solution.

CONSTRUCTION OF CALIBRATION CURVE

Into five 50-ml volumetric flasks are introduced 1, 2, 3, 4 and 5 ml of standard niobium solution; 5 ml of potassium pyrosulfate solution, 0.5 ml of Complexone III, and 0.25 ml of hydrogen peroxide are added to each flask, the volume of the solutions is brought to 45 ml with acetate buffer solution, 0.5 ml of 1-(2-pyridylazo)-resorcinol solution are added to each flask and the solutions are diluted to mark with the acetate buffer solution. After 30 minutes the optical densities of the solutions are measured in an electrophotocolorimeter with a filter which has the maximum light transmission at 536 mμ relative to a solution which contains the same amounts of the reagent, Complexone III, hydrogen peroxide and buffer solution.

The data thus obtained serve to plot a calibration curve.

PROCEDURE

A 0.1–0.2-g alloy sample is placed in a heat-resistant 100-ml beaker, 1–2 g of potassium sulfate and 1–2 ml of sulfuric acid ($d = 1.84$ g/cm^3) are added, the mixture heated until the alloy is fully dissolved and evaporated almost to dryness. To the residue are added 0.5–1 ml of hydrogen peroxide, 20 ml of 1 *N* sulfuric acid, the solution is transferred to a 50-ml or 100-ml volumetric flask and diluted to mark with water.

An aliquot of the solution (containing up to 10 mg of zirconium or up to 5 mg of titanium and 5–50 μg of Nb) is placed in a 50-ml volumetric flask and the determination is continued as described for the preparation of standard solutions. The optical density of the solution is determined and the niobium content found from the calibration curve.

Photometric determination of tantalum by pyrogallol [25, 82]

Pyrogallol reacts with tantalum (V) in 4 *N* HCl and 0.0175 *M* oxalic acid to form a soluble complex compound; the molar extinction coefficient of the solution of the complex is 4775 at 325 mμ. Beer's law is obeyed up to a concentration of 40 μg of Ta/ml.

Mo(VI), W(VI), U(VI), Sn(IV) interfere. The effect of Nb, Ti, Zr, Cr, V(V), Bi and Cu is insignificant and may be compensated for by introducing these ions into the blank solution. Fluorides and platinum ions interfere. The fusion of the samples must not be effected in platinum vessels.

REAGENTS

Standard solution of tantalum pentoxide, 0.2 mg Ta/ml: 0.2000 g of spectroscopically pure tantalum pentoxide is fused with 10 g of $K_2S_2O_7$ in a quartz crucible; the melt is leached with 50 ml of the solution containing 50 g of ammonium oxalate in 1 liter and is heated on a water bath. The solution is cooled, transferred to a 1-liter volumetric flask and diluted to mark with water.

Pyrogallol solution, 200 g pyrogallol are dissolved in 100 ml HCl ($d = 1.19$ g/cm^3), 10 ml of 2 *M* stannous chloride solution are added and the solution is diluted to 1 liter with water; it is stable for one month.

Diluting solution, one liter of the solution contains 12.5 g of ammonium oxalate and 50 g of potassium pyrosulfate. The solution is stable for 2 weeks.

Hydrochloric acid, 8 *N* solution.

CONSTRUCTION OF CALIBRATION CURVE

Into five 50-ml volumetric flasks are placed 1,2,3,4 and 5 ml of the standard solution of tantalum pentoxide; 25 ml of hydrochloric acid solution and 10 ml of pyrogallol solution are added to each flask and the solutions in the flasks are made up to mark with the diluting solution. The optical density of the solutions is measured in SF-4 spectrophotometer at 325 mμ and the calibration curve is plotted.

PROCEDURE

To the sample solution are added all the reagents employed in the construction of the calibration curve, the optical density is determined and the tantalum content found from the calibration curve.

Photometric determination of tantalum in titanium tetrachloride [64]

Tantalum is separated from the bulk of titanium by precipitation with tannin, after reduction of titanium to the trivalent state. The precipitate is dissolved and tantalum is extracted with cyclohexanone and is separated from the coprecipitated titanium. The determination of tantalum is terminated by the photometric pyrogallol method.

REAGENTS

Tannin, 10% solution in 1:9 hydrochloric acid.
Cadmium, metallic.
Tin, metallic.
Potassium pyrosulfate, crystalline.
Acid mixture, 0.4 *M* in HF and 2 *M* in H_2SO_4.
Cyclohexanone.
Ammonium sulfate, crystalline.
Hydrofluoric acid, 0.1 *N* solution.
Ammonium oxalate, 4% solution, containing boric acid.
Hydrochloric acid, dilute (1:3) solution.

PROCEDURE

The titanium tetrachloride sample (2–5 ml) is introduced into 80 ml of hydrochloric acid and the mixture is heated until clear. Tin (scavenger) is added to the solution, then 3 g of metallic cadmium, and the mixture is heated to a permanent violet coloration. Ten ml of tannin solution are added and the mixture is boiled and left to stand for 1 hour. The precipitate is filtered off, ashed together with the filter paper and the residue is fused with potassium pyrosulfate. The melt is dissolved in 25 ml of the acid mixture, with heating. The solution is transferred to a separating funnel, 8 ml of cyclohexanone and 2 g of ammonium sulfate are added and the solution is vigorously shaken for 1 minute. The phases are allowed to separate, the lower layer is drained into another separating funnel and tantalum is again extracted with 8 ml of cyclohexanone. The extraction of tantalum is repeated once more. The combined extracts are washed with the solution of hydrofluoric acid and shaken with a little cyclohexanone, which is combined with the main extracts. Three reextractions are performed, each time with 7 ml of 4% ammonium oxalate solution containing boric acid. The reextracts are diluted to 25 ml with water.

To determine tantalum by the pyrogallol method, an aliquot part of the solution is withdrawn and evaporated with sulfuric acid in the presence of hydrogen peroxide to decompose organic matter. The vitreous residue is dissolved in 4% ammonium oxalate solution and the solution is transferred to a 50-ml volumetric flask, twenty-five ml of 8 *N* HCl and 10 ml of pyrogallol solution are added to the flask and the solution is made up to the mark with ammonium oxalate solution. The optical density is measured; the tantalum content is found from the calibration curve (p. 191).

Extraction-photometric determination of tantalum in technical grade niobium [54]

The optical density of the benzene extract of Methyl Violet fluorotantalate is determined. The optimum concentration range of hydrofluoric acid is 0.25–0.5 *N*. Zirconium, hafnium and niobium ions in solution are not extracted.

REAGENTS

Standard solution of tantalum salt, 50 μg of Ta/ml.
Hydrofluoric acid, 40%.
Methyl Violet, 0.2% solution.
Benzene.
Washing solution, 0.2–0.3 *N* hydrofluoric acid containing 0.2 g Methyl Violet in 100 ml.

CONSTRUCTION OF CALIBRATION CURVE

Into each of four paraffin-coated 50-ml volumetric flasks are introduced 40–45 ml of water, 1.5 ml hydrofluoric acid and, respectively, 0.5,1.0,1.5 and 2.0 ml of the standard solution of titanium salt; the solutions are then diluted to mark with water. Five ml of each solution are placed in a separating funnel, and 4 ml of water, 10 ml of benzene and 1 ml of Methyl Violet solution are added to each funnel, after which each mixture is gently shaken three or four times. The phases are allowed to separate, the water layer decanted, the benzene layers washed with 5 ml of washing solution, transferred to centrifuge tubes and centrifuged for 1–2 minutes. The clear solutions are poured into 1-cm cells and the optical density is measured using a green-orange filter (at 570 mμ). A calibration curve is then plotted.

PROCEDURE

Ten mg of metallic niobium are dissolved in hydrofluoric acid by heating on a water bath while adding nitric acid drop by drop. The solution is cautiously evaporated to dryness on a hotplate, the residue is treated with 2–3 ml hydrofluoric acid and is again evaporated to dryness, taking care not to overheat the residue. The residue is dissolved in 1.5 ml of hydrofluoric acid and 5–6 ml water, the solution transferred to a paraffin-coated 50-ml volumetric flask and diluted to mark with water. A 5-ml aliquot is withdrawn into a separating funnel and the determination continued as in the construction of the calibration curve.

Photometric determination of tantalum in ores by dimethylfluorone [37]

Dimethylfluorone reacts with tantalum (V) ions to form a red-colored precipitate. At low concentrations of tantalum (below 50 μg/10 ml), the precipitate forms a colloidal suspension, especially so in the presence of gelatin.

The colorimetric reaction of niobium is rapidly suppressed in a weakly acid medium in the presence of oxalic acid. Titanium ions are masked with hydrogen peroxide.

The light absorption of the colored tantalum solutions obeys Beer's law at concentrations between 5 and 50 μg of Ta/10 ml. Under the experimental conditions employed the following amounts of accompanying elements (μg/10 ml solution) do not interfere:

Nb	400	Sb(V)	50
Ti	100	Sb(III)	3
Zr	500	Sn(IV)	5
W	150	Ge	3
Mo	100	Fe	1000

In the analysis of lean tantalum ores tantalum must be preliminarily separated from the bulk of the interfering elements.

REAGENTS

Hydrofluoric acid, concentrated.
Sulfuric acid, 1:1 and 6% solution.
Potassium pyrosulfate, crystalline.
Hydrogen peroxide, 3% solution.
Potassium metabisulfite, crystalline.
Tannin, 2% solution.
Gelatin, 1% solution.
Washing solution, 0.6% sulfuric acid containing 0.2% tannin.
Acid mixture, 16 ml of 40% HF are diluted to 200 ml with water and 100 ml sulfuric acid (d = 1.84 g/cm^3) are slowly introduced with cooling.
Isobutanol.
Acetone.
Ammonium sulfate, saturated solution.

Washing solution for extracts. Five ml of hydrofluoric–sulfuric acid mixture, 5 ml of saturated ammonium sulfate solution, 5 ml of acetone and 5 ml of isobutanol are shaken in a separating funnel, the lower layer is decanted and is used for washing.

Boric acid, 5% solution.

Oxalic acid, 4% solution.

Potassium hydroxide, 1 *N* solution.

α-Dinitrophenol, indicator solution.

Hydrochloric acid, 2 *N* solution.

Diluting solution, 1% in potassium pyrosulfate, 2% in boric acid, 0.2% in oxalic acid and 0.1 *N* in hydrochloric acid.

Dimethylfluorone. The reagent (50 mg) is ground with 0.5 ml of 6 *N* sulfuric acid, 50 ml of 96% ethanol are added and the mixture is heated to 60°C. The solution is filtered into a 100-ml volumetric flask, the filter paper washed with ethanol and the solution is diluted to the mark with ethanol.

Standard solution of tantalum, 5 μg of Ta/ml. Tantalum sheet (20 mg) is dissolved in a mixture of 5 ml of 40% HF and 2 ml of HNO_3 (d = 1.42 g/cm^3), and the solution is evaporated to dryness; 4 g of boric acid and 5 ml of water are added to the residue. The mixture is evaporated to dryness and is fused at 700°C in a muffle furnace with 4 g of potassium pyrosulfate, until a clear melt is obtained. The melt is dissolved with heating in 100 ml of 4% solution of oxalic acid, cooled and made up to 200 ml in a volumetric flask with the oxalic acid solution. One ml of the resulting solution contains 0.1 mg Ta. A 2.5-ml portion of this solution is evaporated to dryness on a water bath, the dry residue is ignited over a burner and treated with 2 ml of HF, after adding 20 ml of 5% boric acid solution, the solution is evaporated to dryness, the residue transferred to a muffle furnace and fused with 0.5 g of potassium pyrosulfate. The melt is dissolved with heating in 2.5 ml of 4% solution of oxalic acid, transferred to a 50-ml flask with the aid of water and neutralized with 1 *N* KOH to α-dinitrophenol until the solution becomes pale yellow. Hydrochloric acid (2 *N*; 2.5 ml) is added and the solution is diluted to mark with water. One ml of this solution contains 5 μg of Ta.

CONSTRUCTION OF CALIBRATION CURVE

Into five colorimetric tubes are introduced portions of standard solution of tantalum corresponding to 5, 10, 15, 20 and 25 μg of Ta respectively, and the work is continued exactly as described in the procedure.

PROCEDURE

The ore (1–3 g, depending on the content of tantalum) is placed in a platinum dish and treated with 15 ml of concentrated HF by heating on a water bath. Five ml of 1:1 sulfuric acid are added and the treatment is continued on a sand bath. When the evolution of sulfuric acid fumes has ceased, the residue is fused with 5–7 g of potassium pyrosulfate in a muffle furnace. The clear melt is transferred to the beaker with the aid of 50 ml of 6% sulfuric acid which is poured into the dish and heated on a hot plate. Five ml of 3% hydrogen peroxide are then added and the mixture is heated.

The insoluble residue is separated and washed on the filter with 50 ml water (the total volume of the filtrate and wash waters should be 100 ml, and the concentration of sulfuric acid should be 3%). The solution is heated on the plate and hydrogen peroxide is decomposed by cautiously adding potassium metabisulfite to the disappearance of the yellow color of the peroxy compound of titanium. In the absence of titanium 2 g potassium metabisulfite are added.

In the analysis of titanium-rich ores, the melt is dissolved in 100 ml of 6% sulfuric acid.

The solution is brought to boil, 20 ml hot 2% solution of tannin and 2 ml of 1% solution of gelatin are added and the mixture is left to stand for at least 6 hours. The precipitate is filtered through a double layer of filter paper and washed with 0.6% sulfuric acid containing 0.2% tannin. The filter with the precipitate is placed in a platinum dish, ashed and ignited in a muffle furnace. The precipitate is dissolved in 2 ml of hydrofluoric acid and the solution is evaporated to dryness on a water bath. The dry fluorides are dissolved in 2 ml of sulfuric acid–hydrofluoric acid mixture, the solution transferred to a 50-ml separating funnel, the dish rinsed with 5 ml of isobutanol and 5 ml of acetone which are then added to the funnel, and 2 ml of saturated ammonium sulfate solution are added to the funnel. The mixture is shaken for one minute, then the layers are allowed to separate for 5 minutes. The upper layer is left in the funnel while the lower layer is decanted into another separating funnel, 2.5 ml of acetone and 2.5 ml of isobutanol are added and the extraction is repeated. The upper layer is poured into the first funnel. The second funnel is washed with 1 ml of the washing liquid for the extract, the liquid is poured into the first funnel together with the extracts and shaken for half a minute. The layers are allowed to separate and the lower layer is discarded; the upper layer is again washed with 1 ml of the washing solution. The washed extract is shaken with 5 ml of boric acid solution and the contents of the funnel are introduced into the platinum dish; the funnel is rinsed with three 5-ml portions of boric acid solution which are also placed in the platinum dish.

The platinum dish with the extract is placed on a water bath, and the solution evaporated to dryness, taking care to prevent spattering. The dish with the residue is placed on a sand bath and, without expelling all of the sulfuric acid, fused with 0.5 g of potassium pyrosulfate at 600–700°C for 30–40 minutes. The clear melt is dissolved in 2.5 ml of oxalic acid solution by heating on a hot plate, while adding small portions of water to prevent crystallization of oxalic acid. The solution is transferred from the dish to a 50-ml volumetric flask, rinsing the dish several times with water and pouring the washings into the flask; care should be taken to leave a 10-ml vacant space below the mark in the flask; 2 drops of α-dinitrophenol solution are added to the flask and then potassium hydroxide solution drop by drop until the solution becomes just perceptibly yellow. Hydrochloric acid (2.5 ml) is added to the sample and the contents of the flask are diluted to mark with water.

To a 2–10 ml aliquot of the solution in a tube are added 10 ml of diluting solution, 1 ml of gelatin solution and the mixture is stirred. Dimethylfluorone (0.4 ml) is added and the solution is again stirred, and then heated for 3 minutes in a boiling water bath and the solution left in the bath until the water has cooled (half an hour), after which it is held $1\frac{1}{2}$ hours longer at room temperature to allow the coloration to develop. Hydrogen peroxide (0.5 ml) is added, the solution shaken and the optical density is measured after 15 minutes at 530 mμ, using a photometer or an electrophotocolorimeter.

Polarographic determination of niobium, titanium and iron in technical grade tantalum and tantalum compounds [73]

Niobium, titanium and iron can be polarographically determined with pyrophosphoric acid as the supporting electrolyte without separating tantalum and other elements. The large differences in the half-wave potentials make it possible to determine all the impurities on a single polarogram. The half-wave potentials are 0.2 V for iron, 0.4 V for titanium and 0.8 V for niobium relative to the mercury anode. One determination lasts 5–10 minutes. Any suitable polarograph model may be employed.

The sensitivity of determination of niobium is 10^{-2}%; the relative error is 3–5%.

REAGENTS

Nitric acid, $d = 1.4$ g/cm^3.
Hydrofluoric acid, 40% solution.
Pyrophosphoric acid, $d = 1.9$ g/cm^3.
Standard solutions of niobium and titanium. Samples of oxides Nb_2O_5 and TiO_2 are dissolved in hydrofluoric acid, after which the hydrofluoric acid is removed by heating the solution with pyrophosphoric acid.

CONSTRUCTION OF CALIBRATION CURVE

Polarograms are obtained for standard solutions in which the concentrations of the elements are kept within the limits of those expected to be present in the samples. From these data the curve is plotted.

PROCEDURE

A 0.5–1-g sample of metallic tantalum in a platinum dish is treated with a mixture of hydrofluoric and nitric acids. When the dissolution is complete, 20–30 ml of pyrophosphoric acid are introduced and the solution is heated until water vapor, nitrous oxides and hydrogen fluoride are no longer evolved. The residue is transferred to a 50–100-ml volumetric flask, and diluted to mark with pyrophosphoric acid and water so as to have a 2:1 ratio between $H_4P_2O_7$ and H_2O in the final solution.

A polarogram is taken of an aliquot of the solution, using internal mercury anode.

BIBLIOGRAPHY

1. KANZELMEYER, J. H., O. J. RYAN, and H. FREUND.—*J. Am. Chem. Soc.*, **78**, 3020. 1956.
2. KANZELMEYER, J. H. and H. FREUND.—*Anal. Chem.*, **25**, 1807. 1953.
3. ALIMARIN., I. P., I. M. GIBALO and CHIN KUANG-JUNG.—*ZhAKh*, **17**, 60. 1962.
4. ALIMARIN, I. P. and R. L. PODVAL'NAYA.—*ZhAKh*, **1**, 30. 1946.
5. MUNDY, R. J.—*Anal. Chem.*, **27**, 1408. 1955.
6. BACON, A. and G. W. C.—*Anal. chim. acta*, **15**, 129. 1956.
7. BUKHSH, M. N. and D. N. HUME.—*Anal. Chem.*, **27**, 116. 1955.
8. WARD, F. N. and A. P. MARRANZINO.—*Ibid.*, **27**, 1325. 1955.
9. SAVCHENKO, G. S. and I. V. TANANAEV.—*ZhPKh*, **19**, 1093. 1946; **20**, 385. 1947.
10. ALIMARIN, I. P. and E. I. STEPANYUK.—*Zav. Lab.*, **22**, 1149. 1956; **24**, 1064. 1958.
11. GRIMALDI, F. S. and M. M. SCHNEPFE.—*Anal. Chem.*, **30**, 2046. 1958.
12. ALIMARIN, I. P. and T. A. BUROVA.—*ZhPKh*, **18**, 289. 1945.
13. TELEP, G. and D. F. BOLTZ.—*Anal. Chem.*, **24**, 163. 1952.
14. PAPILLA, F. C., N. ADLER, and C. F. HISKEY.—*Ibid.*, **25**, 926. 1953.
15. BANKS, C. V., K. E. BURKE, J. W. O'LAUGHLIN, and J. A. THOMPSON.—*Ibid.*, **29**, 995. 1957.
16. ALIMARIN, I. P. and B. I. FRID.—*Trudy Vsesoyuznoi Konferentsii po Analiticheskoi Khimii*, **2**, 333. 1943; *Zav. Lab.*, **7**, 913. 1938.
17. MAJUMDAR, A. K. and A. K. MUKHERJEE.—*Anal. chim. acta*, **21**, 330. 1959.
18. MAJUMDAR, A. K. and J. B. RAY CHOWDHURY.—*Ibid.*, **19**, 18. 1958.
19. MAJUMDAR, A. K. and A. K. Mukherjee.—*Ibid.*, **19**, 23. 1958.
20. MAJUMDAR, A. K. and A. K. MUKHERJEE.—*Ibid.*, **21**, 245. 1959.
21. LANGMYHR, F. J. and T. HONGSLO.—*Ibid.*, **22**, 301. 1960.
22. ALIMARIN, I. P., O. M. PETRUKHIN, and TSÊ YÜN-HSIANG.—*Doklady AN SSSR*, **136**, 1073. 1961.
23. ALIMARIN, I. P. and B. I. FRID.—*Zav. Lab.*, **7**, 1109. 1938.
24. PLATONOV, M. S., N. F. KRIVOSHLYKOV, and A. A. MAKAROV.—*ZhOKh*, **6**, 1814. 1936.
25. KRIVOSHLYKOV, N. F. and M. S. PLATONOV.—*ZhPKh*, **10**, 184. 1937.
26. ALEKSEEVSKAYA, N. V. and M. S. PLATONOV.—*ZhPKh*, **10**, 2139. 1937.
27. HUNT, E. C. and R. A. WELLS.—*Analyst*, **79**, 345. 1954.
28. NORWITZ, G., M. CODELL, and J. MIKULA.—*Anal. chim. acta*, **11**, 173. 1954.
29. DOBKINA, B. M. and T. M. MALYUTINA.—*Zav. Lab.*, **24**, 1336. 1958.
30. SCHOELLER, W. R. *The Analytical Chemistry of Tantalum and Niobium.*—London, Chapman and Hall. 1937.
31. KNIPOVICH, YU. N. and YU. V. MORACHEVSKII, EDITORS. *Analiz mineral'nogo syr'ya (Analysis of Mineral Raw Materials)*, 3rd ed., p. 678.—Goskhimizdat. 1959.
32. PONOMAREV, A. I. and A. YA. SHESKOL'SKAYA.—*ZhAKh*, **12**, 355. 1957.
33. DAS, M. S., CH. VENKATESWARLY, and V. T. ATHAVALE.—*Analyst*, **81**, 239. 1956.
34. BYKOVA, V. S.—*Doklady AN SSSR*, **18**, 655. 1938.
35. PATRORSKY, V.—*Coll. Czech. Chem. Comm.*, **23**, 1774. 1958.
36. ZAIKOVSKII, F. V.—*ZhAKh*, **11**, 269. 1956.
37. NAZARENKO, V. A. and M. B. SHCHUSTOVA.—*Zav. Lab.*, **23**, 1283. 1957; *Trudy GIREDMET*, **2**, 52. Metallurgizdat. 1959.
38. LUKE, C. L.—*Anal. Chem.*, **31**, 904. 1959.
39. NIKITINA, E. I.—*ZhAKh*, **13**, 72. 1958.
40. LASSNER, E. and R. PÜSCHEL.—*Mikrochim. acta*, 950. 1963.
41. LASSNER, E.—*Chemist-Analyst*, **51**, 14. 1962.
42. LASSNER, E.—*Talanta*, **10**, 1229. 1963.
43. PÜSCHEL, R. and E. LASSNER.—*Z. anorg. Chem.*, **326**, 317. 1964.
44. ELINSON, S. V. and L. I. POBEDINA.—*ZhAKh*, **18**, 189. 1963.

45. ELINSON, S. V. and A. T. REZOVA.—*ZhAKh,* **19**, 1078. 1964.
46. BABKO, A. K. and M. I. SHTOKALO.—*ZhAKh,* **17**, 1068. 1958.
47. ELINSON, S. V. and L. I. POBEDINA.—*ZhAKh,* **18**, 734. 1963.
48. ALIMARIN, I. P. and KHAN HSI-I.—*Vestnik. MGU,* chem. ser. II. No. 1, 65. 1964.
49. ALIMARIN, I. P. and KHAN HSI-I.—*Vestnik. MGU,* chem. ser. II, No. 2, 41. 1964.
50. BELEKAR, G. K. and V. T. ATHARALE.—*Analyst,* **82**, 630. 1957.
51. POLUEKTOV, N. S., L. I. KONONENKO, and R. S. LAUER.—*ZhAKh,* **13**, 396. 1958.
52. PAVLOVA, N. N. and I. A. BLYUM.—*Zav. Lab.,* **20**, 1305. 1962.
53. POLUEKTOV, N. S. and L. I. KONONENKO.—*Trudy GIREDMET,* **2**, 31. Metallurgizdat. 1959.
54. LAUER, R. S. and N. S. POLUEKTOV.—*Zav. Lab.,* **25**, 903. 1959.
55. STEVENSON, P. C. and H. G. HICKS.—*Anal. Chem.,* **25**, 1517. 1953.
56. HICKS, H. G. and GILBERT, R. S.—*Ibid.,* **26**, 1205. 1954.
57. LEDDICOTTE, G. W. and F. L. MOORE.—*J. Am. Chem. Soc.,* **74**, 1018. 1952.
58. ELLENBERG, J. Y., G. W. LEDDICOTTE, and F. L. MOORE.—*Anal. Chem.,* **26**, 1045. 1954.
59. SCADDEN, E. M. and N. E. BALLOU.—*Anal. Chem.,* **25**, 1602. 1953.
60. MORRIS, D. F. and D. SCARGILL.—*Anal. chim. acta.,* **14**, 57. 1956.
61. RYABCHIKOV, D. I. and M. P. VOLYNETS.—*ZhAKh,* **14**, 700. 1959.
62. MOORE, F. L.—*Anal. Chem.,* **27**, 70. 1955.
63. CASEY, A. T. and A. G. MADDOCK.—*J. Inorg. Nucl. Chem.,* **10**, 289. 1959.
64. CHERNIKHOV, YU. A., R. S. TRAMM, and K. S. PEVZNER.—*Zav. Lab.,* **22**, 637. 1956; *Trudy GIREDMET,* **2**, 22. Metallurgizdat. 1959.
65. ELINSON, S. V., K. I. PETROV, and A. T. REZOVA.—*ZhAKh,* **13**, 576. 1958.
66. SCOTT, J. A. P. and R. J. MAGEE.—*Talanta,* **1**, 329. 1958.
67. WILLIAMS, A. F.—*J. Chem. Soc.,* 3155. 1952.
68. BURSTALL, F. H. and A. F. WILLIAMS.—*Analyst,* **77**, 921. 1952.
69. BLASIUS, E. and A. CZEKAY.—*Z. anal. Chem.,* **156**, 81. 1957.
70. RUDENKO, N. P. and O. M. KALINKINA.—*ZhNKh,* **2**, 959. 1957.
71. AL'TSHULER, O. V., E. A. SUBBOTINA, and A. F. AFANAS'EVA.—*ZhNKh,* **3**, 1192. 1958.
72. MERCER, R. A. and R. A. WELLS.—*Analyst,* **79**, 339. 1954.
73. KURBATOV, D. I.—*ZhAKh,* **14**, 743. 1959; **16**, 36. 1964.
74. ZAKHARIYA, N. F. *Trudy GIREDMET,* **2**, 351. Metallurgizdat. 1959.
75. FRESENIUS, W. and G. JANDER.—In: *Handbuch der analytischen Chemie,* **Vb**, part 3, p. 520, ed. G. J. KOLMESCHATE, Springer-Verlag, Berlin, Göttingen, Heidelberg. 1957.
76. COCKBILE, M. H,—*Analyst,* **87**, 611. 1962.
77. YAKOVLEV, P. YA. and E. F. YAKOVLEVA. *Tekhnicheskii analiz v metallurgii (Technical Analysis in Metallurgy).*—Metallurgizdat. 1963.
78. PONOMAREV, A. I. and A. YA. SHESKOL'SKAYA.—*ZhAKh,* **14**, 67. 1959.
79. KOTLYAR, E. E. and T. N. NAZARCHUK.—*ZhAKh,* **18**, 474. 1963.
80. DOROSH, V. M.—*ZhAKh,* **16**, 250. 1961.
81. ZABOEVA, M. I. and V. F. BARKOVSKII.—*ZhAKh,* **17**, 955. 1962.
82. DINNIN, J.—*Anal. Chem.,* **25**, 1803. 1953.

Molybdenum Mo

Molybdenum in its compounds displays valencies of +2, +3, +4, +5, and +6. Compounds of hexavalent, pentavalent and trivalent molybdenum are the most important.

The standard electrode potentials in aqueous medium at 25°C relative to the standard hydrogen electrode are:

$Mo \rightleftharpoons Mo^{3+} + 3e$	−0.200 V
$Mo + 4H_2O \rightleftharpoons MoO_4^{2-} + 8H^+ + 6e$	+0.154 V
$MoO_2 + 2H_2O \rightleftharpoons MoO_4^{2-} + 4H^+ + 2e$	+0.606 V
$[Mo(CN)_8]^{4-} \rightleftharpoons [Mo(CN)_8]^{3-} + e$	+0.73 V

In aqueous solutions the compounds of molybdenum (VI) are the most stable.

In hydrochloric acid solutions (4 *N* HCl) the compounds of molybdenum (V) are also perfectly stable in the air (Figure 12). As the acidity decreases, the tendency of molybdenum (V) compounds to be oxidized by atmospheric oxygen increases. In alkaline solutions molybdenum (V) ions are very readily oxidized.

Molybdenum (IV) ions are retained in the highly stable cyanide complex $[Mo(CN)_8]^{4-}$.

Compounds of molybdenum (III) in solution are fairly resistant to oxidation by atmospheric oxygen at high hydrochloric acid concentrations (9 *N* HCl) owing to the formation of the brick-red complex ions $[MoCl_6]^{3-}$, $[MoCl_5]^{2-}$, etc. In 2.5–4 *N* HCl hydrated green Mo^{3+} ions are in equilibrium with halide complexes of Mo (III)

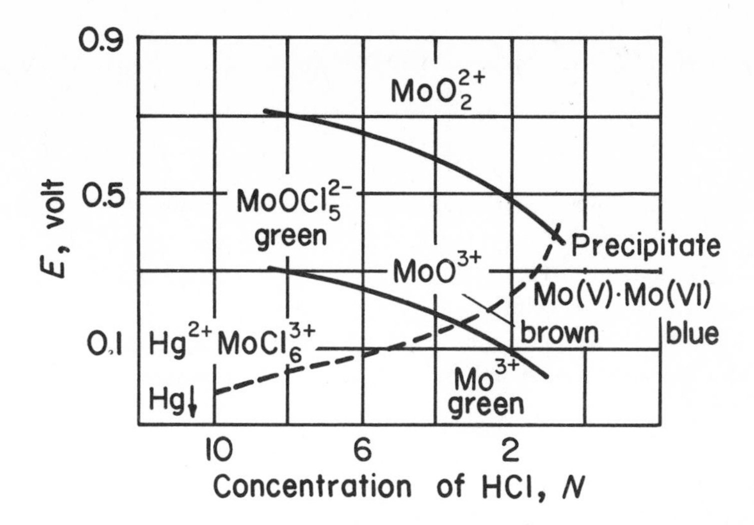

Figure 12

Approximate formal potentials of the systems Mo(VI)/Mo(V), Mo(V)/Mo(III) and Hg(II)/Hg in hydrochloric acid solutions

(this equilibrium is established slowly). Mo^{3+} ions are always oxidized quite readily.

Molybdenum (III) ions are much more easily oxidized in sulfuric acid than in hydrochloric acid. Sulfuric acid solutions contain only the green molybdenum (III) ions.

Molybdenum (III) ions form complexes with oxalate and thiocyanate ions; the yellow $[Mo(SCN)_6]^{3-}$ ion is an example.

Compounds of molybdenum (II) are unstable in solution.

When molybdenum (VI) ions are reduced in a weakly acid solution (less than 2 *N* HCl), intense blue compounds are formed as colloidal solutions or precipitates, in which molybdenum is partly pentavalent and partly hexavalent.

In relatively concentrated hydrochloric acid (e.g., in 8 *N* HCl) molybdenum (VI) ions are reduced to the green $[MoOCl_5]^{2-}$ ions, while in dilute hydrochloric acid (2 *N* HCl) they are reduced to yellow-brown or intense brown ions of molybdenum (V).

In alkaline medium, molybdenum (VI) ions are reduced very slowly and incompletely.

Metallic silver and mercury in strongly acid solutions reduce molybdenum (VI) ions to molybdenum (V) and molybdenum (III) ions, depending on the concentration of HCl (Figure 12). Metallic zinc in dilute acid reduces molybdenum (VI) ions to molybdenum (III) ions.

Solutions of chromium (II) salts reduce molybdenum (VI) ions in hydrochloric acid and sulfuric acid media first to pentavalent and then to trivalent molybdenum [1]. This reaction forms the basis of potentiometric titration methods of molybdenum ions with tungsten indicator electrode. Under certain conditions molybdenum (III) ions quantitatively reduce molybdenum (VI) to molybdenum (V) ions:

$$2Mo(VI) + Mo(III) \rightleftharpoons 3Mo(V).$$

Solutions of molybdenum (III) salts are used in the potentiometric titration of molybdenum (VI) ions in the presence of a number of elements [2, 3].

Reducing agents (Zn, Sn (II), Fe (II), etc.) in weakly acid solutions reduce molybdates to molybdenum blue [4] which contains both Mo (V) and Mo (VI).

Alkaline solutions (pH $\geqq$ 7) contain molybdate ions MoO_4^{2-}; weakly acid solutions contain $Mo_7O_{24}^{6-}$ ions. Other polymerized anions of hexavalent molybdenum may also be formed:

	pK
$MoO_4^{2-} + H^+ \rightleftharpoons HMoO_4^-$	4.08
$7MoO_4^{2-} + 8H^+ \rightleftharpoons Mo_7O_{24}^{6-} + 4H_2O$	57.7
$Mo_7O_{24}^{6-} + H^+ \rightleftharpoons HMo_7O_{24}^{5-}$	4.33
$HMo_7O_{24}^{5-} + H^+ \rightleftharpoons H_2Mo_7O_{24}^{4-}$	~3.7

When the solution is made alkaline, polymolybdate ions are converted to MoO_4^{2-} ions. The hydrolysis of molybdate ions has been described in [5].

Acid solutions also contain different polymerized cations of hexavalent molybdenum.

Molybdic acid in the form of H_2MoO_4 is colorless; the monohydrate $H_2MoO_4 \cdot H_2O$ is yellow. Molybdic acid precipitates out when concentrated molybdate solutions are acidified; on the addition of more acid the precipitate redissolves, unlike tungstic acid. Molybdic acid and paramolybdic acids are readily soluble in alkalis.

Molybdenum (VI) forms complexes with chlorides, fluorides, oxalates, tartrates and citrates. Numerous complexes with o-diphenols and with organic substances containing sulfhydryl groups are known.

The molybdates of most metals ($PbMoO_4$, Hg_2MoO_4, Ag_2MoO_4, $CaMoO_4$, etc.) are sparingly soluble in water. Molybdates of alkali metals alone are soluble. Silver nitrate and silver acetate react with molybdate ions in dilute acetic acid solutions, forming white precipitates ($pL_{Ag_2MoO_4} = 10.5$). The precipitates are not formed in the presence of excess hydrogen peroxide.

Benzidine reacts with molybdate, tungstate and sulfate ions to form sparingly soluble white crystalline salts.

Molybdate ions are partly reduced by hydrogen sulfide in acid solution with the appearance of a blue coloration and separation of the brown precipitate of $MoS_2 + MoS_3$.

Ammonium sulfide reacts with molybdates to form a red-brown solution of the sulfo salt MoS_4^{2-}; when the solution is acidified to pH 6–7, the black sulfide MoS_3 precipitates out. This precipitate is sparingly soluble in dilute hydrochloric acid, but dissolves in ammonium sulfide solutions.

Hydrogen peroxide produces pink or red stains on solid ammonium molybdate, previously moistened with dilute solution of ammonia; this is due to formation of the peroxymolybdate [6]. Titanium (IV) and vanadium (V) interfere.

Potassium thiocyanate reacts with molybdenum (V) ions in hydrochloric acid solution to form a carmine-red compound which can be extracted with ether and other organic solvents [7]. When ammonia is added to the solution, the color disappears, and appears again when the solution is acidified. Molybdenum in different specimens can be detected and photometrically determined in the form of thiocyanate complexes.

Under certain conditions molybdic acid forms heteropoly acids with phosphoric, arsenic, vanadic, silicic, germanic and other acids. Ammonium phosphate gives a yellow precipitate of ammonium phosphomolybdate with molybdate ions in nitric acid solution. The precipitate is sparingly soluble in nitric acid and dissolves in alkalis with decomposition.

In acid solution phosphomolybdates are readily reduced by, say, stannous chloride, to give the intensely colored molybdenum blue.

Germanic, phosphoric, arsenic and silicic acids are detected and photometrically determined as the molybdic heteropoly acids.

The reduced molybdate reacts with 2,2′-dipyridyl to form intensely red solutions [8]. This fact is taken advantage of in highly sensitive detections of molybdenum. The sensitivity of the detection decreases in the presence of tungstate ions WO_4^{2-} owing to the formation of tungsten blue. Fe^{2+}, Fe^{3+}, SeO_3^{2-}, TeO_3^{2-} and TeO_4^{2-} interfere.

α-Benzoinoxime precipitates molybdenum (VI), tungsten (VI), and vanadium (V) from strongly acid solutions; the molybdenum precipitate is insoluble in 20% sulfuric acid [9]. This reagent is used for the detection and quantitative separation of molybdenum from a large number of elements (p. 208). If no precipitation or turbidity is noted on the addition of 2% solution of reagent to the acid sample solution, it may be taken that molybdate and tungstate ions are absent.

Complexone III reacts with molybdenum (VI) and molybdenum (V) ions in the molar ratio of 1:2 [10–12]. The compound of Complexone III with molybdenum (VI) is strongly dissociated in solution and exists only in weakly acid medium. The compound of molybdenum (V) is more stable; it exists both in acid and in weakly alkaline media (between 0.5 *M* HCl and pH 10) and is used in the complexometric determination of molybdenum (p. 209).

o-Diphenols, pyrocatechol, Tiron, Stilbazo, derivatives of 2,3,7-trihydroxy-6-fluorone etc. react with molybdenum (VI) ions in neutral or weakly acid solutions with the formation of yellow or orange-colored compounds, some of which are sparingly soluble. Tiron reacts with molybdate ions to form two soluble yellow compounds: one at pH 3.5–4.5 (molar ratio 1:1, absorption maximum 322 mμ) and the second at pH 6–8 (molar ratio 1:2, absorption maximum at 390 mμ). The molar extinction coefficient of the latter compound in the presence of 500-fold excess of the reagent is 6500. The coloration appears instantly and is stable for a long time [13, 15]. The reagent also forms soluble colored compounds with molybdenum (V); these have an absorption maximum at 332 mμ and a molar extinction coefficient [15] of 4200. The reagent is employed in the detection and photometric determination of molybdenum.

Stilbazo reacts with Mo(VI) and Mo(V) at pH 2 to form intense violet-colored compounds. The reagent is employed in the detection and photometric determination of molybdenum [16].

8-Hydroxyquinoline forms two compounds with Mo(V) [17]: at about pH 3 at room temperature the greenish yellow precipitate of $H_2Mo_4O_{11} \cdot (C_9H_7ON)_7 \cdot 11H_2O$ precipitates out; when the solution is boiled (pH 3.8–4.4), the fine-grained black precipitate of $Mo_2O_3 \cdot (C_9H_6ON)_4 \cdot H_2O$ is obtained. Molybdenum in 50% acetone can be photometrically determined by this reagent. The solutions of $Mo_2O_3(C_9H_6ON)_4 \cdot H_2O$ in chloroform have two absorption maxima, one at 397.5 mμ (E_M = 16,500) and the other at 550 mμ (E_M = 9500). The solution of $H_2Mo_4O_{11} \cdot (C_9H_7ON)_7 \cdot 11H_2O$ has one absorption maximum at 395 mμ (E_M = 18,200).

8-Hydroxyquinoline-5-sulfonic acid reacts with Mo(V) at pH 3–4.5 to form a stable soluble red-colored compound with an absorption maximum at 540 mμ and a molar extinction coefficient of 5200. The reagent is suitable for the photometric determination of molybdenum in the presence of tungsten and other elements [18].

Azo compounds and azomethine compounds which contain the groupings

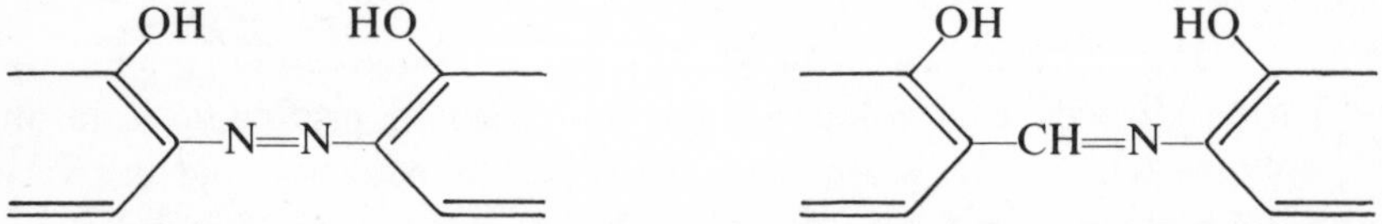

OH HO — N=N — (azo grouping)   OH HO — CH=N — (azomethine grouping)

form typical colored compounds with molybdenum (VI). The reagents may be used in the detection and photometric determination of molybdenum [19]. Thus, lumogallion (2,2′,4′-trihydroxy-5-chloro-(1-azo-1′)benzene-3-sulfonic acid) reacts with molybdate ions in the molar ratio of 1:1, forming an intensely colored stable chelate compound [20].

The true molar extinction coefficient of the solutions of this compound at 510 mμ is $(1.13 \pm 0.19) \cdot 10^4$. The sensitivity of detection [20] of molybdenum is 0.05 of μg/ml. Lumogallion is used in the photometric determination of traces of molybdenum. Ca, Mg, Co, Zn, Cd, Ni, Mn(II), Cr(III), Ce(IV) do not interfere.

Phenylhydrazine $C_6H_5NHNH_2$ reacts with molybdates in acetic acid solutions to give an intense red compound, probably owing to the oxidation of the reagent to the diazonium salt, which forms an azo compound with the excess phenylhydrazine [21]. The reagent is used in the detection and photometric determination of molybdenum. Strong oxidizing and reducing agents must be absent. Tungstate ions interfere if present in concentrations higher than 1%.

Sulfur-containing organic compounds of various types react with Mo(VI) and Mo(V) with formation of colored compounds.

Most sulfur-containing organic compounds with the groupings

$$\begin{array}{c} =\mathrm{C}-\mathrm{SH} \\ | \\ =\mathrm{C}-\mathrm{A} \end{array} \quad \text{or} \quad \begin{array}{c} =\mathrm{C}-\mathrm{SH} \\ | \\ \mathrm{A} \end{array}$$

, where A is COOH, SH, OH,

are valuable analytical reagents for hexavalent and pentavalent molybdenum.

These groupings may form part of both aromatic as well as aliphatic compounds [22].

Thioglycolic acid $HSCH_2COOH$ reacts with Mo(VI) at pH 3–6 and with Mo(V) at about pH 2 in hydrochloric acid medium in the molar ratio of 1:2, giving typically colored compounds [23–25].

At pH < 3 the reagent reduces Mo(VI) to Mo(V) with the formation of the colored compound $Mo_2O_3(SCH_2COOH)_4$. At pH 4 solutions of the compound formed by the reagent with Mo(VI) at pH 4 have an absorption maximum at 365 mμ (molar extinction coefficient 4400); the absorption maximum of the compound of Mo(V) at pH 0.6 and 1.8 is at 352 mμ (molar extinction coefficient 2400).

Thiomalic (mercaptosuccinic) acid [26, 27], $HOOC—CH_2—CH(SH)—COOH$, 2,3-dimercaptopropionic acid [28] $CH_2(SH)CH(SH)COOH$, sodium 2,3-dimercaptopropanesulfonate [29] $CH_2(SH)CH(SH)CH_2SO_3Na$ and sodium p-(mercaptoacetamide)-benzenesulfonate [30] $HSCH_2CONH—C_6H_4—SO_3Na$ react with Mo^{5+} and Mo^{6+} similarly to thioglycolic acid $HSCH_2COOH$.

All these reagents are employed in the detection and photometric determination of molybdenum in the presence of a number of elements.

Anionic complexes of pentavalent and hexavalent molybdenum with thioglycolic and thiomalic acids in the form of diphenylguanidinium salts are extracted with a number of organic solvents, best of all with a mixture of isoamyl alcohol with chloroform. One complex anion of Mo(V) with thioglycolic or thiomalic acid binds one, while Mo(VI) binds two diphenylguanidinium cations [31].

An extraction-photometric method for the determination of molybdenum as the diphenylguanidinium salt of the compound of Mo(V) with thioglycolic acid has been developed.

Toluene-3,4-dithiol reacts with Mo(VI) (and also with W(VI)) with the formation of typically colored compounds which are sparingly soluble in water, but are soluble in some organic solvents. The reagent is used in the detection and photometric determination of molybdenum.

Dialkyl- and diaryldithiophosphoric acids with the atomic grouping

$$\begin{matrix} -O & & S \\ & \rangle P \langle & \\ -O & & SH \end{matrix}$$

react with Mo(VI) ions in acid medium with formation of intensely colored compounds, the composition of which is unknown [23]. The reagents are used to detect Mo(VI) in the presence of tungsten and other elements [32, 33]. Pentavalent molybdenum does not form colored compounds with these reagents.

Various xanthates ROC(S)SNa and dithiocarbamates $R_2NC(S)SNa$ react with Mo(VI) ions in acid solutions, forming intensely colored compounds which can be extracted with organic solvents [34]. These reagents are used in the detection and photometric and titrimetric determination of molybdenum.

Small concentrations of molybdenum can be determined by the polarographic technique. Molybdenum gives a catalytic wave in 1 *M* $HClO_4$ + 0.75 *M* H_2SO_4 at $E_{1/2} = -0.18$ volt. The height of the wave is proportional to the concentration between 0.2 and 2 μg of Mo/ml. Molybdenum is first separated from the accompanying elements as the citrate complex by the method of ion exchange chromatography on KU-2 cation exchanger in the hydrogen form. The polarographic method is used to determine molybdenum in ores and in other materials.

For a survey of the methods of detection and determination of molybdenum the reader is referred to [34, 35].

Separation of traces of molybdenum by coprecipitation [36, 37]

Molybdenum ions together with tungsten and a number of other elements are separated together with the precipitate formed by tannin and Methyl Violet. The precipitate thus obtained is ashed, molybdenum ions are solubilized and then separated together with Methyl Violet thiocyanate precipitate from a solution 0.2 *M* in hydrochloric acid and 0.05 *M* in tartaric acid. In the second coprecipitation only 10–20% of tungsten is coprecipitated together with molybdenum. Small amounts of tin, antimony and bismuth are coprecipitated with molybdenum, in addition to tungsten. In this way 95–100% of molybdenum can be separated at a dilution of $1:10^9$.

The isolated molybdenum is determined by the photometric method.

REAGENTS

Sodium bicarbonate, crystalline.
Sodium carbonate, 1% solution.
Hydrochloric acid, 6 *N* solution.

Tannin, 2% solution.
Methyl Violet, 2% solution.
Methyl orange, 0.1% solution.
Sulfuric acid, dilute (1:1) solution.
Sodium hydroxide, 1 *N* solution.
Tartaric acid, crystalline.
Ammonium thiocyanate, crystalline.
Washing solution I. To 1 liter water are added 10 ml of 6 *N* HCl, 1 ml of 2% solution of tannin and 5 ml of 2% solution of Methyl Violet.
Washing solution II. To 1 liter of water are added 10 ml of 6 *N* HCl, 5 g of ammonium thiocyamate, 5 g of tartaric acid and 5 ml of 2% Methyl Violet solution.

PROCEDURE

Solubilization of rock (granite)

The finely ground sample (1–2 g) of granite is fused in the muffle furnace or over a blowtorch flame with 5 times the amount of sodium bicarbonate. The melt is leached with 200–250 ml of hot water, the residue is filtered off and washed with 50–100 ml sodium carbonate solution. The solution is neutralized with hydrochloric acid to methyl orange and 10 ml of acid are added in excess.

If silicic acid precipitates out in the course of analysis, it is filtered off, the precipitate with the filter paper is ashed and the residue is fused with sodium bicarbonate.

Coprecipitation with Methyl Violet tannate

To the sample solution are added 7.5 ml of tannin solution, and 15 ml of Methyl Violet solution are then introduced drop by drop, with stirring. After 30 minutes the precipitate is filtered through filter paper previously wetted with washing solution I. The first portions of the filtrate are passed through the same filter paper for a second time. The residue is washed 3–4 times with washing solution I, transferred into a crucible together with the filter paper, wetted with a few drops of sulfuric acid, the paper carbonized under infrared lamp and ashed in a muffle furnace at a temperature not above 500°C.

Coprecipitation with Methyl Violet thiocyanate

The residue from the ashing is wetted in the crucible with 3–5 drops of sodium hydroxide solution, 2–3 ml of water are added, the mixture is heated on the water bath for 20–30 minutes, and transferred to a 500-ml beaker while washing the crucible with 250 ml of water. A solution of 7 g tartaric acid in 30 ml of water and a solution of 5 g ammonium thiocyanate in 10 ml of water are added and then 10 ml hydrochloric acid. If a turbidity appears, the solution is filtered. Twenty-five ml of Methyl Violet solution are then slowly added drop by drop, with stirring, and after 30 to 40 minutes the precipitate is filtered through filter paper and wetted with washing solution II; the precipitate is washed 3–4 times with this solution, transferred to a crucible and ashed. The residue is dissolved in a solution of sodium hydroxide and molybdenum is determined photometrically, e.g., as thiocyanate.

Gravimetric determination of molybdenum by α-benzoinoxime [9, 34, 38]

α-Benzoinoxime (Cupron) quantitatively precipitates Mo(VI) ions from acetate–acetic acid solutions, or from cold solutions containing 5 vol.% acetic, phosphoric, nitric, hydrochloric or sulfuric acids. The precipitation is conducted in the presence of bromine water in order to prevent Mo(VI) from being reduced to pentavalent molybdenum which is not precipitated by the reagent.

In order to prevent the excess reagent from precipitating, 20% acetone is introduced into the solution. The precipitate is dried at 105°C and weighed as $MoO_2(C_{14}H_{12}O_2N)_2$.

Ag, Pb, Hg(II), Bi, Cu(II), Cd, As(V), Sb(III), Sn(II), Al, Fe(II), Fe(III), Ti, Zr, Cr(III), V(IV), Ce, U, Ni, Co, Mn, Zn do not interfere; Te, Se, Re, Ru, Rh, Ir, Os, Pt, probably do not interfere. The ions Nb, Si, Pd, W and Ta interfere by contaminating the precipitate and must be separated. The effect of vanadium(V) and chromium(VI) is eliminated by reduction with sulfurous acid or Mohr's salt. Tartaric and hydrofluoric acids interfere with complete precipitation of molybdenum. This method is suitable for the determination of molybdenum in steels not containing tungsten.

α-Benzoinoxime is one of the best reagents for the separation of molybdenum from the accompanying elements.

REAGENTS

Mohr's salt, 5% solution in 5% sulfuric acid.
α-Benzoinoxime, 0.05 *M* solution in 1:1 mixture of acetone and water.
Mixture of acetone and water, 1:1.
Sulfuric acid, $d = 1.84$ g/cm^3 and 1% solution.

PROCEDURE

To the sample solution, which should contain 8–20 mg of Mo(VI), is added 1–1.5 ml of sulfuric acid ($d = 1.84$ g/cm^3), the solution is diluted to 25 ml with water and Mohr's salt is added in excess to reduce chromium(VI) and vanadium(V). Ten ml α-benzoinoxime solution are added drop by drop while stirring, and the mixture is left to stand for 10 minutes. The solution over the precipitate is decanted through a tared No. 2 or No. 3 sintered glass crucible; the precipitate is washed by decantation with 5 ml of 1% sulfuric acid and transferred onto the filter with the aid of water. The analyst should make sure that the precipitate on the filter is covered with the liquid. The walls of the crucible and the precipitate are then washed with two 5-ml portions of acetone-water mixture, dried at 105°C to constant weight and weighed as $MoO_2(C_{14}H_{12}O_2N)_2$.

Note. The precipitate may also be ignited in a porcelain crucible at 500–550°C in a muffle furnace and weighed as MoO_3.

Complexometric determination of molybdenum [11, 39, 40]

Complexone III reacts with Mo(V) to form a stable complex with a component ratio of 1:2 (formation constant $(1.75 \pm 0.52) \cdot 10^{11}$). Molybdenum(VI) is reduced to molybdenum (V) by hydrazine in sulfuric acid medium in the presence of excess Complexone III, which is titrated against a standard solution of zinc salt in the presence of Eriochrome Black T.

The aliquot should not contain more than 20 mg of molybdenum because the solutions of molybdenum complexonate are intensely colored, so that the equivalence point is difficult to establish.

Manganese, alkali metal and alkaline earth metal ions do not interfere. In the presence of tartaric acid and potassium fluoride which are introduced after the reduction of molybdenum (VI), the ions Ti, Nb, Ta, W, Th, Al, Ce(III), La, U(IV) do not interfere. The determination can be carried out in the presence of chlorides, sulfates, nitrates, phosphates, acetates, tartrates and citrates.

The method will determine more than 0.5 mg of molybdenum with an error of ±0.2 mg.

The method is applicable to the analysis of alloys containing bismuth, cadmium, cobalt, zinc, nickel, copper, mercury, vanadium, chromium and lead. A preliminary titration of all the accompanying elements is carried out on one portion of the solution, without reducing molybdenum (VI); the reduced molybdenum and the accompanying elements are determined together in the second portion of the solution.

REAGENTS

Complexone III, 0.01 *M* solution.
Zinc sulfate, 0.01 *M* solution.
Hydrazine sulfate, crystalline.
Sulfuric acid, 6 *N* solution and dilute (1:1) solution.
Ammonia, 25% solution.
Ammonium chloride buffer solution, pH 10: 54 g of ammonium chloride are dissolved in water, 350 ml of 25% ammonia solution are added and the solution is diluted to one liter with water.
Eriochrome Black T. A 1:150 mixture with sodium chloride.
Tartaric acid, crystalline.

Sodium fluoride, crystalline.

Methyl orange, 0.2 g of indicator is dissolved in 60 ml of ethanol and 40 ml of water are added.

Zinc sulfate, 0.005 M solution.

PROCEDURE

Determination of molybdenum in solutions

The sample solution, which should contain 25–50 mg of molybdenum as molybdate, is diluted to the mark with water in a 50-ml volumetric flask, thoroughly mixed, and 10.0 ml of this solution are withdrawn with a pipet into a heat-resistant 300-ml conical flask. To the flask are added 10 ml of Complexone III solution, 3 g of hydrazine sulfate, 10 ml of 6 N sulfuric acid, the mixture is diluted to 100 ml with water, heated and boiled for 5 minutes. The solution is cooled, 5 ml of ammonia solution, 20 ml of ammonium chloride buffer solution and as much Eriochrome Black T as will cover the tip of a spatula (20–30 mg) are added. The blue mixture is titrated against zinc sulfate until it turns lilac-colored. One ml of 0.0100 M Complexone III solution is equivalent to 1.92 mg of molybdenum.

Determination of molybdenum in alloys

The alloy sample, which should contain 10–50 mg of molybdenum, is treated with 10 ml of 1:1 sulfuric acid in a heat-resistant 50-ml beaker by heating on a sand bath. The solution is quantitatively transferred to a 50-ml volumetric flask, diluted to the mark with water and thoroughly mixed.

Ten ml of the resulting solution are withdrawn with a pipet into a 300-ml conical flask, 20 ml of Complexone III solution are added and the solution is heated on a plate to 70–80°C. The solution is cooled, 2 g of tartaric acid, 5 ml of ammonia solution and 20 ml of ammonium chloride buffer solution are added and the mixture is stirred. Eriochrome Black T (20–30 mg) on the tip of a spatula is added and the mixture is titrated against zinc sulfate solution until the blue color turns lilac. Let the volume of Complexone III solution consumed in titrating the components of the alloy Bi, Cd, Co, Ni, Zn, V(IV), Cr(III), Pb be V_1.

Another 10-ml portion of the solution of the alloy is withdrawn with a pipet into another 300-ml conical flask, 20 ml of Complexone III solution, 5 ml of 6 N sulfuric acid, 3 g of hydrazine sulfate, and 2 g of tartaric acid are added, the mixture is diluted with 60–70 ml of water, heated and boiled for 5 minutes. The solution is cooled, neutralized with 5 ml of ammonia solution, 20 ml of ammonium chloride buffer solution are added, then Eriochrome Black T on the tip of a spatula, and the excess Complexone III is titrated against zinc sulfate solution until the blue solution turns lilac. Let the volume of Complexone III solution used in titrating molybdenum together with the other components of the alloy be V_2.

The difference $V_2 - V_1$ corresponds to the amount of Complexone III which has reacted with molybdenum.

Note. The amounts of reagents shown above are for a sample size of 0.025–0.05 g containing 8–40% molybdenum; if less molybdenum is present, it must be preliminarily separated, since the amount of Complexone III as given above will then be insufficient to bind all the components of the alloy. Small amounts of molybdenum can be determined complexometrically by titrating the excess Complexone III against a solution of zirconium sulfate in 0.3 *N* sulfuric acid in the presence of Xylenol Orange.

Complexometric determination of molybdenum ions in acid medium

The excess of the added Complexone III is titrated against a standard solution of zirconium sulfate in 0.3 *N* sulfuric acid in the presence of Xylenol Orange as indicator.

Equal amounts of Al, Mg, Zn, Cd, Mn, Co, Pb and REE do not interfere.

The ions Fe(III), Zr, Hf, which form complexonates stable in acid medium, must be absent. Phosphates, fluorides, oxalates, citrates and tartrates, which react with zirconium in acid medium to form stable complexes or precipitates, must also be separated.

The method is suitable for the determination of molybdenum in solutions of pure salts and in alloys.

REAGENTS

Complexone III, 0.01 *M* solution.
Zirconium sulfate, 0.01 *M* solution in 1 *N* sulfuric acid.
Sulfuric acid, 1 *N* solution.
Hydrazine sulfate, 0.5% solution in 1 *N* sulfuric acid.
Xylenol Orange, 0.05% solution.
Sodium sulfate, crystalline.

PROCEDURE

Fifteen ml of sample solution, which should contain 2–12 mg of molybdenum, are placed with a pipet in a 300-ml conical flask, 25 ml of sulfuric acid are added, the solution is heated to boiling, 15 ml of hydrazine sulfate and 3 g of sodium sulfate are added, the mixture is boiled for another 2–3 minutes and 10 ml of Complexone III solution are introduced into the hot solution.

The solution is diluted with 100 ml of hot distilled water, heated to boiling, 1 ml of Xylenol Orange solution added and the excess Complexone III is titrated against zirconium sulfate until the yellow solution assumes a distinctly red color, persisting for about 30 seconds.

Photometric determination of molybdenum by pyrocatechol [34]

Pyrocatechol reacts with molybdenum ions in neutral or weakly alkaline medium (0.1–3% NaOH) to form an orange-red water-soluble compound with a component ratio of 2:1.

The reaction may be represented by the equation:

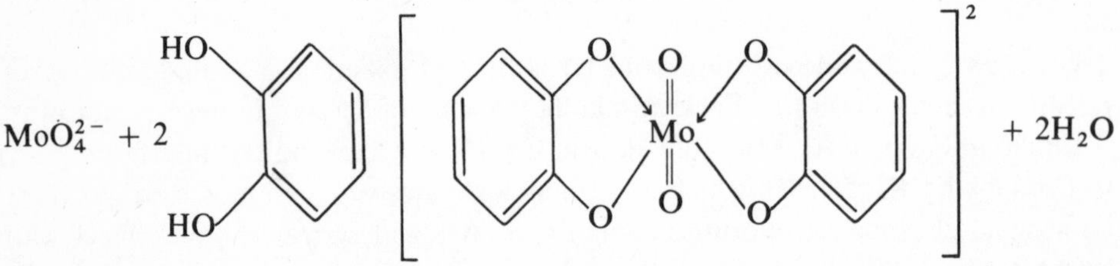

MoO_4^{2-} + 2 (HO, HO) [O O O / Mo / O O O]^2 + $2H_2O$

The absorption maximum of the solutions is at 412 mμ, the molar extinction coefficient is 3400 and the dissociation constant of the compound is $(1.88 \pm 0.03) \cdot 10^{-5}$. Beer's law is obeyed in a wide range of molybdenum concentrations.

In the presence of sodium bisulfite, which prevents the oxidation of pyrocatechol by atmospheric oxygen, the colored solutions are stable for 2 days.

The determination may be carried out in the presence of up to 4 mg cobalt and manganese chlorides; ferric chloride interferes.

REAGENTS

Standard solution of sodium molybdate, 0.1 mg of Mo/ml.

Pyrocatechol, 1% solution. Sodium hydroxide (2 g) is dissolved in 500 ml of water, and 15 g of sodium metabisulfite $Na_2S_2O_5$ and 10 g of pyrocatechol are added, and the solution is diluted with water to one liter. It is stable for several months.

CONSTRUCTION OF CALIBRATION CURVE

Into five 50-ml volumetric flasks are introduced 1, 2, 3, 4 and 5 ml of standard solution of sodium molybdate, 10 ml of pyrocatechol solution are added to each flask, the solutions are made up to the mark with water and thoroughly mixed. The optical densities are measured relative to water in an electrophotocolorimeter or a photometer and the calibration curve is plotted.

PROCEDURE

The reagents used in the preparation of the standard scale are added to the sample solution, the optical density is determined and the content of molybdenum is read off the calibration curve.

Photometric determination of molybdenum in steels by the thiocyanate method [34]

Thiocyanates react with molybdenum (V) in acid solution to form colored compounds, the composition of which depends on the thiocyanate concentration. Hexavalent molybdenum is reduced to the pentavalent state by stannous chloride, potassium iodide, ascorbic acid or thiourea in the presence of cupric salts, and also by other reducing agents.

Depending on the concentration of the thiocyanate, the molar ratio Mo: SCN in the compounds formed may vary between 1:1 and 1:6. The most intense colorations are given by compounds with molar ratio Mo: SCN = 1:5 (molar extinction coefficient 15,000) and Mo:SCN = 1:6 (molar extinction coefficient 12,600).

Thiocyanate compounds of Mo(V) are extracted with ether and with other extractants.

Al, Co, U, Ta, Na, K, Si, Ca, Mg, Ti, V, Cr, Mn, Ni, Zn, As, Ag, Sn, Sb, Hg do not interfere. Ferric and cupric ions enhance the intensity of the coloration, probably owing to the formation of polynuclear thiocyanate complexes containing molybdenum and iron (or copper) in the molar ratio of 1:1. The interfering effect of tungsten is removed by introducing tartaric acid, which prevents the formation of its thiocyanate complexes.

REAGENTS

Standard solution of sodium (ammonium) molybdate, 0.1 mg of Mo/ml.
Sulfuric acid, dilute (1:4 and 1:1) solutions.
Hydrogen peroxide, 30% solution.
Tartaric (citric) acid, crystalline.
Sodium hydroxide, 10% solution.
Potassium thiocyanate, 5% solution.
Stannous chloride, 35% solution. 350 g $SnCl_2 \cdot 2H_2O$ are dissolved with heating in 300 ml hydrochloric acid (d = 1.19 g/cm^3), the clear solution is diluted with 800 ml of water and a few pieces of metallic tin are added.
Diethyl ether.
Ferric sulfate, 8% solution in 1 *N* sulfuric acid.

CONSTRUCTION OF CALIBRATION CURVE

Into each of five 50-ml separation funnels are introduced 3 ml of ferric sulfate solution, 0.5, 1.0, 1.5, 2.0 and 2.5 ml of standard sodium molybdate solution, 5 ml of 1:1 sulfuric acid, 5 ml of potassium thiocyanate solution and the mixtures are vigorously shaken for 2–3 minutes. To each funnel are then added 5 ml of stannous chloride solution and the funnels are again strongly shaken for 1–2 minutes. The solutions gradually become amber-colored or reddish-brown.

To each colored solution are added 10–12 ml ether, the molybdenum compound is extracted for

2–3 minutes, the ether layers are decanted into dry 25-ml volumetric flasks and the ether extraction is repeated. The ether extracts are diluted with ether to 25 ml and the optical densities are measured in an electrophotocolorimeter in cells provided with lids to prevent the evaporation of ether. From the data thus obtained the calibration curve is constructed.

PROCEDURE

A 0.05–0.25-g steel sample, which should contain 0.05–0.1% molybdenum is placed in a 100-ml beaker, 10 ml of 1:4 sulfuric acid are added and the mixture is heated to 60–70°C. Two ml of hydrogen peroxide are added, the solution is boiled several minutes to oxidize the carbides, filtered through filter paper to remove the carbon, and the insoluble residue is washed with water and discarded. The filtrate is combined with the wash waters and evaporated in a beaker to a residual volume of 10–15 ml. The residue is not filtered off if the sample consists of carbon steel or tungsten-containing steel. To the solution is added 0.5 g of tartaric acid if the steel contains tungsten. The solution is then neutralized with sodium hydroxide to pH 8–9 (universal indicator paper), 5 ml of 1:1 sulfuric acid added and the mixture transferred to a 50-ml separating tunnel. Five ml of potassium thiocyanate solution are added and the determination is continued as described for the construction of the calibration curve. The amount of molybdenum is found from the calibration curve.

Photometric determination of molybdenum in steels by p-phenetidide-1-mercaptopropionic acid [41]

p-Phenetidide-1-mercaptopropionic acid reacts with Mo(V) and Mo(VI) to form yellow compounds which are insoluble in water, but soluble in organic solvents. The compound with Mo(V) is formed in the acidity range between 0.2 *N* HCl and pH 6; the maximum optical density of the extracts is noted when the pH of the aqueous phase is 1.6–3.8 (molybdenum concentration $2 \cdot 10^{-4}$ *M*, concentration of reagent $2 \cdot 10^{-2}$ *M*). Mo(VI) ions react with p-phenetidide-1-mercaptopropionic acid between 1 *N* HCl and pH 6, the maximum optical density of the extracts being noted between 0.3 *N* HCl and pH 4 (molybdenum concentration $2 \cdot 10^{-4}$ *M*, reagent concentration $2 \cdot 10^{-2}$ *M*). Molybdenum (VI) ions are not reduced by excess reagent in 0.3 *N* HCl; gradual oxidation of molybdenum (V) compound commences at about pH 4.

A 1:1 mixture of isoamyl alcohol and benzene is the best extractant. The absorption maximum of the extract is 355–360 mμ, the molar extinction coefficients are 2700 and 4550 for Mo(V) and Mo(VI) compounds respectively, and the molar ratio molybdenum: reagent is 1:2, irrespective of the valency of molybdenum. The optical density of the extracts is proportional to the concentration of molybdenum between 10 and 200 μg of Mo/5 ml solvent.

The ions (Fe(III), Cr, Ti, Co, Ni, Zn, Al) and 100-fold amounts of W do not interfere; copper ions interfere.

The method is suitable for the determination of molybdenum in steels.

REAGENTS

Standard solution of sodium paramolybdate, 25 μg of Mo/ml.
p-Phenetidide-1-mercaptopropionic acid, 1% solution in 1:1 isoamyl alcohol–benzene mixture.
Hydrochloric acid, dilute (1:1) solution.
Nitric acid, d = 1.4 g/cm^3.

CONSTRUCTION OF CALIBRATION CURVE

Into eight 50-ml separating funnels are introduced 1, 2, 3, 4, 5, 6, 7 and 8 ml of standard sodium paramolybdate solution; 0.5 ml of hydrochloric acid is added to each funnel and water to make up the volume to 10 ml. Then 5 ml of reagent solution are added to each funnel.

The mixtures are shaken for 30 minutes, the organic phase decanted into dry cells and the optical densities measured at 360 mμ in a spectrophotometer or an electrophotocolorimeter relative to a blank solution prepared in the same manner, except that the molybdenum solution is not introduced.

PROCEDURE

A 0.1-g steel sample is placed in a 50-ml beaker, 5–6 ml of HCl are added and the mixture is gently heated on a sand bath. If the steel is not completely dissolved, 1–2 ml of nitric acid are added. The solution is evaporated to dryness, the residue dissolved in 5 ml of hydrochloric acid, the solution transferred to a 500-ml volumetric flask and diluted to mark with water. Between 1 and 10 ml of the resulting solution are withdrawn into a 50-ml separating funnel, diluted with water to 10 ml, hydrochloric acid added to a final concentration of about 0.3 N and the determination is continued as described for the construction of the calibration curve.

The content of molybdenum is found from the calibration curve.

BIBLIOGRAPHY

1. BUSEV, A. I. and LI KÊNG.—*ZhAKh,* **13**, 519. 1958.
2. BUSEV, A. I. and LI KÊNG.—*ZhAKh,* **14**, 668. 1959.
3. BUSEV, A. I. and LI KÊNG.—*ZhAKh,* **15**, 191. 1960.
4. HAHN, F. L. and R. LUKHANS.—*Z. anal. Chem.,* **149**, 172. 1956.
5. SILLEN, L. G.—*Quart. Rev.,* **13**, 163. 1959.
6. KOMAROVSKY, J.—*Chem. Ztg.,* **37**, 957. 1913.
7. BRAUN, C. D.—*Z. anal. Chem.,* **2**, 36. 1863.
8. KOMAROVSKII, A. S. and N. S. POLUEKTOV.—*ZhPKh,* **10**, 565. 1937; *Michrochim. acta,* **1**, 264. 1937.
9. KNOWLES, H. B.—*Bur. Stand. J. Res.,* Paper 9, 1. 1932.

10. PECSOK, R. L. and D. T. SAWYER.—*J. Am. Chem. Soc.*, **78**, 5496. 1956.
11. BUSEV, A. I. and CHANG FAN.—*ZhAKh*, **14**, 445. 1959.
12. BUSEV, A.I. and CHANG FAN.—*Vestnik MGU*, math., mech., astron., phys., chem., ser., No. 2, 203. 1959.
13. YOE, J. H. and A. R. ARMSTRONG.—*Anal. Chem.*, **19**, 101. 1947.
14. SOMMER, L.—*Coll.*, **22**, 414. 1957.
15. BUSEV, A. I. and CHANG FAN.—*Izvestiya Vuzov, Khimiya i Khimicheskaya Tekhnologiya*, **4**, 905. 1961.
16. BUSEV, A. I. and CHANG FAN.—*Vestnik MGU*, chem. ser. II, No. 4, 55. 1961.
17. BUSEV, A. I. and CHANG FAN.—*ZhAKh*, **15**, 455. 1960.
18. BUSEV, A. I. and CHANG FAN.—*Vestnik MGU*, chem. ser. II, No. 2, 36. 1961.
19. BUSEV, A. I. and CHANG FAN.—*Vestnik MGU*, chem. ser. II, No. 3, 66. 1962.
20. BUSEV, A. I. and CHANG FAN.—*ZhNKh*, **6**, 1308. 1961.
21. ZHAROVSKII, F. G. and E. F. GAVRILOVA.—*Zav. Lab.*, **23**, 143. 1957.
22. BUSEV, A. I., CHANG FAN, and Z. P. KUZYAEVA.—*Izvestiya Vuzov, Khimiya i Khimicheskaya Tekhnologiya*, **5**, 17. 1962.
23. RICHTER, F.—*Chem. Techn.*, **1**, 31. 1949.
24. WILL, F. and J. H. YOE.—*Anal. Chem.*, **25**, 1363. 1953.
25. BUSEV, A. I. and CHANG FAN.—*ZhAKh*, **16**, 39. 1961.
26. SAMARA, R. N. S.—*Sci. and Culture (India)*, **23**, 434. 1958.
27. BUSEV, A. I. and CHANG FAN.—*ZhAKh*, **16**, 169. 1961.
28. BUSEV, A. I., CHANG FAN, and Z. P. KUZYAEVA.—*ZhAKh*, **16**, 695. 1961.
29. BUSEV, A. I., CHANG FAN, and Z. P. KUZYAEVA.—*Zhurnal VKhO im. D. I. Mendeleeva*, **6**, 237. 1961.
30. BUSEV, A. I., CHANG FAN, and Z. P. KUZYAEVA.—*Vestnik MGU*, chem. ser. II, No. 4, 43. 1962.
31. BUSEV, A. I. and G. P. RUDZIT.—*ZhAKh*, **18**, 840. 1963.
32. BUSEV, A. I.—*Doklady AN SSSR*, **66**, 1093. 1949.
33. BUSEV, A. I.—*ZhAKh*, **4**, 234. 1949.
34. BUSEV, A. I. *Analiticheskaya khimiya molibdena (Analytical Chemistry of Molybdenum)*.—Izd. AN SSSR. 1962. [English translation available as part of the *Analytical Chemistry of the Elements* series, Ann Arbor–Humphrey Sci. Pub. 1969.]
35. BUSEV, A. I.—In: *Metody opredeleniya i analiza redkikh elementov*, p. 537. Izd. AN SSSR. 1961.
36. KUZNETSOV, V. I.—In: *Sessiya AN SSSR po mirnomu ispol'zovaniyu atomnoi energii* 1–5 *iyulya* 1955. *Zasedaniya otdeleniya khimicheskikh nauk (Conference of the Academy of Sciences of the USSR on Peaceful Uses of Atomic Energy*, 1–5 *July* 1955. *Meeting of the Section of Chemical Sciences)*, p. 301.—Izd. AN SSSR. 1955.
37. KUZNETSOV, V. I. and G. V. MYASOEDOVA.—*Trudy Komissii po Analiticheskoi Khimii*, **9**, No. 12, 89. 1958.
38. HOENES, H. J. and K. G. STONE.—*Talanta*, **4**, 250. 1960.
39. LASSNER, E. and R. SCHARF.—*Z. anal. Chem.*, **167**, 114. 1959; **168**, 30, 429. 1959.
40. KLYGIN, A. E., N. S. KOLYADA, and D. M. ZAVRAZHNOVA.—*ZhAKh*, **16**, 442. 1961.
41. BUSEV, A. I., G. P. RUDZIT, and A. NAKU.—*ZhAKh*, **19**, 767. 1964.

Tungsten W

In its compounds tungsten displays valencies of + 2, + 3, + 4, + 5 and + 6. Compounds of hexavalent and pentavalent tungsten are the most important.

Standard electrode potentials relative to standard hydrogen electrode in aqueous medium at 25°C are:

$W + 2H_2O \rightleftharpoons WO_2 + 4H^+ + 4e$	-0.119 V
$2WO_2 + H_2O \rightleftharpoons W_2O_5 + 2H^+ + 2e$	-0.031 V
$W_2O_5 + H_2O \rightleftharpoons 2WO_3 + 2H^+ + 2e$	-0.029 V
$W + 4H_2O \rightleftharpoons WO_4^{2-} + 8H^+ + 6e$	$+0.049$ V
$[W(CN)_8]^{4-} \rightleftharpoons [W(CN)_8]^{3-} + e$	$+0.457$ V

Stannous chloride reduces tungstates to tungsten blue which, unlike molybdenum blue, is stable in acid solutions. A number of methods for the detection of tungstate WO_4^{2-} ions are based on the formation of tungsten blue.

Strong reducing agents (zinc, zinc amalgam) in concentrated hydrochloric acid solutions reduce W(VI) to the brown ions of W(III) which are unstable in the air. In moderately concentrated hydrochloric acid solutions tungsten blue is formed as a result.

Metallic mercury also reduces W(VI) to W(V). Ferrous sulfate does not reduce tungstate ions.

When tungstates are fused with lithium carbonate and sodium formate, lithium–

tungsten bronze is formed. This reaction is used in the detection of tungsten by identification of crystals under the microscope.

The red-colored trivalent tungsten ions WCl_5^{2-} are a strong reducing agent. They are converted into the yellow or yellowish-green $W_2Cl_9^{3-}$ ions. Trivalent tungsten in solutions is oxidized by atmospheric oxygen.

Compounds of tungsten(II) are unstable in solution.

Metallic tungsten occurs in the passive state, in which it is inert to hydrochloric acid. The effect of concentrated nitric acid on bulk tungsten is slow.

Hot concentrated phosphoric and hydrofluoric acids dissolve tungsten with complex formation. Fused alkalis, especially in the presence of oxidizing agents, react with metallic tungsten.

When solutions of alkali tungstates are acidified, a precipitate of tungstic acid separates out; it is insoluble in excess HNO_3, $HClO_4$, H_2SO_4, unlike molybdnic acid. The precipitate of tungstic acid is soluble in 9–12 *N* hydrochloric acid. In the cold, the white amorphous precipitate $H_2WO_4 \cdot nH_2O$ separates out, while a less hydrated yellow precipitate separates out of hot solutions.

Tungstic acid has a tendency to form colloidal solutions. In order to prevent peptization, the tungstic acid precipitate is washed with a solution of an electrolyte. Tungstic acid becomes fully insoluble only when the acid solutions are evaporated to dryness.

If a solution of the melt of the sample material with alkali hydroxides or carbonates is acidified with hydrochloric acid, tungstic acid and silicic acid separate (acid hydrolysis). Evaporation with HCl yields a less hydrated and less soluble modification. In this way tungsten can be separated from numerous elements.

Tungstic acid in acid solutions forms different polytungstic acids, in analogy to polymolybdic and polyvanadic acids, depending on its concentration and on the pH of the solution. In alkaline medium polytungstic acids readily decompose with the formation of tungstate ions WO_4^{2-}.

Tungstic acid is readily soluble in solutions of ammonia and alkali hydroxides with the formation of tungstates:

$$\left.\begin{array}{l} H_3WO_5^- \rightleftharpoons H_2WO_5^{2-} + H^+ \\ H_2WO_5^{2-} \rightleftharpoons HWO_5^{3-} + H^+ \\ HWO_5^{3-} \rightleftharpoons WO_5^- + H^+ \end{array}\right\} \quad pK = 6-8$$

Silver, mercury (I), lead and alkaline earth tungstates are sparingly soluble in water; thus, for example, $pL_{Ag_2WO_4} = 9.3$.

Chlorides, fluorides, oxalates, tartrates, and citrates form different complexes with W(VI).

Hydrogen sulfide does not precipitate sulfides from solutions containing WO_4^{2-} ions. These ions react with ammonium sulfide to form the orange-yellow thiosalt

WS_4^{2-}. When the solution is acidified, the light brown WS_3 precipitates out. The thiosalt is fully decomposed below pH4.

Orthophosphoric acid reacts with W(VI) ions to form a heteropoly acid. Thiocyanates form colored complexes with W(V). Thiocyanates are used in photometric determination of tungsten.

Nitrogen-containing organic bases and certain triphenylmethane dyes precipitate tungstic acid ions and are used in the quantitative isolation of W(VI) from solution. The resulting precipitates, the composition of which is usually variable, are ignited to WO_3. Tungstic acids are precipitated by alkaloids (quinine, cinchonine), benzidine, 1-naphthylamine, Nitron, vanillylidenebenzidine, β-naphthoquinoline (see below), sulfamido-2,4-diaminoazobenzene (red streptocide), dimethylaminoantipyrine (pyramidone), Rhodamine B etc.

Pyramidon [1] is an important reagent in the isolation of tungstic acid; first hydrochloric acid and then pyramidon are added to a dilute solution of sodium tungstate. The reagent is successfully employed in the determination of tungsten in ferrotungsten.

Rhodamine B precipitates tungstic acid from acidified solutions [2,3]. The precipitation is most complete from solutions 0.10–0.15 *M* in HCl. In neutral medium the precipitate is not formed. The solubility product of the precipitate is $2 \cdot 10^{-18}$ in 0.12 *M* HCl. The reagent is used in the detection of tungsten in minerals and alloys and also in its gravimetric determination.

Numerous derivatives of 2,3,7-trihydroxy-6-fluorone [4] react with tungstate ions in weakly acid media (pH 2.0–3.5) to form red-colored compounds, in which the molar ratio is 1:1. If the concentration of the tungstate ions is sufficiently high, these compounds precipitate out. In the presence of gelatin, the colored solutions remain clear. Citric acid is added to the tungstate solution prior to acidification in order to prevent the separation of tungstic acid; ethanol is added to prevent the reagent from precipitating out.

The best reagents for the photometric determination of tungsten are 9-(2′-hydroxyphenyl)-trihydroxyfluorone and 9-(9′-anthracenyl)-2,3,7-trihydroxyfluorone.

8-Hydroxyquinoline precipitates tungsten(VI) ions and is used in the quantitative determination of tungsten.

The most valuable photometric methods for the determination of tungsten are based on the capacity of W(V) to form fairly stable colored complexes with thiocyanates and with toluene-3,4-dithiol under certain conditions (p. 224).

Various titrimetric methods for the determination of tungsten are based on redox reactions.

For a survey of quantitative methods of determination of tungsten the reader is referred to [5].

Gravimetric determination of tungsten in ferrotungsten [6]

Tungsten is precipitated by β-naphthoquinoline at pH 1–6.

Ni, Zn, Mn, Al, Ti, Pb, Be, Zr, V, Cr and small amounts of Fe do not interfere. If much iron is present, the precipitate obtained is treated with ammonia and the tungsten is again precipitated by β-naphthoquinoline. If molybdenum is present, tungsten is precipitated in a more acid medium.

The method is used for the analysis of cast irons, steels, alloys, slags and ores.

REAGENTS

β-Naphthoquinoline, 2% solution. Two grams of the reagent are dissolved in 100 ml of water acidified with a few drops of concentrated sulfuric acid.
Oxalic acid, saturated solution.
Hydrogen peroxide, 30% solution.
Sulfuric acid, dilute (1:1) solution.
Hydrochloric acid, 2% solution.
Ammonia, 10% and 2.5% solutions.
Ammonium nitrate, 2% solution.
Potassium ferrocyanide, crystalline.

PROCEDURE

A 0.25–0.50-g sample of ferrotungsten is placed in a 500-ml beaker and dissolved in 30 ml of oxalic acid solution and 5 ml of hydrogen peroxide solution by heating on an electric hot plate covered with a sheet of asbestos. The solution is cautiously evaporated to a small volume, the residue cooled and 15 ml of sulfuric acid added drop by drop; the content of the beaker is heated until the sulfuric acid fumes have evolved for 3 minutes. After cooling the solution is diluted with 400 ml of cold water and is gradually heated until the salts have dissolved. Thirty ml of β-naphthoquinoline solution are added and the mixture is left to stand at room temperature for 1–2 hours or overnight.

The precipitate is separated on Blue Ribbon filter paper, and washed with hydrochloric acid solution to a negative reaction for iron (ferrocyanide test). The washed precipitate together with the unfolded filter paper is placed in the beaker which was used for the precipitation and treated with 50 ml of 10% solution of ammonia. When the precipitate has dissolved, the filter paper is disintegrated with a glass rod. The beaker with the solution is heated for 20–30 minutes on a water bath and the macerated paper filtered off and washed several times with 2.5% solution of ammonia.

After cooling the filtrate is acidified to pH 3–4 with hydrochloric acid and 20 ml of β-naphthoquinoline solution are added. The precipitate is filtered off, washed twice with hydrochloric acid and twice with ammonium nitrate solution, placed in a porcelain dish, dried on asbestos-covered electric plate and cautiously ashed in a muffle furnace, first at gentle heat to avoid sublimation of tungsten, and then at 600–650°C. After cooling the residue is weighed as WO_3.

The conversion factor to tungsten is 0.7931.

Potentiometric determination of tungsten by chromium (II) salts [7, 8]

Tungsten(VI) is reduced to tungsten(V) in acid medium by salts of bivalent chromium:

$$WO_2Cl_3^- + 2H^+ + 2Cl^- + e \rightleftharpoons WOCl_5^{2-} + H_2O.$$

The equivalence point is established by the potentiometric method. At the end of the reduction there is a sharp jump of the potential of the platinum electrode. The potential of the platinum electrode is established immediately after the addition of each new portion of the solution of chromium(II) salt. The titration is conducted in concentrated hydrochloric acid (100 or 150 ml of acid are added for each 5 or 10 ml of 0.1 *M* solution of sodium tungstate, respectively). In the presence of Fe(III), Cu(II), Cr(VI), Mo(VI), two potential jumps are observed: the first jump corresponds to the completed reduction of these elements, while the second corresponds to the completed reduction of tungsten (VI) and also to the reduction of molybdenum (V) to molybdenum (III). For this reason tungsten (VI) and other elements are first reduced with a solution of chromium (II) salt, and tungsten (V) is then titrated against a solution of potassium dichromate. Citric, tartaric, oxalic and formic acids are without effect on the course of the titration. The method is used in the determination of tungsten in scheelite concentrate.

REAGENTS

Hydrochloric acid, $d = 1.19$ g/cm^3.

Chromium (II) *sulfate*, 0.1 *N* solution: 29.421 g of dried $K_2Cr_2O_7$(A.R.) are dissolved in 500 ml of water in a 2-liter flask and the solution is acidified with 27.8 ml of 36 *N* sulfuric acid; 75 ml of 30% hydrogen peroxide are slowly added with stirring, to reduce the dichromate to Cr^{3+}. The solution is heated to boiling, cooled to room temperature, transferred to a 2-liter volumetric flask and diluted to mark with water. The bottle in which the solution is to be stored is two-thirds filled with zinc amalgam and connected to the titration buret and to a Kipp apparatus serving as a source of hydrogen. The bottle with the zinc amalgam is rinsed with two small portions of the solution, after which the solution is poured into it. The reduction of chromium(III) to chromium(II) is complete within 24 hours. The evolving hydrogen should be allowed to escape through the Kipp apparatus. The titer of the solution of $CrSO_4$ is established by titrating against a standard solution of copper sulfate $CuSO_4 \cdot 5H_2O$. To do this, 3 ml of concentrated hydrochloric acid are added to 2 ml of 0.1 *M* solution of copper sulfate and the mixture is titrated potentiometrically, using a platinum indicator electrode. The solution of $CrSO_4$, when stored over zinc amalgam in an atmosphere of hydrogen, remains stable for several weeks.

Sodium hydroxide, 20% solution.
Oxalic acid, saturated solution.
Potassium dichromate, 0.1 *N* solution.

PROCEDURE

A 0.25-g sample of finely ground scheelite concentrate in a small porcelain dish is treated with 4–5 ml of hydrochloric acid while heating. The excess acid is evaporated and the almost dry residue is wetted with 3 ml of sodium hydroxide solution. Ten ml of oxalic acid solution are added to the resulting solution and the contents of the dish are washed into a titration beaker containing 100 ml of hydrochloric acid. The beaker is closed with a rubber stopper provided with holes serving as inlet and outlet for carbon dioxide, inlet of buret, salt bridge and platinum indicator electrode. In order to remove oxygen, carbon dioxide is passed for 30 minutes through the solution from a Kipp apparatus. The solution is titrated against chromium(II) sulfate without interrupting the stream of carbon dioxide and with magnetic stirring until the potential jump corresponding to the reduction of tungsten(VI) is observed. Then tungsten(V) is titrated against potassium dichromate solution.

Dichromatometric determination of tungsten [9]

Tungsten(VI) is reduced to tungsten(III) in concentrated hydrochloric acid by the action of granulated lead in the presence of ammonium chloride. To ensure quantitative reduction, the solution is passed through a lead reductor. The solution containing tungsten(III) ions is added to a solution of ferric salt. The ferrous ions, which are formed in an equivalent amount, are titrated against potassium dichromate in the presence of diphenylaminosulfonic acid as indicator.

REAGENTS

Ferric ammonium alum: 10 g of $Fe(NH_4)(SO_4)_2 \cdot 12H_2O$ are dissolved in 20 ml of concentrated hydrochloric acid and 80 ml of water are added.
Potassium dichromate, 0.01 *N* and 0.05 *N* solutions.
Diphenylaminosulfonic acid, 0.005 *M* solution.
Granulated lead. The reagent is placed in a reductor 18 cm tall and 2 cm in diameter, washed with a solution containing 15 g of NH_4Cl in 150 ml hot hydrochloric acid (2:1) and the reductor is then filled with 1:1 solution of HCl.
Hydrochloric acid, $d = 1.19$ g/cm^3 and dilute (1:1) solution.
Ammonium chloride, crystalline.

PROCEDURE

The sample solution, which should contain 5–50 mg tungsten, is diluted to 25 ml, 5–8 g of ammonium chloride are added and the mixture is heated until the salt has dissolved. Hydrochloric acid (25 ml, d = 1.19 g/cm^3) is added to the solution and the mixture is boiled, 10 g of lead is introduced, and the flask is covered with a watch glass and boiled for 2 minutes. The hot solution is transferred to the reductor, washing the walls of the flask with three 8-ml portions of concentrated hydrochloric acid. The solution is passed through the reductor at a rate of 20 ml/min. The solution issuing from the reductor is collected in a flask containing an excess of the solution of ferric ammonium alum in hydrochloric acid. The reductor is washed with 10 ml of 1:1 hydrochloric acid, with two 15-ml portions of water and again with 15 ml of hydrochloric acid. Two drops of the indicator are added, and the warm solution is titrated against 0.05 N solution of potassium dichromate to the appearance of a violet coloration persisting for 15 seconds.

The expended volume of potassium dichromate solution must be corrected for the indicator; the correction is found by performing the titration on the blank solution.

Titrimetric determination of tungsten in steels [10]

Tungstic acid is isolated by acid hydrolysis, dissolved in excess sodium hydroxide solution and the excess of sodium hydroxide is titrated against acid.

This method is suitable for the analysis of steels.

REAGENTS

Hydrochloric acid, dilute (1:1 and 5:95) solutions.

Nitric acid, d = 1.4 g/cm^3.

Potassium nitrate, 1% solution.

Silver nitrate, 1% solution.

Standard solution of alkali. To 16 ml of concentrated solution of sodium hydroxide (200 g/liter) are added 1.8 liter of freshly boiled water and 0.5 g of $BaCl_2 \cdot 2H_2O$; the solution is stored in a two-liter bottle fitted with a calcium chloride tube. The solution is standardized against accurately weighed $H_2C_2O_4 \cdot 2H_2O$.

Standard solution of acid. Nitric acid (65 ml, d = 1.4 g/cm^3) is diluted with water to 2 liters. The titer of the solution is determined against a standard solution of sodium hydroxide to phenolphthalein.

Phenolphthalein, 0.1% solution in ethanol.

PROCEDURE

A 0.5–1-g sample of steel in a beaker is dissolved in hot 1:1 hydrochloric acid and nitric acid is added. The solution is boiled, 60 ml of hot water added, then 5 ml of 1:1

hydrochloric acid and the mixture is boiled for 5 more minutes. The precipitate of tungstic acid is filtered through a double thickness of Blue Ribbon filter paper, and washed with hot 5:95 hydrochloric acid and with water, until the test for chlorides is negative. The beaker in which the steel was dissolved is washed with a solution of potassium nitrate, the filter paper with the precipitate is placed in the beaker, 60 ml of hot water are added, the filter paper disintegrated by shaking, 2 drops of phenolphthalein are added and the tungstic acid is dissolved in excess standard solution of alkali. The excess alkali is titrated against nitric acid solution until colorless. A blank determination on the filter paper is carried out at the same time.

The content of tungsten in the steel is calculated from the results of the titration.

Photometric determination of tungsten by the thiocyanate method [11, 12]

Tungsten(V) reacts with thiocyanate ions to form a greenish-yellow complex compound. The molar extinction coefficient of the solution of the complex compound is 17,600 at 398 mμ.

Tungsten(VI) is reduced to tungsten(V) by titanium trichloride. The excess of the reducing agent is violet-colored, but is practically without effect on the determination of the optical density with a blue filter.

Molybdenum(VI) does not interfere with the determination, since it is reduced to molybdenum(III) and forms a weakly colored complex compound with thiocyanate ions; Cr, V, Se and Te interfere; the effect of As and Sb is eliminated by adding hypophosphite. Ti, Nb, Ta, Cu, platinum group metals, F and P do not interfere.

This method will determine between 0.003% and 1.5% tungsten in mineral specimens in the presence of up to 10% arsenic, up to 3% antimony, up to 0.5% molybdenum, up to 0.3% chromium, and up to 0.1% each of vanadium, selenium and tellurium.

REAGENTS

Standard solution, 0.1 mg of W/ml.
Potassium thiocyanate, 25% solution.
Sodium hydroxide, 2% solution.
Titanium trichloride, 0.1 *N* solution. To 1 volume of 15% solution of $TiCl_3$ are added 9 volumes of 1:1 hydrochloric acid. The solution must be freshly prepared.
Hydrochloric acid, $d = 1.19$ g/cm^3.

CONSTRUCTION OF CALIBRATION CURVE

Into five 50-ml volumetric flasks are placed 1, 2, 3, 4 and 5 ml of the standard tungsten solution, the solutions are diluted to 10 ml with sodium hydroxide solution, and 2 ml of potassium thiocyanate solution, 12 ml of hydrochloric acid solution and 0.2 ml of titanium trichloride solution are added to each flask. After 5 minutes the solutions are diluted to the mark with water and their optical densities measured in an electrophotocolorimeter with a blue filter. From the results obtained the calibration curve is plotted.

PROCEDURE

The content of tungsten in the sample solution is determined as described in the preparation of the standard scale.

Photometric determination of tungsten by toluene-3,4-dithiol [13]

Toluene-3,4-dithiol reacts with tungsten (V) ions to give a blue-colored compound (Figure 13) which can be extracted with butyl, amyl or isoamyl acetate. The molar extinction coefficient of the extract is 20,000 at 640 mμ. The reducing agent is usually metallic titanium, and the rather unstable toluene-3,4-dithiol is replaced by its zinc complex.

More than 1 mg of lead and more than 10–20 μg of molybdenum interfere. Bismuth, antimony and tin do not interfere.

If the sample contains molybdenum in amounts several times higher than those of tungsten, the two elements are first isolated and are determined in succession. The determination is based on the fact that molybdenum(VI), unlike tungsten(VI), forms a green-colored complex with the dithiol. The absorption maximum of the

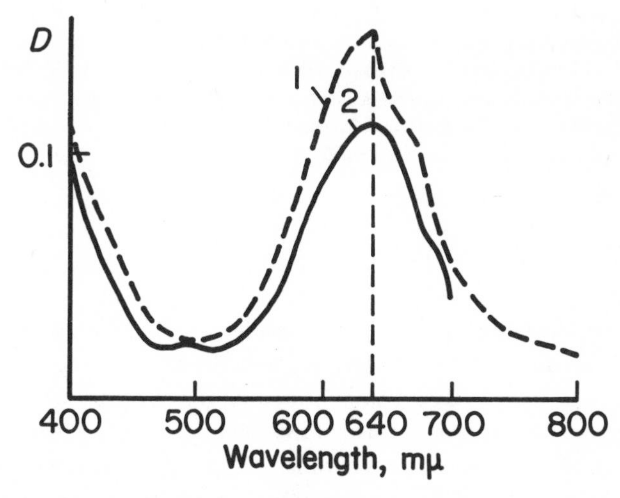

Figure 13

Light absorption by solutions of complexes of tungsten with dithiol in isoamyl acetate:

1) 5 μg of W, reduced by metallic titanium (in 5 ml of isoamyl acetate); 2) 90 μg of W(VI) (in 5 ml of isoamyl acetate).

complex is at 690 mμ; in this way molybdenum may be determined in the presence of more than 0.4 mg of tungsten. After removal of molybdenum by dithiol and isoamyl acetate the tungsten in the aqueous phase is determined.

REAGENTS

Hydrochloric acid, $d = 1.19$ g/cm^3.
Toluene-3,4-dithiol, zinc salt, alcoholic suspension. The reagent (0.2 g) is ground in a porcelain mortar with a minimum amount of alcohol, then ethanol is added to make the volume up to 50 ml. The mixture is shaken before use.
Titanium, metallic, shavings.
Isoamyl acetate.
Standard solution of sodium tungstate, 2 μg of W/ml.
Sulfuric acid, $d = 1.84$ g/cm^3 and dilute (1:1) solution.
Hydrofluoric acid, 40% solution.
Hydrogen peroxide, 30% solution.
Sodium carbonate, or *potassium pyrosulfate*.
Ammonia, solution, $d = 0.9$ g/cm^3.
α-Benzoinoxime, 2% alcoholic solution.
Chloroform.

PREPARATION OF SCALE OF STANDARDS

Into 5 tubes are placed portions of standard tungstate solution corresponding to 1.0, 2.0, 3.0, 4.0 and 5.0 μg of W, respectively, the solutions are diluted to 3 ml with water, and 7 ml of hydrochloric acid and 100–200 mg metallic titanium are added to each tube. The solutions are heated until the appearance of a weakly violet color, transferred to glass-stoppered tubes, and a suspension of zinc dithiol is added to the appearance of a faint turbidity; the tubes are loosely stoppered and placed in a beaker with boiling water for 20 minutes. The solutions are transferred to separating funnels, cooled to room temperature and extracted with 2 ml of isoamyl acetate for 1½–2 minutes. The extracts are transferred to tubes with ground glass stoppers containing 10 ml of redistilled hydrochloric acid. The scale is stable for 10–12 days if kept in a refrigerator.

Oxidizing acid mixture: 2–3 ml of redistilled nitric acid, 5 drops of sulfuric acid ($d = 1.84$ g/cm^3) and 5 drops of concentrated perchloric acid.

PROCEDURE

In the analysis of silicate rocks a 0.5–1.0-g sample is wetted with a small amount of water in a platinum dish. To the mixture are added 1–2 ml of 1:1 sulfuric acid and 8–10 ml of hydrofluoric acid. The dish is heated on a hot plate, stirring periodically with a platinum rod. The hydrofluoric acid treatment is repeated twice. The contents of the dish are evaporated to the appearance of sulfuric acid fumes. Hot water is added to the cooled residue and the solution is again evaporated to a small volume. Hydrogen peroxide is then added drop by drop to clarify the solution, the solution is diluted with hot water and boiled until the evolution of gas bubbles has ceased. It is filtered through filter paper into a 50-ml or 100-ml volumetric flask. If there is much undecomposed residue, the filter paper is ashed in a platinum crucible, the residue is fused with sodium carbonate or potassium pyrosulfate, dissolved and the solution is combined with the main filtrate.

A 10–15-ml aliquot of the resulting solution is neutralized with ammonia to the appearance of the precipitate of hydroxides, which is dissolved by adding 1–2 drops of 1:1 sulfuric acid; the solution is transferred to a separating funnel, the acid added to a concentration of 2–4% by volume, 2 ml α-benzoinoxime solution are added, the solution being stirred after the addition of each successive reagent; it is extracted after 5–10 minutes with three 5-ml portions of chloroform, shaking for 1 minute each time. The extracts are collected in a beaker and the chloroform is distilled off. To the dry residue is added the oxidizing acid mixture and the beaker is covered with a watch glass. The contents of the beaker are heated until sulfuric acid fumes are evolved, the mixture is left to stand for 15–20 minutes, the watch glass is removed, the solution is evaporated to dryness and the residue is ignited. Three ml of water and 10 drops of concentrated solution of ammonia are added to the colorless residue. The solution is boiled until the odor of ammonia disappears, 7 ml of concentrated hydrochloric acid are added and the determination is continued as described in the preparation of the scale of standards. The content of tungsten in the sample is found by visual matching of the color of the sample solution against the standards.

Note. In the analysis of polymetallic ores, tungsten (and molybdenum) is isolated as the complex with α-benzoinoxime by chloroform extraction from sulfuric acid solution.

Determination of small amounts of tungsten in molybdenum and molybdenum compounds [14]

Molybdenum(VI) is reduced by hydrazine in sulfuric acid medium and the resulting molybdenum(V) is bound by Complexone III; tungsten(VI) is then extracted with chloroform as the 8-hydroxyquinolate. The extraction of molybdenum(V) is insignificant under these conditions. After removal of chloroform and decomposition of the 8-hydroxyquinolate, tungsten is determined by the thiocyanate method, using titanium trichloride as the reducing agent. The effect of small amounts of molybdenum is thus eliminated.

REAGENTS

Sulfuric acid, dilute (1:1) and 8 *N* solutions.
Hydrochloric acid, $d = 1.19$ g/cm^3 and dilute (1:1) solution.
Nitric acid, $d = 1.40$ g/cm^3.
Ammonia, 25% solution.
Aqua regia.
Complexone III, 10% solution.
Hydrazine hydrochloride, 20% solution.

Mixture of butanol with chloroform, (1:2).

8-*Hydroxyquinoline*, 1% solution. One g of the preparation is dissolved in 100 ml of a 1:2 mixture of butanol with chloroform.

Potassium thiocyanate or *ammonium thiocyanate*, 50% solution.

Titanium trichloride, 1.5% solution, obtained by diluting a 15% solution.

Standard solution of tungsten. 1 ml corresponds to 0.01 mg of tungsten. Tungsten trioxide WO_3 (0.126 g) is dissolved in a few ml of 10% solution of sodium hydroxide and is diluted to one liter with water. One ml of this solution contains 0.1 mg of tungsten. A solution containing 0.01 mg of W/ml is obtained by dilution.

PREPARATION OF A SCALE OF STANDARDS

Into ten 25-ml volumetric flasks are introduced portions of standard tungstate solution corresponding to 10, 15, 20, 25, 30, 35, 40, 45 and 50 μg of W (or to 2, 4, 6, 8, 10, 12, 14, 16, 18 and 20 μg of W), 12 ml of 1:1 hydrochloric acid are added to each solution, the solutions are diluted with water to about 22 ml and all the reagents listed in the procedure are added.

PROCEDURE

Molybdenum or molybdenum concentrate (0.1–0.2 g) is dissolved in 10–15 ml of aqua regia and 5 ml of 1:1 sulfuric acid. The solution is evaporated to the appearance of sulfuric acid fumes, the residue is wetted with water and the evaporation is repeated to the evolution of sulfuric acid fumes. To the residue are added 5 ml of water, the mixture heated until the precipitate is dissolved, the solution neutralized with ammonia to a neutral or weakly acid reaction and transferred to a 100-ml volumetric flask.

An aliquot part of the solution, containing not more than 50 μg of W, is withdrawn, 5 ml of 8 *N* sulfuric acid, 10–20 ml (depending on the molybdenum content) of Complexone III solution are added, the solution is diluted with water to 30 ml, heated to boiling, 10 ml of hydrazine hydrochloride solution added and the mixture is boiled for 2–3 minutes. The solution is rapidly cooled, neutralized with ammonia solution until the precipitate disappears and is there transferred to a 100–150-ml separating funnel.

The pH of the solution is adjusted to not more than pH 2–3, 10 ml of 8-hydroxyquinoline solution are added and the mixture is shaken for 1–2 minutes. The layers are allowed to separate, and the lower (organic) layer is decanted into another separating funnel. The extraction is performed two more times, each time with 5 ml of butanol–chloroform mixture. The combined extracts are washed with water (5–10 ml), the organic layer is decanted into a 50-ml beaker, evaporated on a water bath or a sand bath and the residue is heated with a mixture of concentrated sulfuric and nitric acids until a dry white residue is obtained. Twelve ml of 1:1 hydrochloric acid are added to the beaker, the contents of the beaker are heated to dissolve the precipitate and washed with water into a 25-ml volumetric flask, bringing the volume up to approximately 22 ml. The solution is cooled, mixed, 1 ml of potassium thiocyanate (or ammonium thicoyanate) solution is added and titanium trichloride solution is introduced drop by drop with stirring until the brown coloration disappears and then 5 drops in excess. The solution is made up to the mark with water and its color is matched against the scale of standards after 30 minutes.

Note. In the analysis of ammonium molybdate, 0.2 g of the salt is dissolved in 5–10 ml of 1 *N* sulfuric acid, 20 ml of Complexone III solution are added and the determination is continued as above.

BIBLIOGRAPHY

1. GUSEV, S. I. and V. I. KUMOV.—*ZhAKh,* **3**, 373. 1948.
2. KUZNETSOV, V. I.—*Uspekhi Khimii,* **18**, 75. 1949.
3. ADAMOVICH, L. P. and T. F. NEVALENOVA.—*Uchenye Zapisky KhGU,* **4**, *Trudy Khimicheskogo Fakul'teta i Nauchno-Issledovatel'skogo Instituta Khimii KhGU,* **12**, 179. 1954.
4. POLUEKTOVA, E. N. and V. A. NAZARENKO.—*ZhAKh,* **19**. 1964.
5. BUSEV, A. I., V. G. TIPTSOVA, and A. D. KHLYSTOVA.—*Zav. Lab.,* **18**, 1414. 1962.
6. GOLUBTSOVA, R. B.—*ZhAKh,* **3**, 118. 1948; **6**, 357. 1951; *Zav. Lab.,* **18**, 412. 1952.
7. BUSEV, A. I. *Primenenie soedinenii dvukhvalentnogo khroma v analiticheskoi khimii (Utilization of Divalent Chromium Compounds in Analytical Chemistry),* p. 103.—Moscow, Izd. VINITI. 1960.
8. CHERNIKHOV, YU. A. and V. G. GORYUSHINA.—*Zav. Lab.,* **12**, 397. 1946.
9. LUKE, C. L.—*Anal. Chem.,* **33**, 1365. 1961.
10. YAKOVLEV, P. YA. and E. F. YAKOVLEVA. *Tekhnicheskii analiz v metallurgii (Technical Analysis in Metallurgy),* p. 94.—Metallurgizdat. 1963.
11. FINKEL'SHTEIN, D. N.—*Zav. Lab.,* **22**, 911. 1956.
12. GOTTSCHALK, G.—*Z. anal. Chem.,* **187**, 164. 1962.
13. STEPANOVA, N. A. and G. A. YAKUNINA.—*ZhAKh,* **17**, 858. 1962.
14. VINOGRADOV, A. V. and M. I. DRONOVA.—*ZhAKh,* **20**, 343. 1965.

Rhenium Re

Rhenium displays valencies of + 3, + 4, + 5, + 6 and + 7. In analytical chemistry the compounds of heptavalent rhenium (perrhenates) and of tetravalent rhenium are the most important. In many compounds formed by the reaction between potassium perrhenate and analytical reagents the valency of rhenium is not known with certainty.

Chemical properties of rhenium resemble those of molybdenum. The compounds of heptavalent and tetravalent rhenium also have some resemblance to the corresponding compounds of manganese.

The standard electrode potentials relative to standard hydrogen electrode in aqueous medium at 25°C are as follows:

$$Re^3 + 2H_2O \rightleftharpoons ReO_2 + 4H^+ + e \qquad + 0.157\ V$$

$$Re \rightleftharpoons Re^3 + 3e \qquad + 0.300\ V$$

$$Re^3 + 4H_2O \rightleftharpoons ReO_4^- + 8H^+ + 4e \qquad + 0.422\ V$$

$$ReO_2 + 2H_2O \rightleftharpoons ReO_4^- + 4H^+ + 3e \qquad + 0.510\ V$$

$$Re^{3+} + 4H_2O \rightleftharpoons ReO_4^{2-} + 8H^+ + 3e \qquad + 0.795\ V$$

Metallic rhenium is insoluble in hydrochloric acid, hydrofluoric acid and dilute sulfuric acid; it is readily dissolved by nitric acid with oxidation to $HReO_4$. When fused with alkalis, it forms the brown-green rhenate ions ReO_4^{2-} (in analogy to MnO_4^{2-}) which immediately disproportionate in aqueous solution to form the colorless perrhenate ions ReO_4^- and the black dioxide ReO_2:

$$3ReO_4^{2-} + 2H_2O = 2ReO_4^- + ReO_2 + 4OH^-.$$

Rhenium dioxide ReO_2 is stable and resembles MnO_2; unlike the latter, the former dissolves in concentrated hydrochloric acid with the formation of the chlororhenate(IV) complex, $H_2[ReCl_6]$.

Compounds of Re(V) slowly disproportionate in solution with formation of Re(VII) and Re(IV) compounds. Oxalate and tartrate complexes of rhenium(V) are known.

Perrhenates are much weaker oxidizing agents than permanganates.

The ions ReO_4^- undergo typical reduction reactions. The relatively weak reducing agents such as sulfur dioxide or hydrazine probably reduce Re(VII) to Re(VI), the solutions becoming yellow in the process. Stronger reducing agents, such as stannous chloride or zinc in HCl medium, result in a more far-reaching reduction of perrhenate ions; the solutions become brown and then green, after which a black turbidity $ReO_2 \cdot n\ H_2O$ appears or else the color of the solution disappears.

Liquid amalgams of zinc, cadmium, lead and bismuth reduce perrhenate ions in sulfuric acid or hydrochloric acid solutions to rhenium ions of different valencies [1].

Bismuth amalgam reduces heptavalent rhenium to the pentavalent state in 7–18 *N* sulfuric acid and to a valency state close to four in solutions containing more than 2 moles HCl/liter [1].

In the titrimetric determination [1, 2] heptavalent rhenium ions are reduced to Re(V) within 10 minutes by 2–3% bismuth amalgam in 18 *N* sulfuric acid in an atmosphere of CO_2, after which the amalgam is separated and Re(V) ions titrated against a solution of potassium dichromate in the presence of N-phenylanthranilic acid or against a solution of ceric sulfate. Satisfactory results are obtained in determining 4–11 mg of Re. Best results are obtained if ferric sulfate solution in excess is added to the sulfuric acid solution of rhenium(V) and the ferrous ions thus formed are titrated against ceric sulfate.

The perrhenate ions are quantitatively reduced [3–5] to Re(IV) by $CrSO_4$ solutions in 4–12 *N* sulfuric acid at 60–70°C or in 8 *N* HCl.

Solutions of chromium(II) salts are used in the potentiometric titration of heptavalent rhenium (p. 236).

Mercurous nitrate and perchlorate reduce [6] perrhenate ions to rhenium(IV) ions in hydrochloric acid medium. The reduction does not take place in 0.5–1 *N* sulfuric acid in the presence of thiocyanate ions (but only in 5 *N* sulfuric acid); molybdate ions, on the contrary, are reduced under these conditions. The resulting thiocyanate complex of molybdenum is extracted with ether. In this way molybdenum can be separated from rhenium.

Perrhenate ions are reduced by stannous chloride in the presence of potassium thiocyanate to form a red-colored compound, which is extractable with ether (analogy with molybdates). The valency of rhenium in this compound is not known with certainty. The formation of the thiocyanate complex of rhenium forms the principle of the different photometric methods of determination of this element. The thiocyanate method is used to determine 0.0001–0.01% Re in ores, molybdenites and concentrates [7] (pp. 237 and 238). The sample material is sintered with CaO in the presence of $KMnO_4$ or else is fused with sodium hydroxide and sodium peroxide. Rhenium is then separated from the interfering elements by extraction with methyl ethyl ketone from 5 *N* NaOH solution. The thiocyanate compound of rhenium is extracted with n-butanol. The sensitivity is 2 μg of Re in 20 ml of n-butanol.

Hydrogen sulfide precipitates the heptasulfide Re_2S_7 mixed with lower valency sulfides out of hydrochloric acid solutions (4 *N* HCl) of Re(VII) salts, especially in the heat. The precipitate is black, and is readily soluble in HNO_3 with formation of $HReO_4$. Rhenium sulfide is precipitated by hydrogen sulfide out of alkaline solutions only slowly and after prolonged standing. The precipitate of rhenium sulfide is sparingly soluble in solutions of ammonium polysulfide.

A pure precipitate of Re_2S_7 is formed if sodium or ammonium sulfide in excess is added to the solution of a perrhenate and is followed by a large excess (20–25 g) of ammonium sulfate, hydrochloric acid is added to a concentration of 6 *N* and the mixture is heated to boiling [8]. After washing and drying, the precipitate is weighed as Re_2S_7; this gravimetric determination gives satisfactory results.

Perrhenic acid $HReO_4$ is a strong acid. The ReO_4^- ions are colorless. Perrhenates of thallium(I), silver, cesium, rubidium and potassium are sparingly soluble in water. The solubility of $KReO_4$ is 0.95 g/liter. Sodium and ammonium perrhenates are quite soluble (similarly to perchlorates).

Perrhenate anions form compounds with large organic cations (tetraphenylarsonium, nitron, 2,2′,2″-tripyridyl methylene blue, Methyl Violet, Rhodamine 6G, Butylrhodamine B, etc.) which are sparingly soluble in water. The resulting compounds are soluble in different organic solvents and are used in the separation and in gravimetric and especially photometric determination of rhenium.

Tetraphenylarsonium chloride quantitatively precipitates perrhenate ions at pH 8–9. The precipitate is extractable with chloroform [9, 10]. Molybdate ions do not precipitate out under these conditions and remain behind in the aqueous phase. The chloroform extract is shaken with 6 *N* HCl, when ReO_4^- ions pass into the aqueous phase. Without separating the chloroform layer, the colored thiocyanate complex of rhenium is obtained by successively adding NaSCN and $SnCl_2 \cdot 2H_2O$. Isoamyl alcohol is then added, the thiocyanate complex of rhenium is extracted and the optical density of the extract is measured at 430 mμ. Satisfactory results are obtained when 0.01–0.1 mg of Re are determined in a solution containing 0.07 g of molybdenum. A determination takes about 2 hours.

It is possible to determine 0.1–10% Re in alloys containing molybdenum and tungsten by extracting tetraphenylarsonium perrhenate with chloroform from citrate-containing solution at pH 8–9 and measuring the optical density at 255 mμ

relative to chloroform [11]. Tin, permanganate, perchlorate, bromide and fluoride ions interfere.

Tetraphenylarsonium chloride is also used in the gravimetric determination of rhenium (p. 235).

Perrhenate ions are extracted with ethyl acetate in the presence of methylene blue; in the presence of Methyl Violet they are extracted with ethyl acetate from neutral or ammoniacal solutions or else with chloroform and ethyl acetate from sulfuric acid solutions [12].

In the extraction-photometric method the compound of perrhenate with Methyl Violet is extracted from neutral solutions with ethyl acetate [12], benzene or toluene [13] (p. 240).

Rhodamine 6G reacts with perrhenate ions in 0.5–1.5 *N* sulfuric acid to give a compound which is extractable with benzene. The extract emits an orange-colored fluorescence [14]. The fluorescence spectrum has the form of an amorphous band with a maximum at 550–560 mμ. If rhenium compounds are extracted from 10–25 ml of aqueous solution with 6 ml of benzene, 1–30 μg of Re can be determined. Major amounts of mercuric ions, which also form an extractable fluorescing compound, interfere. The fluorescence intensity of the rhenium compound decreases in the presence of auric gold, chromate, permanganate and tungstate ions. Antimony and uranium ions in the amount of 5–10 mg enhance the fluorescence of 5 μg Re by 30–40%. Molybdenum (25–30 mg) does not interfere. In the presence of halide ions gallium, indium and thallium react with Rhodamine 6G to form complexes which are extractable with benzene and capable of fluorescing. For this reason halide ions must not be present in the sample solution of rhenium.

Butylrhodamine B reacts with ions of rhenic acid in 5 *N* sulfuric acid or 3 *M* phosphoric acid to give a colored compound extractable with benzene [15]. The optical density of the resulting extract is measured and the amount of rhenium found. The method is suitable for the determination of rhenium in ores. The ore sample is sintered with magnesium oxide in the presence of potassium permanganate. The interfering vanadium, tungsten and mercury remain behind in the insoluble residue and do not affect the results. The sensitivity of the method is 0.0002% for a 2 g sample.

Dimethylglyoxime, α-furildioxime and other oximes react with $SnCl_2$-reduced perrhenate to form red-colored compounds, the composition of which has not been determined. The colored compound yielded by α-furildioxime is extractable with cyclohexanone. The oximes are used in the detection and photometric determination of rhenium.

Thiourea reacts with lower valency rhenium compounds to give a colored compound [16, 17]. The compound is formed if thiourea and then a solution of Cu, $2H_2O$ in HCl are added to rhenic acid dissolved in hydrochloric acid. Thiourea is employed in the photometric determination of rhenium (p. 242).

8-Mercaptoquinoline (thioxine) reacts with perrhenate ions to give differently colored compounds [18] at different acidities. The valency of rhenium and the compositions of the resulting compounds have not yet been established. The very stable compound of rhenium with the reagent which is formed in 5–12 *N* HCl makes it possible to determine small amounts of rhenium by the extraction-photometric method. This reagent is also used in the extraction-photometric determination of 0.05–7% Re in titanium alloys [19].

Diphenylcarbazide reacts with the ions of rhenic acid in 8 *N* HCl; the resulting colored compound is extractable with chloroform [20]. The optical density of the extract is measured at 540 mμ. The ions Cd, Ag, Bi, Zn, Mn, Al, Fe, Au(I), Cr(III), W, Ti, Co, Hi, Zr, Nb do not interfere. Cu, Se, V and Mo and oxidizing agents interfere. In some of the alloys rhenium may be determined by diphenylcarbazide without a preliminary separation. The sensitivity is 0.1 μg of Re/ml.

When 2,4-diphenylthiosemicarbazide is added to a warm (80°C) solution of potassium perrhenate in approximately 6 *N* HCl or 14–16 *N* H_2SO_4, an intense red coloration appears [21]. The colored compound is readily extracted with chloroform. The absorption maximum of the extract is at 510 mμ. The color intensities of the water solution and of the chloroform extract remain practically unchanged for several hours. The nature of the compound has not been established. Tetravalent rhenium compounds (K_2ReCl_6) do not give a color reaction with the reagent.

Less than 50 μg of Re in 10 ml of solution can be photometrically determined by 2,4-diphenylthiosemicarbazide. Under certain conditions molybdate ions in moderate amounts do not interfere.

Rhenium compounds display catalytic properties. Traces of rhenium catalyze the reduction of tellurates by stannous chloride.

$$Na_2TeO_4 + 3SnCl_2 + 8HCl = Te + 3SnCl_4 + 2NaCl + 4H_2O.$$

In practice, the reaction fails to take place in the absence of rhenium compounds in the solution. The black elementary tellurium separates out; it remains in solution in the presence of a protective colloid. The photometric method will determine 0.1–0.001 μg of Re in 1.5 ml of solution [22, 23] (p. 244). The optical density of the resulting colloidal solution depends on the concentration of the reactants, the duration of the reaction, and the presence of other electrolytes (if any). All these factors are taken into account in the analytical procedure. The sample and the standard solution must be prepared in the same manner. If fairly large amounts of molybdenum ions are present, they catalyze the reaction and must be removed. Rhenium is separated from the bulk of molybdenum by sintering the sample with calcium oxide and calcium nitrate at 700–800°C, and subsequent treatment with water; practically all the rhenium passes into solution, while molybdenum remains in the precipitate as $CaMoO_4$. The remaining molybdenum is separated from rhenium by chloroform extraction as molybdenyl 8-hydroxyquinolate. The method will determine rhenium contents of the order of 10^{-4}% in molybdenite [23].

Heptavalent rhenium can be determined polarographically (p. 245). The mechanism of the reduction of such compounds on mercury dropping electrode is complex [24]. Satisfactory results are obtained [25] when perrhenate ions are polarogrammed with 4 *N* HCl or 4 *N* perchloric acid in the presence of 0.005% gelatin or 5 *N* sulfuric acid as supporting electrolyte.

The determination of rhenium in most objects must be preceded by preliminary concentration of the element, e.g., by distillation from acid solutions, extraction with various organic solvents, or hydrogen sulfide precipitation in acid medium in the presence of copper, osmium, mercury and arsenic salts. Molybdenum(VI) ions are removed by coprecipitation with ferric hydroxide at pH 5–7.5.

Satisfactory results are obtained by concentrating rhenium in the form of sulfide with the aid of thioacetamide. When sulfuric acid solution of potassium perrhenate which contains thioacetamide is heated, a black precipitate separates out [26], with a composition close to Re_2S_7.

The precipitate is readily filtered. At concentrations above 10 mg/liter rhenium is quantitatively precipitated from solutions in 2–6 *N* sulfuric acid. Smaller amounts of rhenium are separated by coprecipitation with copper and mercury sulfides from 3 *N* sulfuric acid. In the presence of 100 mg of copper sulfate, 5 μg of Re can be precipitated from 500 ml of solution.

The methods for the separation of rhenium from molybdenum are very important. These may be based on the volatility of Re_2O_7 in a stream of HCl, on the precipitation of molybdenum by α-benzoinoxime and other organic reagents, on the extraction of molybdenum compounds (e.g., with ethylxanthate or thiocyanate ions) or on the extraction of rhenium compounds. A large number of methods for the chromatographic separation of rhenium from molybdenum have been proposed. All these methods are time-consuming, difficult to handle and not always reliable.

Sintering with calcium oxide is employed to separate the bulk of the molybdenum during the decomposition of molybdenum minerals. This is accompanied by the formation of the sparingly soluble calcium molybdate, which during the subsequent leaching is partly solubilized together with the perrhenate. Rhenium may be determined in the resulting solution directly by the thiocyanate method without separation of the small amounts of molybdenum present if the HCl concentration is 5 *N*, since the thiocyanate complex of molybdenum is unstable at high acid concentrations.

For a survey of the analytical methods concerning rhenium the reader is referred to [27].

Gravimetric determination of rhenium as tetraphenylarsonium perrhenate [28]

Perrhenate ions are quantitatively precipitated by tetraphenylarsonium chloride in a wide acidity range (between 5 *M* HCl and 6 *M* NH_4OH) Hg^{2+}, Bi^{3+}, Pb^{2+}

Ag^+, Sn(II), VO^{2+} and MnO_4^-, ClO_4^-, IO_4^-, Br^-, F^-, SCN^- interfere. Tungstates and vanadates do not interfere. Molybdates do not interfere if the precipitation is carried out from ammoniacal solutions or in the presence of 0.6 *M* tartaric acid.

REAGENTS

Tetraphenylarsonium chloride, 1% solution.
Sodium chloride, crystalline.

PROCEDURE

Sodium chloride is added to a concentration of 0.5 *N*, to 5–25 ml of the hot solution of perrhenate, which should contain between 0.5 and 100 mg Re; the reagent is then introduced with stirring until no more precipitate is formed. The solution is stirred, cooled and left to stand for several hours. The precipitate is filtered through a No. 3 sintered glass crucible, washed a few times with ice water, dried at 110°C and weighed as $(C_6H_5)_4AsReO_4$. The conversion factor to ReO_4 is 0.3952.

Potentiometric determination of rhenium in alloys [3–5]

The method is based on the reduction of perrhenate ions to rhenium(IV) by a standard solution of chromium(II) salt:

$$ReO_4^- + 4H^+ + 3e \rightarrow ReO_2 + 2H_2O.$$

Rhenium(VII) is titrated in hot (60–70°C) sulfuric acid solution (4 *N*) in the presence of small amounts of potassium iodide as a catalyst. The equivalence point is established with the aid of an indicator electrode (plantinum leaf); a saturated calomel half-cell serves as the comparison electrode.

Equal amounts of molybdenum do not interfere. The ions Fe^{3+}, Ti^{4+}, Cr^{3+}, V, Ni^{2+}, Co^{2+}, Nb and Cu^{2+}, which interfere with the determination, are readily removed by precipitation with ammonia or alkali.

The method determines more than 0.5% Re in alloys; the relative error is 1–3%.

REAGENTS

Chromium (II) sulfate. For preparation see p. 221.
Hydrochloric acid, $d = 1.19$ g/cm^3.
Nitric acid, $d = 1.4$ g/cm^3.
Sulfuric acid, $d = 1.84$ g/cm^3.
Ammonia solution, $d = 0.9$ g/cm^3.
Sodium hydroxide, 5% solution.
Potassium iodide, crystalline.

PROCEDURE

A 0.5-g sample of the alloy is dissolved with heating in a mixture of 5 ml hydrochloric and 5 ml nitric acids. The solution is evaporated on the water bath to a residual volume of 2–3 ml, a little water is added, then 2 ml of sulfuric acid and the solution is again evaporated, first on a water bath and then on a hot plate with gentle heating to the appearance of sulfuric acid fumes. The residue is dissolved with heating in 50 ml of water, the hydroxides are precipitated by the solution of ammonia (if the sample consists of an alloy of rhenium with iron, chromium, vanadium, titanium, niobium) or by sodium hydroxide (for alloys of rhenium with cobalt, nickel, copper). The filtrate is evaporated to a small volume, neutralized to litmus with sulfuric acid, sulfuric acid is added to a final concentration of 5–6 *N* (7 ml acid for each 50 ml of the final volume of the solution) and the solution is diluted with water to 50 ml in a volumetric flask. An aliquot (5, 10 or 20 ml, depending on the content of rhenium) is placed in a sealed potentiometric titration cell and a stream of nitrogen is passed through. The solution is heated to 80°C, 1 mg potassium iodide is introduced into the cell, and the contents of the cell are titrated against the solution of chromium sulfate, which is added first in 1-ml portions, and then in 0.1-ml portions as the equivalence point is approached. In the presence of molybdenum ions the titration is continued until a second potential jump is noted. The volume of the solution consumed in the titration of rhenium corresponds to the difference between the volumes of $CrSO_4$ between the second and first potential jumps. One ml of 0.100 *N* solution of $CrSO_4$ is equivalent to 6.210 mg of Re.

Photometric determination of rhenium by the thiocyanate method in molybdenum-containing and tungsten-containing alloys [9, 29, 30]

The method is based on the measurement of the optical density of the solutions of the yellow thiocyanate complex compound of rhenium, probably $K_3[ReO_2(SCN)_4]$. The heptavalent rhenium is reduced by stannous chloride in hydrochloric acid medium. The absorption maximum of the solutions is at 420 mμ.

Less than 2 mg of V, Ga, Ge, In, Ir, Co, Ni, Os, Ru, Pb, Tl, Cr, U, Ce do not interfere. Molybdenum(VI) and tungsten(VI) interfere. They are separated by a chromatographic method based on the sorption of phosphomolybdic and phosphotungstic heteropoly acid anions from 2 *M* H_3PO_4 on EDE-10 anion exchanger in the phosphate form; heptavalent rhenium is not sorbed under these conditions.

REAGENTS

Standard solution of $KReO_4$, 10 μg at Re/ml.
Potassium thiocyanate, 20% solution.

Stannous chloride, 35% solution: 350 g $SnCl_2 \cdot 2H_2O$ are dissolved in 200 ml 1:1 hydrochloric acid, the solution is cooled and diluted to 1 liter with freshly boiled water.
Hydrochloric acid, $d = 1.19$ g/cm^3.
Mixture of concentrated nitric and hydrochloric acids (1:3).
Ammonia, concentrated solution.
Phosphoric acid, concentrated and 2 *M* solution.
Hydrogen peroxide, 30% solution.

CONSTRUCTION OF CALIBRATION CURVE

Into five 25-ml volumetric flasks are placed 1, 2, 3, 4 and 5 ml of the standard rhenium solution; 10 ml of hydrochloric acid and 4 ml of 2 *M* phosphoric acid are added to each flask and the solutions are diluted to 20 ml with water. The solutions are cooled in a stream of tap water and 2 ml of potassium thiocyanate solution and 1 ml of stannous chloride solution are added to each flask, with thorough mixing after the addition of each reagent. The solutions are then made up to mark with water and their optical densities measured in a FEK-M electrophotocolorimeter with a blue filter, after which the calibration curve is constructed.

PROCEDURE

A 1-g sample of the alloy is dissolved in 30 ml of hydrochloric acid–nitric acid mixture and is evaporated twice on a water bath, once after the addition of 10–15 ml of hydrochloric acid and another time after the addition of water. The dry residue is dissolved in 10–15 ml ammonia solution, and concentrated phosphoric acid is added to an acid reaction. In order to ensure complete solubilization of tungsten, 30% hydrogen peroxide is added to the solution and the solution is boiled to decompose the excess peroxide. The solution is transferred to a 200-ml volumetric flask and diluted to mark with water.

Chromatographic separation is effected by passing 5 ml of the resulting solution through a 20 × 6 cm column filled with EDE-10 anion exchanger in the phosphate form (50 mesh). The column is then eluted with 2 *M* phosphoric acid. The eluate is passed through at the rate of 1 drop in 15 seconds and 40 ml are collected. The rhenium-containing eluates are placed in a 50-or 100-ml volumetric flask and the volume is made up to mark with 2 *M* phosphoric acid.

Five ml of the solution are taken for the photometric determination, which is performed as described for the construction of the calibration curve.

Determination of rhenium in molybdenite by the thiocyanate method [6, 31]

Unlike the color of the rhenium–thiocyanate complex, the coloration of the molybdenum–thiocyanate in 5 *M* hydrochloric acid is impermanent: within 10 minutes the light absorption by the molybdenum complex becomes so low that 4–175 μg of Re can be determined in the presence of less than 100 μg of Mo in 25 ml of solution. Ferric chloride is added to enhance the sensitivity of the determination.

REAGENTS

Potassium permanganate, crystalline.
Calcium oxide, freshly ignited.
Hydrochloric acid, $d = 1.19$ g/cm^3.
Ferric chloride, 6% solution.
Potassium thiocyanate, 20% solution.
Stannous chloride, $SnCl_2 \cdot 2H_2O$, 35% solution.

PROCEDURE

A 0.3–3-g sample of molybdenite is mixed with 0.2 g of $KMnO_4$ in a porcelain crucible. To the mixture are added 2–4 g of freshly ignited calcium oxide and the mixture is again stirred. The contents of the crucible are covered with a layer of 1.0–2.0 g calcium oxide and the crucible is placed in a cold muffle furnace, the temperature of which is slowly raised to 650–700°C. This temperature is maintained for 2 hours. The crucible is cooled and the cake is leached for 1 hour, stirring periodically, in 40–50 ml of water. The solution is filtered into a 100-ml volumetric flask. The precipitate on the filter is washed 3–4 times with water. The filtrate is made up to the mark with water.

Into a 25-ml volumetric flask are introduced 10 ml of the resulting solution, and 10 ml of hydrochloric acid, 0.5 ml of ferric chloride solution and 2 ml of potassium thiocyanate solution are added. The contents of the flask are cooled, 1 ml of stannous chloride solution is added, the solution diluted to mark with water and left to stand for 30 minutes; its optical density is measured, using a blue filter, and the amount of rhenium is found from a calibration curve established under identical conditions.

Note. When the cake is leached with water, 93–96% Re passes into solution, while the bulk of molybdenum remains behind in the precipitate as calcium molybdate. The filtrate contains between 0.6 and 1.0 mg of Mo/100 ml. The content of molybdenum in the aliquot is as low as 100 μg. However, it is not expedient to take an aliquot in the analysis of materials containing about 10^{-4}% rhenium. It is therefore necessary to reduce the molybdenum content by adding barium chloride solution to the filtrate. The solution contains 30 μg of residual molybdenum and less than 3 μg of residual tungsten in 100 ml. Some 5–6 ml of a 10% solution of $BaCl_2 \cdot 2H_2O$ are added to the filtrate and the solution is evaporated to a residual volume of 35–40 ml. The solution is filtered through a dense filter paper and washed with two small portions of hot water, the filtrate and the washings being collected together in a 50-ml volumetric flask. The solution is diluted to mark, 25 ml are withdrawn, evaporated to a residual volume of 7–8 ml, transferred to a 25-ml volumetric flask, and the determination is continued as above.

Photometric determination of rhenium by Methyl Violet [13]

The compound of Methyl Violet with perrhenate ions at pH 3.5-5 is extracted with toluene. The extract has absorption maxima at 540 and 600 mμ (Figure 14). The molar absorption coefficients are 28,000 and 39,500 respectively.

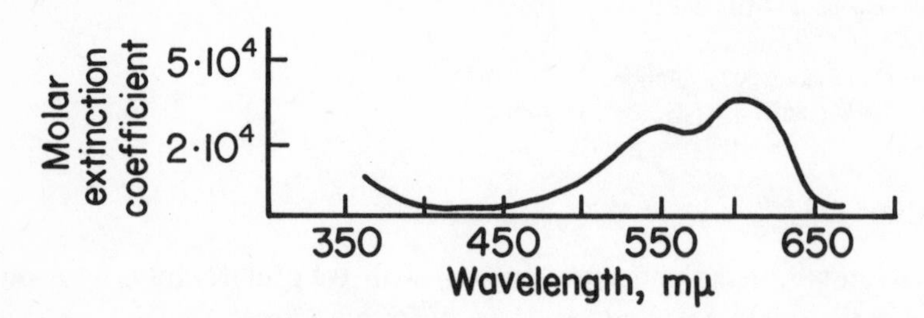

Figure 14

Light absorption by solutions of the perrhenate complex of Methyl Violet in toluene

Titanium, bismuth and antimony and also 40-fold amounts of molybdenum do not interfere. In the presence of major concentrations of Mo, W, Ta, Nb sodium tartrate is added to the solution. The following anions must be absent: NO_3^-, ClO_4^-, ClO_3^-, CH_3COO^-, SCN^-, I^-, Br^-, MnO_4^-, $PtCl_6^{2-}$, $RuCl_5^{2-}$, RuO_4^{2-}, $OsCl_6^{2-}$ $IrCl_6^{2-}$, $AuCl_4^-$, CrO_7^{2-}, CrO_4^{2-}.

The method is suitable for the analysis of specimens with a high content of Mo, W, Ta, Nb, Ti and Zr. The detectable minimum is 0.1 μg of Re/ml.

REAGENTS

Standard solution of $KReO_4$, 10 μg Re/ml.
Sodium tartrate, saturated solution.
Methyl Violet, 30 mg of reagent/ml.
Toluene, pure.
Calcium oxide, solid.
Sodium molybdate solution: 2.522 g $Na_2MoO_4 \cdot 2H_2O$ are dissolved in water and made up to 1 liter in a volumetric flask.
Hydrochloric acid, dilute (1:3) solution.
Hydrogen peroxide, 30% solution.

CONSTRUCTION OF CALIBRATION CURVE

Into five separating funnels are placed 1, 2, 3, 4 and 5 ml of standard perrhenate solution and 40 ml of water, 1.25 ml of sodium molybdate solution (in the case of molybdenum-containing sample) and 5 ml of sodium tartrate are added to each funnel. The solutions are diluted to 50 ml with water, acidified to pH 4–4.6 with hydrochloric acid, 14 drops of Methyl Violet solution are added to each funnel and the colored rhenium compound is extracted 4–5 times with 4–5 ml of toluene each time; the last portion of toluene should remain colorless. The extracts are drained into a 50-ml volumetric flask and the volume is made up to mark with toluene.

The optical densities of the extracts are determined in a FEK-M electrophotocolorimeter with a green filter and a calibration curve is plotted.

PROCEDURE

In the analysis of materials containing 0.01–0.001% Re (molybdenum glance, ferromolybdenum, molybdenum concentrate, etc.), 0.05–0.5 g of the finely ground sample is mixed in a metal crucible with a 3–5-fold amount of calcium oxide and the mixture is sintered in a muffle furnace at 700–750°C during 3.5–4 hours; it is then cooled and leached with water, adding 1 ml of 30% hydrogen peroxide and heating the contents of the beaker for 5–10 minutes on a plate to gentle boiling. The hot solution is filtered through a White Ribbon paper, washing the walls of the beaker and the filter paper several times with hot water. The filtrate is evaporated in a porcelain dish to 30 ml. After cooling, the solution is again filtered, collecting the filtrate in a separating funnel.

The subsequent procedure is as described for the construction of the calibration curve, except that the molybdate solution is not introduced. The content of rhenium is found from the calibration curve.

Photometric determination of rhenium by thiooxine [18, 19]

Thiooxine reacts with perrhenate ions in strongly acid medium (8–11 *N* HCl) to form a compound which is extractable with chloroform. The absorption maximum of the extract is at 438 mμ; the molar extinction coefficient is 8470. Beer's law is obeyed between 1 and 40 μg of Re in 1 ml of chloroform.

More than 0.7 mg of Au and Pt, more than 1 mg of Pd and Ru, more than 0.15 mg of Os, more than 2 mg of Rh and more than 10 mg of Ir interfere with the determination of rhenium in 9–10 *N* HCl. The following do not interfere; 100 mg of Mo, 10 mg of W, 17 mg of As(V), 3–5 mg of Se and Te, 4–5 mg of Cu, up to 5 mg of Sn(II), up to 50 mg of Sn(IV) and large amounts of Cr, Ni, Co, Zn, Cd, In, Sb, Bi, Mn, U, Pb, Tl(I), Ge, Th, Nb, Ta, Ga, Al, Zr and REE. The following anions do not interfere: Br^-, I^-, F^-, SO_4^{2-}, PO_4^{3-} and ClO_4^-; nitrates and arsenites should be absent.

REAGENTS

Standard solution of $KReO_4$, 20 μg Re/ml.
Thiooxine hydrochloride, 6% solution in concentrated HCl.
Chloroform.
Hydrochloric acid, $d = 1.19$ g/cm^3 and 9 *N* solution.

CONSTRUCTION OF CALIBRATION CURVE

Into five 100-ml conical flasks are introduced 1, 2, 3, 4 and 5 ml of standard solution of $KReO_4$, 10 ml of hydrochloric acid (d = 1.19 g/cm^3) and 5 ml of thiooxine hydrochloride solution are added to each flask. The flasks are heated for exactly 3 minutes in a boiling water bath, rapidly cooled in tap water and the solutions are transferred to separating funnels. Ten ml of chloroform are added to each funnel and the mixtures are shaken. The layers are allowed to separate, the chloroform extract is separated and filtered through a dry plug of glass wool into a 1-cm cell. The optical densities are determined in an electrophotocolorimeter with a green filter, and the calibration curve is plotted.

PROCEDURE

The procedure is exactly as described for the construction of the calibration curve, except that in the presence of ferric ions the extract is washed with 9 N HCl (22 ml, containing 2.5 ml of added reagent solution).

Differential-spectrophotometric determination of rhenium by thiourea [16, 32]

Thiourea reacts with pentavalent rhenium to form a soluble colored compound. The solutions obey Beer's law between 5 and 200 μg of Re/ml. The absorption maximum of the solution is at 390 mμ. The optimum conditions of complex formation are: concentration of hydrochloric acid 2.5–5 N, concentration of reducing agent (stannous chloride) 0.8–1%, concentration of thiourea 1%. The coloration develops within 20 minutes and is stable for more than 5 hours.

The determination of rhenium can be effected in the presence of 50 μg of molybdenum and 100 μg tungsten. Cd, Bi, Hg, Se, Te and As interfere. Flouride, citrate, tartrate and oxalate anions do not interfere if present in concentrations not above 0.1 N. The interference by Sb(III) is eliminated by adding fluorides.

In the differential spectrophotometric method the blank solution contains a high concentration of the element to be determined and all the reagents employed. The method is suitable for use with specimens having a high content of rhenium.

REAGENTS

Solution of $KReO_4$, 10 mg of Re/ml. Chemically pure metallic rhenium (1.00 g) is placed in a beaker and dissolved in 25 ml of 5% nitric acid with gentle heating. The solution is transferred to a 100-ml volumetric flask and made up to mark with water.

Thiourea, 5% solution.

Stannous chloride, 20% solution in concentrated HCl.

Hydrochloric acid, d = 1.19 g/cm^3 and dilute (1:1) solution.

Sodium hydroxide, solid.

Hydrofluoric acid, 38–40%.

Boric acid, crystalline.
Hydrogen peroxide, 3%.
Sodium nitrate, crystalline.

PROCEDURE

Alloys containing up to 30% rhenium are dissolved in a mixture of acids. The alloy (50–100 mg) is placed in a 150-ml beaker, 10 ml of 1:1 hydrochloric acid and 4 ml of hydrofluoric acid are added and the dissolution is effected in the cold. When the alloy has dissolved, 4 g of boric acid, 40–60 ml of water, 1–2 drops of hydrogen peroxide are added and the solution is boiled until perfectly colorless, periodically making up the volume to 50–80 ml with water. After cooling the solution is transferred to a 50-ml or 100-ml volumetric flask and made up to mark with water.

Alloys containing 30–40% rhenium are decomposed by alkaline fusion. The alloy (50–100 mg) is placed in a nickel or a silver crucible, as much sodium nitrate as will cover the end of a spatula is added, then 2 g sodium hydroxide, and the mixture is fused at 400–500°C in a muffle furnace until a clear melt is obtained. After cooling 20–40 ml of hot water are added to the cooled melt in a 150 ml beaker, the mixture is heated, 20–30 ml hydrochloric acid are added to clarify the solution, the solution is cooled, transferred to a 50-ml or 100-ml volumetric flask and diluted to mark with water.

An aliquot part of the solution is placed in a 50-ml volumetric flask, 10 ml of concentrated HCl, 10 ml of thiourea solution and 2 ml of stannous chloride solution are added. The solution is made up to the mark with water and mixed. The optical density is measured after 35–40 minutes relative to a blank solution containing all the reagents and 2 mg Re, using a SF-4 spectrophotometer at 390 mμ.

Calculation

The concentration of rhenium C_x is calculated from the formula:

$$C_x = C_0 + D_x F,$$

where D_x is the optical density of the sample solution, C_0 is the concentration of the blank solution, mg/50 ml, and F is a constant.

In order to determine the value of the constant F, optical densities of a series of solutions with concentrations varying between 2.0 and 2.7 mg of Re/50 ml of solution are determined. The value of the constant is given by

$$F = \Delta C/D, \qquad \Delta C = C_1 - C_0,$$

where C_0 is the concentration of the blank solution, mg/50 ml, C_1 is the concentration of a solution containing between 2.1 and 2.7 mg of Re in 50 ml, and D is the optical density.

Photometric determination of rhenium by the catalytic (kinetic) method [22, 33]

The reduction of sodium tellurate to elementary tellurium by stannous chloride is catalyzed in the presence of rhenium compounds. The liberated tellurium colors the solution black-brown in the presence of a protective colloid (gelatin).

Between 0.1 and 0.001 μg of Re may be determined in the presence of more than 100 μg of Cu, Hg, Ge, Sn, Pb, Sb, Bi, As, Ru, and Os. The interference by molybdenum and tungsten is eliminated by complexing them with tartaric acid. The method is suitable for the determination of rhenium in rocks after separation as sulfide.

REAGENTS

Standard solutions of $KReO_4$, 0.05 μg of Re/ml or 0.005 μg of Re/ml.

Reaction mixture. The mixture consists of 5 ml of 0.5% solution of sodium tellurate, 2 ml of tartaric acid solution (45 g in 100 ml, 1.5 ml of 0.5% gelatin solution and a solution of stannous chloride obtained by dissolving 10 g of metallic tin in 25 ml of concentrated HCl with heating. The mixture is prepared shortly before use.

PROCEDURE

One ml of the sample solution is placed in each one of three graduated tubes; 0.05 μg of Re as the standard solution is then added to the first tube, 0.005 μg of Re is added to the second tube, 1 ml of distilled water is added to the third tube and 1.0 ml of reaction mixture is then added to each of the three tubes. The solutions are mixed and left to stand for as long as necessary (between 1–2 and 16–18 hours) until a sufficiently intense coloration appears in the tube containing no added standard solution of rhenium. The optical density of this solution is measured in an electrophotocolorimeter with a blue filter simultaneously with the one of the two standards, the coloration of which is closest to that of the control solution.

Calculation

The content of rhenium is calculated according to the formula:

$$C_x = \frac{C_a \cdot D_x}{D_{x+a} - D_x},$$

where C_a is the amount of rhenium added to the sample solution, D_x is the optical density of the sample solution, and D_{x+a} is the optical density of the sample solution with added rhenium.

Polarographic determination of rhenium alloyed with molybdenum [24]

Rhenium is solubilized and at the same time separated from molybdenum by sintering the sample with calcium oxide. Rhenium in the form of perrhenate is dissolved while the bulk of the molybdenum is removed as the sparingly soluble calcium molybdate. In the resulting solution rhenium is determined polarographically between −0.3 and −0.4 volt with 5 *N* sulfuric acid as the supporting electrolyte.

REAGENTS

Calcium oxide, solid.
Ammonium persulfate, 10% solution.
Sulfuric acid, 5 *N* solution.

PROCEDURE

A 0.1-g alloy sample is mixed with three times the amount of calcium oxide, the mixture is sintered in a muffle furnace at 700–750°C, cooled and transferred to a beaker containing 50 ml of water. The crucible is taken out of the beaker and rinsed, 5 ml of ammoniun persulfate solution are added and the mixture is heated to boiling. The precipitate is filtered through a Buchner funnel, and washed 2–3 times with hot water. The filtrate is then transferred to a 100-ml volumetric flask and made up to mark with water. A 10-ml aliquot is withdrawn into a 100-ml volumetric flask and the volume is made up to mark with sulfuric acid. The solution is placed in an electrolyzer, a stream of nitrogen is passed through to expel the oxygen and a polarogram is taken between −0.3 and −0.4 volt relative to saturated calomel electrode.

BIBLIOGRAPHY

1. LAZAREV, A. I.—*ZhNKh*, **1**, 385. 1956.
2. SPITZY, H., R. J. MAGEE, and C. L. WILSON.—*Mikrochim. acta*, 354. 1957.
3. RYABCHIKOV, D. I., V. A. ZARINSKII, and I. I. NAZARENKO.—*ZhAKh*, **14**, 737. 1959.
4. RYABCHIKOV, D. I., V. A. ZARINSKII, and I. I. NAZARENKO.—*ZhAKh*, **15**, 752. 1960.
5. ZARINSKII, V. A. and V. A. FROLKINA.—*ZhAKh*, **17**, 75. 1962.
6. TARAYAN, V. M., E. N. OVSEPYAN, and L. G. MUSHEGYAN. *Renii (Rhenium).—Trudy Vsesoyuznogo Soveshchaniya po Probleme Reniya* 26–27 *May* 1958, p. 214.—Izd. AN SSSR. 1961.
7. STOLYAROVA, I. A. and G. V. NIKOLAEVA.—In: *Metody khimicheskogo analiza mineral'nogo syr'ya (Vsesoyuznyi Nauchno-Issledovatel'skii Institut Mineral'nogo Syr'ya)*, No. 7, 44. Gosgeoltekhizdat. 1963.
8. TAIMNIL, K. and G. B. S. SALARIA.—*Anal. chim. acta*, **12**, 519. 1955.
9. TRIBALAT, S.—*Ibid.*, **3**, 113. 1949; **5**, 115. 1951.
10. BEESTON, J. M. and J. R. LEWIS.—*Anal. Chem.*, **25**, 651. 1953.

11. ANDREW, T. R. and C. H. R. GENTRY.—*Analyst,* **82**, 372. 1957.
12. POLUEKTOV, N. S., L. I. KONONENKO, and R. S. LAUER.—*ZhAKh,* **13**, 396. 1958.
13. PILIPENKO, A. T. and V. A. OBOLONCHIK.—*Ukrainskii Khimicheskii Zhurnal,* **24**, 506. 1958; **25**, 359. 1959; **26**, 99. 1960.
14. IVANKOVA, A. I. and D. P. SCHERBOV.—*Zav. Lab.,* **29**, 787. 1963.
15. BLYUM, I. A. and T. K. DUSHINA.—*Zav. Lab.,* **28**, 903. 1962.
16. RYABCHIKOV, D. I. and A. I. LAZAREV.—*ZhAKh,* **10**, 228. 1955.
17. RYAZHENTSEVA, M. A. and YU. A. AFANAS'EVA.—*ZhAKh,* **16**, 108. 1961.
18. BANKOVSKII, YU. A., A. F. IEVIN'SH, and E. A. LUKSHA.—*ZhAKh,* **14**, 714. 1959.
19. EGOROVA, K. I. and A. N. GUREVICH.—*Zav. Lab.,* **29**, 789. 1963.
20. RYABCHIKOV, D. I. and L. V. BORISOVA.—*Zav. Lab.,* **29**, 785. 1963.
21. GEILMANN, W. and R. NEEB.—*Z. anal. Chem.,* **151**, 401. 1956.
22. POLUEKTOV, N. S.—*ZhPKh,* **14**. 695. 1941.
23. POLUEKTOV, N. S. and L. I. KONONENKO.—*Zav, Lab.,* **25**, 548. 1959.
24. GEYER, R.—*Z. anorg. Chem.,* **263**, 47. 1950.
25. AREF'EVA, T. V., A. A. POZDNYAKOVA, and R. G. PATS.—*Sbornik Nauchnykh Trudov Gosudarstuennogo Nauchno-Issledovatel'skogo Instituta Tsvetnykh Metallov,* No. 12, 94. Metallurgizdat. 1956.
26. YUDENICH, D. M. *Renii (Rhenium).—Trudy II Vsesoyuznogo Soveshchaniya po Probleme Reniya* 19–21 *November* 1962, p. 236. Izd. "Nauka" 1964.
27. RYABCHIKOV, D. I. and YU. B. GERLIT.—In: *Metody opredeleniya i analiza redkikh elementov,* p. 628. Izd. AN SSSR. 1961.
28. WILLARD, H. H. and G. M. SMITH.—*Ind. Eng. Chem., Anal. Ed.,* **11**, 305. 1939.
29. YAKOVLEV, P. YA. and E. F. YAKOVLEVA. *Tekhnicheskii analiz v metallurgii (Technical Analysis in Metallurgy),* p. 106.—Metallurgizdat. 1963.
30. RYABCHIKOV, D. I., V. A. ZARINSKII, and I. I. NAZARENKO.—*ZhNKh,* **6**, 641. 1961; *ZhAKh,* **19**, 229. 1964; IORDANOV, N. and M. PAVLOVA.—*ZhAKh,* **19**, 221. 1964.
31. TSYVINA, B. S. and N. K. DAVYDOVICH.—*Zav. Lab.,* **26**, 930. 1960.
32. MALYUTINA, T. A., B. M. DOBKINA, and YU.A. CHERNIKHOV.—*Zav. Lab.,* **26**, 259. 1960.
33. LAZAREV, A. I., V. I. LAZAREVA, and V. V. RODZAEVSKII.—*ZhAKh,* **18**, 202. 1963.

Gallium Ga

Gallium mainly displays a valency of +3. Gallium (II) compounds, such as $GaCl_2$, are unstable.

Standard electrode potentials relative to standard hydrogen electrode in aqueous solutions at 25°C are the following:

$Ga \rightleftharpoons Ga^{3+} + 3e$	−0.529 V
$Ga + 3H_2O \rightleftharpoons Ga(OH)_3 + 3H^+ + 3e$	−0.419 V
$Ga + 3H_2O \rightleftharpoons GaO_3^{3-} + 6H^+ + 3e$	+0.319 V

Gallium (III) ions are colorless and their properties resemble those of aluminum (III) and zinc (II) ions.

The hydroxide $Ga(OH)_3$ is colorless and somewhat amphoteric; $pL_{Ga(OH)_3} = 35$–36.5. The hydroxide $Ga(OH)_3$ is much more amphoteric than the indium hydroxide $In(OH)_3$ (Figure 15). As distinct from $Al(OH)_3$ $Ga(OH)_3$ is soluble in concentrated ammonia solution (displays a more acidic character); when the solution is boiled, the hydroxide again precipitates out. A practically pure hydroxide may be obtained by adding 0.1 *M* sodium hydroxide to a 0.01 *M* solution of gallium chloride at room temperature. A colloidal solution is formed first, and the coagulation becomes significant at about pH 6.7, when some 2.9 equivalents of the alkali have been added.

On the contrary, when 0.1 *M* sodium hydroxide is added to 0.005 *M* gallium sulfate solution, a precipitate of basic gallium sulfate is formed, which is fully coagulated at pH 4.5–5.0, after the addition of 2.5–2.6 equivalents of sodium hydroxide.

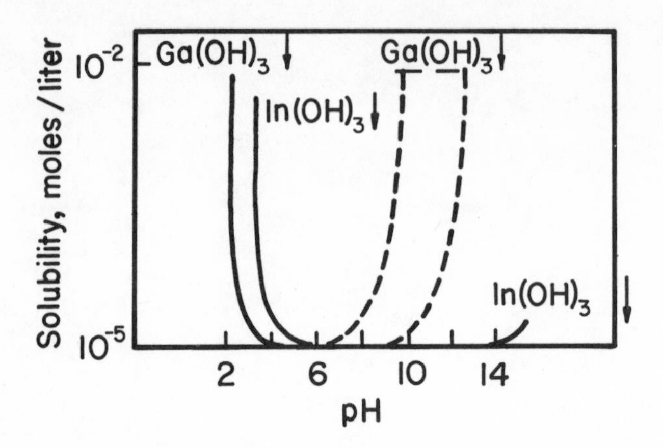

Figure 15

Solubility of gallium and indium hydroxide as a function of the pH of the solution

After four equivalents NaOH have been added (pH 9.7), the precipitates formed by adding sodium hydroxide to solutions of gallium chloride or sulfate, dissolve forming sodium gallate, probably $Na[GaO_2]$ or $Na[Ga\ (OH)_4]$.

Gallium chloride in the form of $HGaCl_4$ is extracted from solutions in 6 *N* HCl with ether. It can be separated from aluminum and from many other elements in this way. Fe(III), Au(III), Sb(III), Tl(III), As(III), Ge(IV), V(V), Mo(VI) are completely or partly extracted together with gallium using ether.

In the presence of water the sulfide Ga_2S_3 is readily converted to $Ga(OH)_3$.

Potassium ferrocyanide $K_4[Fe(CN)_6]$, precipitates the white $Ga_4[Fe(CN)_6]_3$ even from strongly acidic solutions (33% HCl) of gallium salts. Indium reacts in a similar manner, forming the precipitate of $KIn_5[Fe(CN)_6]_4$.

Azo compounds which contain two OH-groups ortho to the azo group react with gallium ions forming colored and frequently also fluorescent compounds. Such compounds also react with In^{3+}, Y^{3+}, Th^{4+}, Zn^{2+}, Ce^{3+}, Al^{3+}, Sc^{3+}, La^{3+}, Fe^{3+} and other ions [2]. 2,2′-Hydroxy-4-sulfonaphthaleneazonaphthalene causes solutions of gallium salts to emit an orange-colored fluorescence in the ultra violet. Lumogallion, which is used in the fluorimetric determination of gallium (p. 254), reacts in a similar manner. The reagent is employed in the analysis of semiconductor materials.

Gallion (0.01% aqueous solution) reacts with gallium ions in weakly acid medium to form a blue compound. The optimum pH is 2.4–3.4. The absorption maximum of the solution is at 600 mμ. At pH 3.2 the sensitivity of the determination is 0.2 μg Ga in 5 ml of solution. The selectivity of the reagent is poor (iron and other elements interfere). The reagent is employed in the photometric determination of gallium after separating the element from most accompanying elements [3, 4].

Sodium diethyldithiocarbamate precipitates gallium ions at pH 1.5–6; the white precipitate is soluble in ethyl acetate [5]. Gallium diethyldithiocarbamate is less stable than indium diethyldithiocarbamate. This fact is used in separating gallium from indium [6].

Gallium 8-hydroxyquinolate is extracted with chloroform at pH 2.6. The chloroform extract fluoresces in the ultraviolet. This is the principle of the highly sensitive methods of determination of gallium, e.g., in bauxites [7]. However, the method is not very selective ($pL_{GaOx_3} = 40.8$).

Morin produces the same fluorescence in solutions of gallium, indium and aluminum salts. The difference consists in the fact that the fluorescence in gallium solution is not quenched on the addition of fluoroborate ions, while in indium solutions it is not quenched on the addition of sodium fluoride. In this way gallium and indium ions can be detected in the presence of aluminum ions, the fluorescence of which is quenched by both these reagents [8]. Morin is also used as the indicator during the complexometric determination of gallium.

Azo derivatives of pyridine are also reagents for gallium. Thus, 1–(2-pyridylazo)-2-naphthol reacts with gallium ions at pH 3.2 to form a raspberry-red compound, in which the molar ration is 1:1. The absorption maximum is at 550 mμ; the compound is sparingly soluble in water, but readily soluble in methanol, ethanol, higher alcohols and carbon tetrachloride. The molar extinction coefficient [9] is 22,000 (in 50% methanol). The reagent is used as complexometric indicators during the titration of gallium ions.

7-(5-Sulfo-2-naphthylazo)-8-hydroxyquinoline-5-sulfonic acid, 7-(4-sulfo-1-naphthylazo)-8-hydroxyquinoline-5-sulfonic acid, 7-(4,8-disulfo-2-naphthylazo-8-hydroxyquinoline-5-sulfonic acid, 7-(5,7-disulfo-2-naphthylazo)-8-hydroxyquinoline-5-sulfonic acid and 7-(1-naphthylazo)-8-hydroxyquinoline-5-sulfonic acid react [10, 11] with gallium ions in weakly acid medium in the molar ratio of 1:1. The molar extinction coefficients of these gallium compounds are 21,000, 19,000, 16,000, 16,000 and 16,000 respectively. The respective equilibrium constants of the formation reaction are 90, 310, 840, 360, and 100.

All these azo derivatives of 8-hydroxyquinoline-5-sulfonic acid and sulfo derivatives of naphthylamine are satisfactory indicators in direct complexometric titration of gallium ions at pH 2.2 and above. The most selective indicator is 7-(4-sulfo-2-naphthylazo)-8-hydroxyquinoline-5-sulfonic acid, which will determine gallium in the presence of major amounts of alkaline earth metals, Zn, Cd, Mg and Al.

Cupferron precipitates gallium (III) from 10–20% sulfuric acid, and also Fe (III), Nb(V), Ta(V), Ti(IV), Zr(IV), Ge(IV), V(V) (p. 251). The reagent will separate gallium from aluminium and zinc. The precipitate is ignited to the oxide Ga_2O_3.

Basic triphenylmethane dyes and their antipyrin analogs react with $GaCl_4^-$ ions in strong hydrochloric acid medium forming intensely colored compounds which are extractable with a number of organic solvents; the excess dye may remain behind in the aqueous phase. This fact made it possible to develop highly sensitive photometric methods for the determination of gallium after its separation from the interfering elements.

Rhodamine B and Rhodamine C are the most frequently employed basic triphenylmethane dyes [12–17]. The more sensitive Butylrhodamine B is used in the photometric determination of gallium in different gallium-containing materials [18]. In 6–6.5 *N* HCl Butylrhodamine B forms a gallium compound which can be extracted with benzene or toluene; the extract is violet-pink and has an absorption maximum at 565 mμ, with a molar absorption coefficient of 90,000. The ions Au(III), Fe(III), Sb(V), Tl(III) are reduced by $TiCl_3$.

Chlorogallate ions $GaCl_4^-$ in strong hydrochloric acid solutions react with various antipyrin dyes [19]. bis-(p-Methylbenzylaminophenyl)-antipyrylcarbinol is sufficiently sensitive and sufficiently selective for use in spectrophotometric determination of gallium.

Gallium in 5.5–6 *N* HCl is quantitatively precipitated [20] by diantipyrylpropylmethane as the compound $C_{26}H_{30}O_2\ N_4 \cdot HGaCl_4$. The compound is a suitable gravimetric form. The determination may be terminated by weighing the precipitate of $C_{26}H_{30}O_2N_4 \cdot HGaCl_4$, or by complexometric titration of the gallium in the precipitate in the presence of 1-(2-pyridylazo)-2-naphthol. This determination may be performed in the presence of a large number of extraneous ions.

Diantipyrylmethane precipitates gallium ions from hydrochloric acid solutions, but the precipitation is not quantitative. The compound obtained was found to have the composition $C_{23}H_{24}O_2N_2 \cdot HGaCl_4$ [21].

Diantipyrylphenylmethane quantitatively precipitates gallium from solutions containing chlorides, bromides and iodides; the precipitates are readily soluble in chloroform. Diantipyrylphenylmethane may be employed in the separation of small amounts of gallium by extraction [21].

Tartrates and oxalates react with gallium ions at pH 9–11 to form the stable complex ions $[Ga(C_4H_4O_6)_3]$ and $[Ga(C_2O_4)_3]^{3-}$; with their aid gallium in ammoniacal solution can be separated from $[Zn(NH_3)_4]^{2+}$ on cation exchangers in the NH_4^+-form [22].

Gallium may be determined spectroscopically by excitation in the spark. The spectrum has two intense violet lines at 294.4 and 287.4 mμ. Volatile gallium salts color the flame violet and the spectrum displays a line at 417.2 mμ.

For a survey of the qualitative and quantitative analytical methods the reader is referred to [23, 24].

Gravimetric determination of gallium by cupferron [25]

Gallium is precipitated from sulfuric acid solution as cupferronate, ignited and weighed as Ga_2O_3. In this way gallium is separated from In, Fe, Al and REE.

REAGENTS

Cupferron, 6% aqueous solution, freshly prepared.
Sulfuric acid, dilute (1:1) and 5% solutions.

PROCEDURE

The cold sample solution in sulfuric acid (which should contain less than 0.1 g Ga) is neutralized with ammonia, acidified with 10 ml of 1:1 sulfuric acid, made up to 200 ml with cold water, and cupferron solution is added drop by drop, while stirring, until the precipitate has coagulated. The precipitate is then separated on Red Ribbon filter paper, and washed with 5% sulfuric acid containing some cupferron. In the presence of chlorides the precipitate must be washed until free from chlorides in order to avoid the loss of gallium as the volatile chloride during the ignition. The filter paper with the precipitate is cautiously ashed over a burner flame, the temperature is then increased to burn up the carbon and finally ignited at 1000–1300°C to constant weight. Gallium oxide ignited at 1000°C is hygroscopic, and for this reason it must be weighed as rapidly as possible in a crucible with a tightly fitting lid.

Separation and complexometric determination of gallium [20]

Gallium is precipitated by diantipyrylpropylmethane from 6 *N* HCl as $C_{26}H_{30}O_2N_4 \cdot \cdot HGaCl_4$. The precipitate is dissolved in water and gallium is titrated complexometrically in the presence of 1-(2-pyridylazo)-2-naphthol.

The method will separate gallium from Zn, Cd, In, Cu, Al, Ni, Mn, Mg, Co, Bi. Fe^{3+}, Tl^{3+}, Sb(V) interfere; the first two elements are reduced before precipitating gallium, while Sb(V) is masked by tartaric acid.

REAGENTS

Diantipyrylpropylmethane, 1% solution in 1:10 acetic acid.
Hydrochloric acid, $d = 1.19$ g/cm^3.
1-(2-*Pyridylazo*)-2-*naphthol,* 0.1% solution in methanol.
Washing solution, contains 0.05% diantipyrylpropylmethane in 6 *M* hydrochloric acid.
Complexone III, 0.05 *M* solution.
Ammonium acetate, crystalline.

PROCEDURE

The diantipyrylpropylmethane solution (1 ml of solution for each 1 mg of gallium) and an equal volume of hydrochloric acid are added to the sample soultion. After cooling the precipitate is separated on the filter and washed several times with the washing solution. The precipitate on the filter is dissolved in 50–70 ml of water and the filter paper is carefully washed with a few portions of water. Ammonium acetate is added to the solution to pH 4–5, the solution is heated to 70–80°C, a few drops of 1-(2-pyridylazo)-2-naphthol are added and the red solution is slowly titrated against Complexone III until the color turns yellow. The solution is left to stand for 5 minutes. When the red color returns, the titration is continued until the solution has turned yellow again. It is left to stand for another 5 minutes with occasional heating, and the titration is continued in this way until the pink color no longer returns.

Complexometric determination of gallium [11]

Complexone III reacts with gallium ions to form a stable complex (pK = 20.26 ± 01); gallium can thus be determined in acid medium (pH 2.0–3.0) in the presence of 7-(1-naphthylazo)-8-hydroxyquinoline-5-sulfonic acid or 7-(4-sulfo-1-naphthylazo)-8-hydroxyquinoline-5-sulfonic acid.

At the equivalence point there is a sharp color change from yellow to violet, which is typical of the indicator in acid medium.

Gallium may be determined in the presence of large amounts of Mg, Ca, Ba, Zn, Mn, Cd, and Al, but the last-named compound must be bound by adding NaF. The ions Fe^{3+}, In^{3+}, Tl^{3+}, Bi^{3+}, Cu^{2+}, Ni^{2+}, Co^{2+} interfere.

REAGENTS

7-(1-*Naphthylazo*)-8-*hydroxyquinoline*-5-*sulfonic acid*, (0.1% solution in dimethylformamide) or 7-(4-*sulfo* 1-*naphthylazo*)-8-*hydroxyquinoline*-5-*sulfonic acid* (0.1% aqueous solution).
Complexone III, 0.01 *M* solution.
Ammonia, 1 *N* solution.
Biphthalate buffer solution, pH 2.2–2.6.

PROCEDURE

The sample solution is neutralized with ammonia solution to pH 2.0–2.6, 15–20 ml biphthalate buffer solution are added, the mixture heated to 70–80°C, a few drops of indicator solution are added and the mixture is titrated against Complexone III solution until the color changes from yellow to violet.

Photometric determination of gallium by Rhodamine C [12–15]

Rhodamine C reacts with gallium ions in 6 *N* HCl to form Rhodamine hydrochloride which is extracted with benzene, benzene:ether (3:1) mixture or benzene: butyl acetate (4:1) mixture or else with a 3:1 mixture of chlorobenzene with carbon tetrachloride. The molar extinction coefficient of the benzene extract of the compound is 60,000 at 565 mμ; when the compound is extracted with 3:1 chlorobenzene: carbon tetrachloride mixture, it is 78,900 at 562 mμ.

Fe(III), Sb(V), As(V), Mo(VI), Tl(III), Te(IV), Se(IV), more than 0.3 mg of Al and more than 2 mg of Cu interfere. Aluminum and copper ions are separated from gallium ions by butyl acetate extraction from 6 *N* HCl. The effect of Fe, Sb, As, Mo, Tl, Te, and Se is eliminated by introducing titanium trichloride. The accuracy of the method is $\pm$ 5%.

REAGENTS

Standard solution of gallium, in 6 *N* HCl, 1 μg Ga/ml.
Hydrochloric acid, 6 *N* solution.
Mixture of benzene and butyl acetate, 4:1 mixture.
Titanium trichloride, 5% solution in hydrochloric acid.
Rhodamine C, 0.5% aqueous solution.

CONSTRUCTION OF CALIBRATION CURVE

In five 25–30-ml separating funnels are placed 1, 2, 3, 4 and 5 ml portions of standard gallium solution; the solutions are diluted to 5 ml with hydrochloric acid. Then 0.2 ml titanium trichloride solution, 10 ml of benzene–butyl acetate mixture and 0.5 ml of Rhodamine C solution are added to each funnel. The funnels are shaken for 2 minutes and the layers are allowed to separate. The lower layer is discarded and the colored extract is filtered through glass wool. The optical density of the extracts is measured relative to benzene and butyl acetate in a photocolorimeter with green filter, in 1-cm cells.

PROCEDURE

Gallium is determined in the sample solution in a similar manner. Its content is read off the calibration curve.

Isolation of gallium in the analysis of technical aluminum [26]

The best method for the isolation of gallium from the accompanying elements for subsequent photometric determination is by ether extraction from 6 *N* HCl or extraction with the less volatile butyl acetate at a phase ratio of 1:1. In butyl acetate extraction, the distribution coefficient is about 400. Gallium is reextracted with water and is determined by the Rhodamine method (p. 252).

REAGENTS

Butyl acetate.
Hydrochloric acid, 6 *N* solution.

Titanium trichloride, 5% solution.
Sodium chloride, crystalline.

PROCEDURE

Metallic aluminum (0.25 g) is dissolved in 20–30 ml hydrochloric acid with gentle heating. Titanium trichloride solution is added drop by drop, to the appearance of a violet color. The solution is left to stand for 2–3 minutes and transferred to a 100-ml separating funnel; the beaker is then rinsed with 3–5 ml of hydrochloric acid. To the solution is added an equal volume of butyl acetate and the solution is vigorously shaken for 1 minute. After layer separation the lower phase is discarded, and the extract is washed twice with hydrochloric acid, taking 2–3 ml each time. Gallium is reextracted with 10–15 ml water for 1 minute. The lower layer is transferred to a porcelain dish. Gallium is again reextracted with water, and the reextract is poured into the original dish. Sodium chloride (0.1 g) is added and the solution is evaporated to dryness in a water bath. The residue is dissolved in hydrochloric acid, the solution transferred to a 25-ml flask and the solution made up to the mark with hydrochloric acid.

Five ml of the solution are withdrawn, transferred to a 25–50 ml separating funnel and the determination is continued as shown on p. 252.

Determination of gallium by lumogallion [27]

Gallium reacts with lumogallion (2,2′,4′-trihydroxy-5-chloro-(1-azo-1′)-benzene-3-sulfonic acid) at pH 1.7–4.0 to give a complex compound which fluoresces orange-red. In the absence of gallium the reagent does not fluoresce under these conditions. The compound is extracted with isoamyl alcohol; the extract emits an intense fluorescence. The fluorescence intensity is a linear function of the gallium concentration up to 0.5 μg/5 ml of solution and between pH 1.7 and pH 4.0 both for the aqueous solution and for the extract. In aqueous solution the sensitivity is 0.01 μg of Ga/5 ml; after the complex has been extracted with isoamyl alcohol, it is 0.005 μg of Ga in 5 ml of extract.

Gallium can be determined between pH 2.0 and 3.0 (0.05 μg of Ga/5 ml) in the presence of thousandfold amounts of Li, Na, K, Cs, NH_4^+, Ag, Zn, Cd, Hg, Be, Mg, Ca, Sr, Ba, Tl, Ge, Pb, As, Sb(III), Sb(V), Bi, Cr, Se, Te, Mn, Ru, Co, Th, Nd, Ce, Pr, Ni, Cl, NO_3^-, SO_4^{2-}, phthalates, citrates and acetates. Less than 5 μg of Sn, Zr, Pd in 5 ml of solution and less than 0.5 μg of Cu, Fe, V, Mo in 5 ml of solution do not interfere.

The reaction of aluminum with the reagent produces a much less intense fluorescense of the solution than its reaction with gallium. If gallium and aluminum are present in equal amounts (pH 1.7–3.5) the presence of aluminum may be neglected. If tenfold amounts of aluminum are present, the pH must be 1.7–2.7, and in the presence of hundredfold amounts of aluminum the pH must be 1.7–2.2, in order to preserve the sensitivity of the reagent to gallium.

REAGENTS

Lumogallion, 0.01% solution in acetone.
Complexone III, 0.05 *M* solution.
Standard solution of gallium salt, 1 μg Ga/ml:0.100 g of metallic gallium is dissolved in 10 ml of 1:1 hydrochloric acid by heating on a water bath, while adding a few drops of perhydrol. The solution is evaporated to dryness with 1 ml of 10% solution of sodium chloride. The residue is dissolved in 2 *N* hydrochloric acid, the solution transferred to a 100-ml volumetric flask and diluted to mark with 2 *N* hydrochloric acid (solution A). Ten ml of solution A are placed in a 100-ml volumetric flask, 10 ml of sodium chloride solution are added and the solution is diluted to mark with water (solution B). On the day of the determination 1 ml of solution B is placed in a 100-ml volumetric flask and the solution is made up to mark with water (solution C). One ml of solution C contains 1.0 μg of Ga. Ten ml of solution C is transferred to a 100-ml volumetric flask and made up to mark with water (solution D). One ml of solution D contains 0.1 μg Ga.
Buffer solution, pH 3.0.

PROCEDURE

The sample solution is neutralized to pH 3.0–3.5 (universal indicator paper), transferred to a 25-ml volumetric flask and made up to mark with the buffer solution. Five ml of the resulting solution are placed in each of four test tubes. Solution C (0.1 ml) is introduced into one of the test tubes, 0.1 ml of solution D into the second test tube, 0.1 ml Complexone III solution into the third tube and 0.1 ml water into the fourth tube. To each tube is added 0.2 ml of reagent solution and the solutions are left to stand for 60–80 minutes. The fluorescence intensities are then measured in a 6-ml cell at 580 mμ.

Calculation

The amount of gallium x is calculated by the formula:

$$x = \frac{\Delta\mathrm{Ga}(I_x - I_m)}{I_{\mathrm{Ga}} - I_x},$$

where ΔGa is the amount of gallium added, μg, I_x is the relative intensity of fluorescence of the sample solution, I_m is the correction of the blank solution with Complexone III and I_{Ga} is the fluorescence intensity of the solution with added gallium.

Determination of gallium in germanium preparations [28]

Gallium (III) forms a complex compound with sulfonaphtholazoresorcinol (4-sulfo-2-naphthol (1-azo-1′)-2′,4′-dihydroxybenzene).

In aqueous ethanolic medium at pH 3 the reagent reacts with gallium (III) to give a bright orange-red luminescence, the intensity of which is proportional to the concentration of gallium. The fluorescence develops gradually to a maximum after about 30 min and then remains constant. The determinable minimum is 0.01 μg of Ga.

The quenching effect of ferric ions is eliminated by adding hydroxylamine. Aluminum reacts similarly to gallium and interferes if present in amounts exceeding 15 mg.

REAGENTS

Sulfonaphtholazoresorcinol, 0.01% ether solution.
Chloroacetate buffer solution (mixture of equal volumes of 0.5 *M* chloroacetic acid and ammonium chloroacetate).
Hydroxylamine hydrochloride, 10% solution.
Ethanol, 96%.
Hydrochloric acid, 7–8 *N* and 0.2 *N* solutions.
Sodium chloride, 10% solution.
Standard solution of gallium salt, 0.2 μg of Ga/ml.

PROCEDURE

The weighed sample of germanium dioxide or metallic germanium is solubilized. Germanium is removed as the tetrachloride by evaporation with 7–8 *N* HCl. To the residue is added 0.3 ml of sodium chloride solution and 2 ml of 7 *N* hydrochloric acid. The solution is evaporated to dryness and the residue ignited for 2 minutes over a Bunsen flame. After cooling the residue is dissolved in 0.2 ml of 0.2 *N* HCl; if necessary, the solution is filtered through a small paper and the filter paper is rinsed with 0.8 ml of water. To 1 ml of the resulting solution in a tube are added 0.2 ml chloroacetate buffer solution, 0.2 ml hydroxylamine hydrochloride, 2 ml alcohol and 0.2 ml sulfonaphtholazoresorcinol. After 30 min the fluorescence of the sample solution is matched against at scale of standards containing 0.05, 0.10, 0.15 and 0.20 μg of Ga.

Polarographic determination of gallium [29, 30]

Gallium (III) in ammoniacal solution, with ammonium sulfate or ammonium chloride as the supporting electrolyte, is reduced on a dropping mercury electrode over a range of −1.4 to −1.8 volt. Variations in the concentration of ammonia between 1.3 and 6.6 *N* do not affect the value of the limiting current. Oxygen is best expelled with hydrogen but if the concentration of gallium is more than 0.01 mg/ml, sodium sulfite may be used. The method will determine 0.01–0.3 mg of Ga/ml.

Small amounts of gallium may be separated from extraneous elements prior to its polarographic determination in polymetallic ore wastes, bauxites and other materials by ion exchange chromatography, using anion exchange and cation exchange in succession. When a solution of the sample in 6 *N* HCl is passed through a column of the strongly basic AV-17 anion exchanger in the Cl^- - form, the elements which do not form complex chloride anions (aluminum, chromium, nickel, cobalt, etc.) are not sorbed, whereas the remaining elements remain behind on the column. If the anion exchanger is now eluted with 4 *N* HCl, the bulk of copper and indium can be separated while all of the gallium remains behind on the anion exchanger, together with tin, lead, cadmium, zinc and other elements which form chloride complexes.

Gallium is extracted with 4 *N* solution of ammonia. The final purification of the ammoniacal solution from the extraneous elements takes place on SBS cation exchanger in the NH_4^+-form which retains all positively charged ammine complexes, while gallium emerges with the filtrate.

REAGENTS

Hydrochloric acid, $d = 1.19$ g/cm³ and 4 *N* solution.
Ammonia, 25% and 4 *N* solution.
AV-17 anion exchanger, in the Cl^- -form.
SBS cation exchanger, in the NH_4^+-form.
Supporting electrolyte, 2.6 *N* solution of ammonia and 1.8 *N* solution of ammonium sulfate or ammonium chloride.
Sodium sulfite, saturated solution.
Gelatin, 0.2% solution.

PROCEDURES

The sample solution is diluted with an equal volume of concentrated hydrochloric acid and passed through a column of AV-17 anion exchanger in the Cl^- -form; the anion exchanger is washed with 100 ml of 4 *N* hydrochloric acid and then with 10 ml of water. The contents of the column are transferred to a separating funnel and gallium is extracted with four 20-ml portions of 4 *N* ammonia solution. The resulting solution is passed through a column with SBS cation exchanger in the NH_4^+-form, the resin is washed with 20 ml of 4 *N* ammonia solution and the eluate is evaporated to a residual volume of 10 ml.

To the resulting solution, which should contain 0.25–2.5 mg Ga, are added 4 drops of sodium sulfite solution and 2 drops of gelatin solution and the volume of the solution is made up to 25 ml with the supporting electrolyte. The polarogram is taken choosing a suitable sensitivity of the galvanometer. The amount of gallium is found from the calibration curve.

BIBLIOGRAPHY

1. Porter, L. E. and P. E. Browning.—*J. Am. Chem. Soc.*, **41**, 1491. 1919.
2. Korenman, I. M., F. P. Sheyanova, and S. D. Kunshin.—*ZhAKh*, **15**, 36. 1960.
3. Lukin, A. M. and G. B. Zavarikhina.—*ZhAKh*, **13**, 66. 1958.
4. Karanovich, G. G., L. A. Ionova, and B. L. Podol'skaya.—*ZhAKh*, **13**, 439. 1958.
5. Delepine, M.—*Ann. chim.*, **6**, 636. 1951.
6. Busev, A. I., T. N. Zholondkovskaya, and Z. M. Kuznetsova.—*ZhAKh*, **15**, 49. 1960.
7. Lacroix, S.—*Anal. chim. acta*, **2**, 167. 1948.
8. Patrovsky, V.—*Chem. Listy*, **47**, 676. 1338. 1953; *Z. anal. Chem.*, **142**, 66. 1954; **143**, 50. 1954.
9. Busev, A. I. and L. M. Skrebkova.—*Izvestiya Sibirskogo Otdeleniya AN SSSR*, No. 7, 1957. 1962.
10. Busev, A. I., L. M. Skrebkova, and L. L. Talipova.—*ZhAKh*, **17**, 831. 1962.
11. Busev, A. I., L. L. Talipova, and L. M. Skrebkova.—*ZhAKh*, **17**, 180. 1962.
12. Onishi, H.—*Anal, Chem.*, **27**, 832. 1955.
13. Onishi, H. and E. B. Sandell.—*Anal. chim. acta*, **13**, 159. 1955.
14. Culkin, F. and I.P. Piley.—*Analyst*, **83**, 208. 1958.
15. Saltykova, V.S. and E.A. Fabrikova.—*ZhAKh*, **13**, 63. 1958.
16. Cherkashina, T.V.—*Materialy Soveshchaniya po Voprosam Proizvodstva i Primeneniya Indiya, Galliya i Talliya* (Gosudarstvennyi Nauchno-Issledovatel'skii i Proektnyi Institut Redkometalliches-koi Promyshlennosti), **2**, No. 13, part 1, 146. 1960.
17. Kuznetsova, V.K. and N.A. Tananaev.—*Nauchnye Doklady Vysshei Shkoly. Khimiya i Khimicheskaya Tekhnologiya* **2**, 289. 1959.
18. Skrebkova, L. M.—*ZhAKh*, **16**, 422. 1961.
19. Busev, A.I., L.M. Skrebkova, and V.P. Zhivopistsev.—*ZhAKh*, **17**, 685. 1962.
20. Busev, A.I. and V.G. Tiptsova.—*ZhAKh*, **15**, 698. 1960.
21. Busev, A.I. and L.M. Skrebkova.—*ZhAKh*, **17**, 56. 1962.
22. Alimarin, I.P. and E.P. Tsintsevich.—*Zav. Lab.*, **22**, 1276. 1956.
23. Busev, A.I. and L.M. Skrebkova.—In: *Metody opredeleniya i analiza redkikh elementov*, p. 201. Izd. AN SSSR. 1961.
24. Weibke, F.—In: *Handbuch der analytischen Chemie*, R. Fresenius and G. Jander, Editors, **III**, part 2, p. 45. Berlin, Springer-Verlag. 1944.
25. Schoeller, V.R. and A.R. Powell. *Analysis of Minerals and the Rarer Elements.*—Hafner Pub. Co. 1955.
26. Kuznetsova, V.K. and N.A. Tananaev.—*Nauchnye Doklady Vysshei Shkoly*, No. 4, 258. Metallurgiya. 1958.
27. Lukin, A.M. and E.A. Bozhevol'nov.—*ZhAKh*, **15**, 43. 1960.
28. Nazarenko, V.A., E.A. Biryuk, G.I. Byk, S.Ya. Vinkovetskaya, N.V. Lebedeva, T.A. Surichan, and M.B. Shustova.—*Trudy GIREDMET*, **2**, 77. Metallurgizdat. 1959.
29. Lysenko, V.N. and P.P. Tsyb.—*Zav. Lab.*, **23**, 794. 1957.
30. Nadezhina, L.S.—*ZhAKh*, **17**, 383. 1962.

Indium In

Indium displays a valency of $+3$. Indium compounds of lower valencies are unstable. The chlorides InCl and $InCl_2$ disproportionate in the presence of water according to the equations

$$3InCl \rightleftharpoons 2In + InCl_3,$$

$$3InCl_2 \rightleftharpoons In + 2InCl_3.$$

The standard electrode potentials in aqueous medium at 25°C have the following values (relative to the standard electrode):

$In^+ \rightleftharpoons In^{3+} + 2e$	-0.443 V
$In + Cl^- \rightleftharpoons InCl + e$	-0.34 V
$In + 3H_2O \rightleftharpoons In(OH)_3 + 3H^+ + 3e$	-0.172 V
$In \rightleftharpoons In^+ + e$	-0.139 V
$In + 2H_2O \rightleftharpoons InO_2^- + 4H^+ + 3e$	$+0.146$ V

Indium (III) ions are colorless.

The reactions of indium (III) ions resemble those of aluminum (III) and cadmium (II) ions.

When a solution of sodium hydroxide is added to a 0.005 *M* solution of $InCl_3$ at pH 3.7, the hydroxide separates out together with a small admixture of the basic salt; the precipitation is terminated at pH 6.70.

The basic sulfate begins to separate out from 0.0025 *M* solution of $In_2(SO_4)_3$ at pH 3.4; the precipitation terminates at pH 3.75. If more sodium hydroxide solution is added, the basic sulfate is fully converted to indium hydroxide [1].

The solubility product of indium hydroxide is $pL = 33.9$. Indium hydroxide is soluble in concentrated solutions of caustic alkalis and is therefore markedly amphoteric (see Figure 15, p. 248).

Hydrogen sulfide precipitates the yellow indium sulfide In_2S_3 from neutral or weakly acid solutions (pH 3–4); it is readily soluble in dilute hydrochloric acid. Indium sulfide In_2S_3 resembles CdS in its properties.

Ammonium sulfide precipitates the white indium hydroxide $In(OH)_3$ from solutions of indium salts (similarity to aluminum ions).

Potassium ferrocyanide precipitates indium ions as the white $KIn_5[Fe(CN)_6]_4$; the compound is more readily soluble in hydrochloric acid than is the corresponding gallium salt.

Indium forms oxalate, fluoride, and bromide complexes $In(C_2O_4)_2^-$, InF_6^{2-}, $InBr_4^-$. The compound $KIn(C_2O_4)_2$ is sparingly soluble. The solubility of InF_3 is 0.5 mole/liter, whereas the solubility of Na_3InF_6 is 0.35 mole/liter.

Indium in the form of $HInBr_4$ is extracted [2] with ether from 3–5 *N* HBr.

Indium is extracted with benzene from 2.5 *N* solutions of hydrobromic acid in the presence of a basic dye—Rhodamine C—to which a little acetone has been added [3]. Fe(II), Ca, Ga, Cr(III), Ti, Th, Bi, Be, Pb, As(III), U(IV), V(V), Ni, Co, Zr do not interfere.

Satisfactory results were obtained [3] in the extraction-photometric determination of 6–22 μg of In. Several variants of photometric and fluorimetric determinations of indium [4–6] are based on this principle. The fluorescence is excited by the light of a mercury lamp.

Sodium diethyldithiocarbamate quantitatively precipitates [7] indium ions at pH 1.5–9 as $[(C_2H_5)_2NC(S)S]_3In$. The precipitate is extracted with ethyl acetate, carbon tetrachloride and with some other solvents. The reagent is employed in the gravimetric determination of indium and its separation from gallium [8] (p. 261).

Alizarin, quinalizarin, stilbazo, morin and other organic reagents react with indium (III) similarly to aluminum (III). The compound of In^{3+} with alizarin is fluorescent [9].

1-(2-Pyridylazo)-2-naphthol and 1-(2-pyridylazo)-resorcinol react with indium in the molar ratio of 1:1 with formation of intensely colored compounds [10]. The solutions of the compound of indium with 1-(2-pyridylazo)-2-naphthol in a mixture of water with ethanol, water and dioxane or water and acetone are raspberry-red. The absorption maximum is at 550 mμ. The molar extinction coefficient is 18,700. The optimum pH value is 4.3–6.

The reagents are employed in the photometric determination of indium (p. 264) and as complexometric indicators [11–13] in the titration of indium ions at pH 2.3–2.5.

8-Hydroxyquinoline azo compounds, containing the hydroxyl group in ortho-position to the azo group, react with indium ions with formation of typical colored compounds [14, 15]. The best indicators in the direct complexometric titration of indium at pH 2.8–3.5 are 7-(1-naphthylazo)-8-hydroxyquinoline-5-sulfonic acid, 7-(4-sulfo-2-naphthylazo)-8-hydroxyquinoline-5-sulfonic acid, 7-(6-sulfo-2-naphthylazo)-8-hydroxyquinoline-5-sulfonic acid, 7-(4,8-disulfo-2-naphthylazo)-8-hydroxyquinoline-5-sulfonic acid, 7-(5,7-disulfo-2-naphthylazo)-8-hydroxyquinoline-5-sulfonic acid, 7-(4,8-disulfo-2-naphthylazo)-8-hydroxyquinoline-5-sulfonic acid and 7-(5,7-disulfo-2-naphthylazo)-8-hydroxyquinoline-5-sulfonic acid, which react with indium ions in the molar ratio of 1:1.

The aqueous solutions of the first-named reagent are raspberry-red (maximum absorption at 520 mμ) in acid medium, and are yellow-orange (absorption maximum at 500 mμ) in alkaline medium. The second-named reagent reacts with indium ions to form a yellow compound which is readily soluble in water; the absorption maximum of the solutions is at 420 mμ. The indium compound is formed at pH 2.8–6.0. The molar extinction coefficient [14] of this compound is 16,000.

Diphenylcarbazone in weakly acid medium (pH below 7) reacts with indium ions with the formation of a violet-colored compound. The absorption maximum of the solutions is approximately at 530 mμ. At high indium concentrations a precipitate of the same color separates out [16]. In the absence of indium the color of the solution remains grayish-yellow. At pH 5–6 the ions Mg, Ca, Al, Cr(III), Ti, Mn(II), U(VI), Cd, Pd, Bi, Sn(IV), Sb, As, Ag, Ge, Tl do not interfere if present in amounts not exceeding 50 μg of each element. The ions Cu(II), Zn, Ga, Fe(III) interfere (the effect of small amounts of ferric iron can be eliminated by adding thiourea at pH 5.6).

The reagent is used to determine more than $2 \cdot 10^{-5}\%$ of indium in germanium (2-g sample) after germanium has been distilled off as $GeCl_4$.

Indium compounds impart to the flame a characteristic violet-blue color. The spectrum contains an intense blue line at 451.1 mμ and a weaker violet line at 410.2 mμ. The spark spectrum contains numerous lines; the main lines at 325.6 and 303.9 mμ are easily confused with those of aluminum, chromium and manganese.

For a review of analytical methods for the determination of indium the reader is referred to [7].

Separation of indium from gallium as diethyldithiocarbamate [8]

The separation of indium from gallium is based on the precipitation (or ethyl acetate extraction) of indium diethyldithiocarbamate in the presence of excess sodium oxalate at pH 3–5. Gallium remains in solution, and is determined by precipitation with 8-hydroxyquinoline.

REAGENTS

Sodium diethyldlthiocarbamate, 1% aqueous solution.
8-*Hydroxyquinoline*, 3% solution in acetic acid.
Sodium oxalate, 2% solution.

PROCEDURES

The sodium oxalate solution is added to the sample which should contain 2–10 mg of indium and 2.5–30 mg of gallium (1–2 ml of solution for each mg of gallium). The solution is diluted to 40 ml, neutralized with ammonia to methyl orange and sodium diethyldithiocarbamate is added, while stirring, until the formation of the precipitate is complete and an excess of the reagent (3 times the amount of In) is present.

The precipitate is separated by filtering through a No. 3 sintered glass crucible, washed with water, dried to constant weight at 105°C and weighed.

The conversion factor to indium is 0.2050.

The filtrate and the washings are combined and gallium is determined by 8-hydroxyquinoline or by Complexone III (see p. 252).

Complexometric determination of indium [13, 14, 17–19]

Indium reacts with Complexone III in a molar ratio of 1:1 with the formation of a stable complex ($\log K_{st} = 24.9 \pm 0.1$). Indium (III) ions are titrated in acid medium (pH 2.3–2.5) against a standard solution of Complexone III in the presence of 1-(2-pyridylazo)-2-naphthol. At the end point there is a sharp color change from the red to pure yellow.

The ions Ga^{3+}, Tl^{3+}, Fe^{3+}, Bi^{3+}, Zr^{4+}, Th^{4+}, interfere, as they form stable complexonates in this pH range. Pb^{2+}, Cd^{2+}, Zn^{2+} and other cations which react with Complexone III in neutral and weakly alkaline medium, do not interfere.

1-(2-Pyridylazo)-resorcinol may also be used as an indicator; the color change is from red to yellow. Various azoxines may also be used, for example: 7-(1-naphthylazo)-8-hydroxyquinoline-5-sulfonic acid, 7-(4-sulfo-1-naphthylazo)-8-hydroxyquinoline-5-sulfonic acid, 7-(5-sulfo-2-naphthylazo)-8-hydroxyquinoline-5-sulfonic acid, 7-(5,7-disulfo-2-naphthylazo)-8-hydroxyquinoline-5-sulfonic acid (color change from yellow to raspberry-red).

REAGENTS

Complexone III, 0.01 *M* solution.
Ammonia, 1 *N* solution.
Acetic acid, $d = 1.05$ g/cm³.
1-(2-*Pyridylazo*)-2-*naphthol*, 0.1% solution in methanol.

PROCEDURE

The sample solution is neutralized with a solution of ammonia to the appearance of a white turbidity, 2 ml of acetic acid are added to dissolve the precipitate, 2 drops of indicator solution are introduced and the solution is titrated against a solution of Complexone III, which is added drop by drop near the end point, with vigorous stirring, until the red solution turns yellow. One ml of 0.0100 *M* solution of complexone is equivalent to 1.148 mg indium.

Photometric determination of small amounts of indium by Rhodamine C [3]

The compound of the bromide complex of indium with Rhodamine C is extracted from 2.5 *N* hydrobromic acid with a 5:1 mixture of benzene with acetone. The absorption maximum of the extract is at 530 mμ.

REAGENTS

Rhodamine C, 0.25% aqueous solution.
Benzene.
Acetone.
Standard solution of indium salt. The solution is prepared in 2.5 *N* hydrobromic acid; 1 ml of solution contains 20 μg of indium.
Hydrobromic acid, 2.5 *N* solution.
Titanium trichloride, 5% solution.

CONSTRUCTION OF CALIBRATION CURVE

Portions of 1, 2, 3, 4 and 5 ml of standard solution of indium salt are introduced into five separating funnels, hydrobromic acid is added to a total volume of 10 ml, 1 ml of Rhodamine C solution, 0.2 ml of titanium trichloride solution, 5 ml of benzene and 1 ml of acetone are added to each funnel. The mixtures are shaken for 1 minute and after layer separation the optical density of the extracts is measured in a FEK-M electrophotocolorimeter with a green filter against a blank solution. The calibration curve is then plotted.

PROCEDURE

If indium is present in the pure form in solution, the procedure is exactly as for the construction of the calibration curve.

Photometric determination of indium by 1-(2-pyridylazo)-resorcinol [10, 20]

1-(2-Pyridylazo)-resorcinol reacts with indium ions forming a red-colored complex with an absorption maximum at 500 mμ and molar extinction coefficient of 32,800 (Figure 16). At pH 6–8 the optical density of the colored solutions is constant. Beer's law is obeyed at concentrations between 5 and 120 μg In/25 ml of solution.

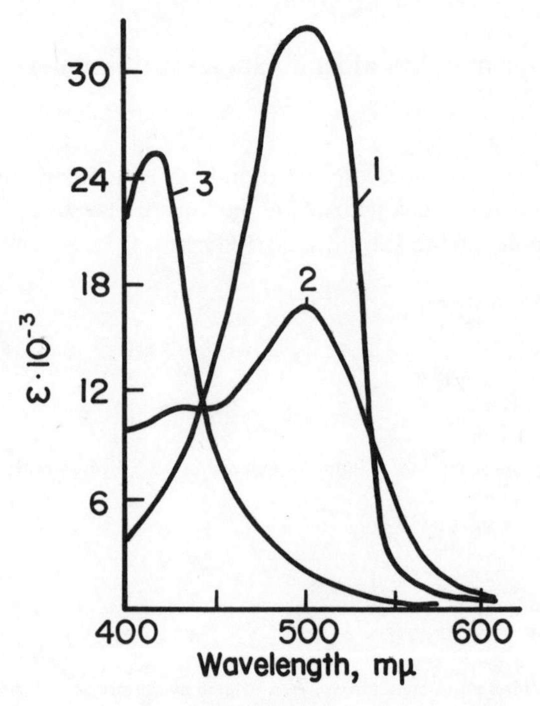

Figure 16.

Light absorption by solutions of indium compounds with 1-(2-pyridylazo)-resorcinol as a function of the wavelength:

1) at pH 6; 2) at pH 3.

For comparison, curve 3 shows the absorption of 1-(2-pyridylazo)-resorcinol solutions at pH 6.

The ions Na^+, K^+, Mg^{2+}, Ca^{2+}, Ba^{2+}, Cl^-, Br^-, I^-, SCN^-, CN^-, SO_4^{2-}, SO_3^{2-}, $S_2O_3^{2-}$, NO_3^-, ClO_3^-, BrO_3^-, IO_3^-, SiO_3^{2-}, BO_2^-, BF_4^-, WO_3^{2-}, CrO_4^{2-}, MoO_4^{2-}, CH_3COO^-, citrates, tartrates and sulfosalicylates do not interfere in 1000–10,000-fold amounts. The ions Fe^{3+}, Fe^{2+}, Co^{2+}, Ni^{2+}, V(V), Zr(IV), Bi^{3+}, Sn(II), F^-, NO_2^-, $P_2O_7^{4-}$, $C_2O_4^{2-}$ and ethylenediaminetetraacetate interfere. Small amounts of Zn^{2+}, Cd^{2+}, Al^{3+}, Mn^{2+}, Cr^{3+}, Pb^{2+}, Sn(IV), Cu^{2+} ions do not interfere if the determination is conducted at pH 3 and cupric ions are masked by thiosulfate.

The method is suitable for the determination of indium in indium concentrates and distillates when indium is separated from the interfering elements by extraction as iodide from 1.5 *N* hydriodic acid.

REAGENTS

Standard solution of indium salt, 20 μg In/ml.
1-(2-*Pyridylazo*)-*resorcinol*, $2 \cdot 10^{-3}$ *M* solution.
Acetate buffer solution, pH 3.
Hydrochloric acid, $d = 1.19$ g/cm^3 and dilute (1:1) solution.
Nitric acid, $d = 1.4$ g/cm^3.
Hydrazine hydrochloride, crystalline.
Potassium bromide, crystalline.
Ammonia, 25% solution.
Ammonium chloride, 1% solution.
Sulfuric acid, 1%, 3 *N* and 3.5 *N* solutions.
Sodium thiosulfate, crystalline.
Potassium iodide, crystalline.
Ether.
Hydriodic acid, 1.5 *N* solution.

CONSTRUCTION OF CALIBRATION CURVE

Into five 25-ml volumetric flasks are introduced 1, 2, 3, 4 and 5 ml of standard indium solution, 2 ml of 1-(2-pyridylazo)-resorcinol are added to each flask and the solutions are diluted to mark with acetate buffer solution. The optical densities of the solutions are measured at 500 mμ, relative to a comparison solution containing the same amount of the reagent in acetate buffer solution.

PROCEDURE

A 0.1–1-g sample of the material (depending on the content of indium) is decomposed by hydrochloric and nitric acids. The solution is evaporated almost to dryness, 5 ml of hydrochloric acid, 0.1 g of hydrazine hydrochloride and 0.1 g of potassium bromide are added and the solution is evaporated almost to dryness. The evaporation with hydrochloric acid is repeated twice more. The residue is dissolved in dilute hydrochloric acid, sulfuric acid solution is added, the solution is evaporated to white fumes, the residue of lead sulfate is filtered off and washed with 1% sulfuric acid.

Two ml of nitric acid are added to the filtrate and, when the solution boils, the hydroxides of trivalent metals are precipitated by adding ammonia solution. The precipitate is separated through a loose fibered filter paper and washed with a solution of ammonium chloride containing a little ammonia. The precipitate is dissolved in dilute 1:1 hydrochloric acid and the precipitation by ammonia is repeated. The hydroxide precipitate is dissolved in hot 3.5 *N* sulfuric acid, the solution is transferred to a 25-ml volumetric flask and made up to the mark with 3 *N* sulfuric acid.

A 10-ml aliquot of the solution is transferred to a separating funnel, 10 ml of water are added, then 0.1 g of sodium thiosulfate and 5 g of potassium iodide, the contents are mixed and indium is extracted with 20 ml of ether. The ether extraction is repeated. The combined extracts are washed with hydriodic acid, 5 ml of water are added and the ether is evaporated on a water bath. The solution is filtered into a 25-ml volumetric flask and the procedure is continued as described for the construction of the calibration curve.

The optical density of the sample thus found serves to determine the indium content by using the calibration curve.

Photometric determination of indium in ores by 8-hydroxyquinoline [21]

The method is based on the extraction of indium 8-hydroxyquinolate with carbon tetrachloride at pH 8.0 ± 0.5. Indium is separated from the bulk of the interfering elements as $HInBr_4$ by ether extraction, while the interfering elements are masked with citric acid and cyanide. Iron is preliminarily reduced by ascorbic acid. In weakly alkaline solution copper, nickel, zinc, cadmium and cobalt are masked by the cyanide.

REAGENTS

Standard solution of indium salt, 15 μg of In/ml.
Ammonium chloride, crystalline.
Hydrochloric acid, 3 *N*, 6 *N* and concentrated solutions.
Hydrobromic acid, concentrated, containing 1% bromine.
Sodium bromide, solution; 100 g of $NaBr \cdot 2H_2O$ are dissolved in 50 ml water.
Sodium sulfite, anhydrous.
Hydrogen peroxide, 30% solution.
Methyl orange, 0.02% solution.
Ether, saturated with sodium bromide. Fifty-six ml of 1:1 sulfuric acid are added to 140 ml solution of sodium bromide (see above). The solution is cooled, 20 g of sodium sulfite are added and the mixture is stirred until dissolved. The solution is transferred to a 1000-ml separating funnel, 600 ml ether are added, the mixture is shaken for 2–3 minutes and the aqueous phase is separated.
Ether, saturated with hydrochloric acid. Equal volumes of ether and 6 *N* hydrochloric acid are shaken in a separating funnel for 2–3 minutes and the aqueous phase is discarded.
Solution of sodium bromide in sulfuric acid. The aqueous phase obtained in the preparation of ether saturated with sodium bromide is employed; the separated crystals of sodium sulfate are filtered off.
Ascorbic acid, 5% aqueous solution.
Sodium citrate, 20% aqueous solution.
Potassium cyanide, 2% solution. To 20 g of the reagent are added 150 ml of ammonia and the solution is diluted to 1000 ml with water.

8-*Hydroxyquinoline*, 0.15% solution in carbon tetrachloride.
Nitric acid, $d = 1.4$ g/cm^3.
Sulfuric acid, 18 *N*.
Ammonia, 6 *N* solution.

CONSTRUCTION OF CALIBRATION CURVE

Into five 50-ml separating funnels are introduced 2, 4, 6, 8 and 10 ml of standard solution of indium salt; 3 ml of 3 *N* hydrochloric acid, 5 g of ammonium chloride, 2 ml of 5% solution of ascorbic acid, 2 ml of 20% solution of sodium citrate and 5 ml of potassium cyanide solution are added to each funnel, the solutions are diluted to 25 ml with water and thoroughly mixed. Five ml of 8-hydroxyquinoline solution are added to each funnel, the mixtures are shaken for 1–2 minutes and left to stand for 30 seconds. The optical densities of the extracts are measured in an electrophotocolorimeter with a blue filter, after which the calibration curve is plotted.

PROCEDURE

A 0.5–1.0-g sample of ore is decomposed in a 250-ml beaker by heating with 15 ml of concentrated hydrochloric acid. When the bulk of the sample has decomposed, 5 ml of nitric acid are added and the solution is evaporated to dryness. To the residue are added 5–10 ml of concentrated hydrochloric acid and 50 ml of hot water. The mixture is boiled until the salts have dissolved, and the solution is filtered through a White Ribbon filter paper, collecting the filtrate in a 250-ml beaker. The residue on the paper is washed with 2% hot hydrochloric acid and then with water. If the sample consists of a copper or a lead–zinc ore, the residue is discarded. In the analysis of tin ores, the residue is transferred to a nickel crucible, ashed, wetted with water, 3–5 ml of hydrofluoric acid are added and the mixture is evaporated to dryness. The precipitate is transferred with the aid of a little water into a small beaker and treated 2–3 times with a little hydrobromic acid containing bromine, evaporating the solution to dryness each time (to drive off tin tetrabromide). The residue is dissolved in hydrochloric acid and combined with the main filtrate.

Ten ml of 18 *N* sulfuric acid are added to the filtrate and the mixture is evaporated until abundant white fumes have been evolving for 5–10 minutes. The residue is cooled, the beaker walls are washed with a little water and the solution is again evaporated to the evolution of sulfuric acid fumes. It is then cooled, the beaker walls are washed down with a little water, 18 ml of hot solution of sodium bromide are added and the solution is evaporated to an approximate residual volume of 20 ml.

Four grams of sodium sulfite are added to the cool solution. It is stirred and left to stand for 30 minutes. It is then transferred to a 125-ml separating funnel, washing the walls of the beaker with the minimum amount of water (not more than 5–10 ml). Twenty-five ml of ether saturated with sodium bromide are added to the solution, and the mixture is shaken for 1–2 minutes, after which the ether layer is allowed to separate. The aqueous layer is decanted into another separating funnel and is then extracted again with 20 ml of ether saturated with sodium bromide. The aqueous solution is discarded, the ether extracts are combined and washed 1–2 times (with a 10-ml portion each time) of the solution of sodium bromide in sulfuric acid. If the washing is conducted twice, both portions of the washing solution are successively extracted with the same portion of ether saturated with sodium bromide (10 ml) and combined with the main ether extract. To the extract are added 25 ml of 6 *N* hydrochloric acid and a few drops of 30% hydrogen peroxide, the mixture is cooled and shaken for 1–2 minutes. The acid solution is transferred to another separating funnel, and the ether extract is again shaken with 10 ml of 6 *N* hydrochloric acid.

Each acid solution is shaken with the same 10-ml portion of ether saturated with hydrochloric acid. If the hydrochloric acid solutions have been combined, they are washed twice, each time with a 10-ml portion of ether saturated with 6 *N* hydrochloric acid.

The ethereal solutions obtained by washing the hydrochloric acid solutions, are successively shaken with the same portion of 6 *N* hydrochloric acid (10 ml). All hydrochloric acid solutions are combined, while the ether solutions are discarded.

The combined hydrochloric acid solutions are evaporated to a small residual volume, a few drops of hydrogen peroxide are added and the solution is evaporated almost to dryness. The residue is dissolved in water. The resulting solution is neutralized to Methyl Orange with 6 *N* ammonia solution until the indicator becomes yellow. Hydrochloric acid (3 *N*) is then added until the solution becomes orange, then 3 ml more are added in excess. Two ml of ascorbic acid and 2 ml of sodium citrate solution are then added. The mixture is transferred to a 50-ml separating funnel, washing the walls of the beaker with a minimum amount of water, 5 g of ammonium chloride and 5 ml of potassium cyanide solution are added and the solution is diluted to 25 ml with water. Five ml of 8-hydroxyquinoline solution are added, the mixture is shaken for 1–2 minutes, left to stand in a dark place for 30 minutes and all or part of the extract is transferred to a cell, the optical density measured in an electrophotocolorimeter and the indium content is read off a calibration curve.

Fluorimetric determination of indium in ores by Rhodamine 6G [22]

The method is based on the reaction of the bromide complexes of indium with Rhodamine 6G. The resulting compound is extracted with benzene from 15 *N* sulfuric acid and the concentration of indium is found from the fluorescence intensity of the extract. The interfering Fe(III), Cu^{2+}, Sn(IV), Sb(III), Tl(III), Au(III) and Hg^{2+} ions are removed.

REAGENTS

Rhodamine 6G, 0.25% solution.
Ammonia, 25% solution.
Ammonium sulfate, 1% solution.
Iron, reduced by hydrogen.
Sulfuric acid, $d = 1.84$ g/cm^3 and 15 *N* solution.
Titanium trichloride, 5% solution.
Potassium bromide, 40% solution.
Standard solution of indium salt, 1 μg In/ml.
Sulfuric acid, 15 *N* solution and dilute (1:1 and 1:2) solutions.
Benzene.

CONSTRUCTION OF CALIBRATION CURVE

Into ten 10-ml tubes are introduced portions of the standard solution corresponding to 0.1, 0.2, 0.3, 0.4, 0.5, 0.6, 0.7, 0.8, 0.9, and 1.0 μg of In, the solutions are diluted to a final volume of 10 ml with 15 *N* sulfuric acid, 100 mg of iron are introduced into each tube and the mixtures are left to stand for 30 minutes. A 5-ml aliquot is withdrawn with a pipet and is transferred into a 25-ml separating funnel through a cotton plug; 0.1 ml of Rhodamine 6G solution, 6 ml of benzene and 0.5 ml of potassium bromide solution are then added to each funnel. Immediately after the addition of the bromide the funnel is shaken for 30 seconds and the layers are allowed to separate. The extract is transferred to a dry tube, the fluorescence intensity is measured in a fluorimeter and the calibration curve is plotted.

PROCEDURE

A 0.3-g sample of ore is dissolved with heating on a sand bath in 25–30 ml of aqua regia. When the sample has decomposed, the solution is evaporated to a small residual volume (5 ml) and the concentrate is diluted to 25–30 ml with water. The hydroxides are precipitated by adding ammonia solution in slight excess. The precipitate is separated on a dense filter, washed 2–3 times with solution of ammonium sulfate and dissolved in hot 1:2 sulfuric acid, washing the filter 2–3 times with water and collecting the filtrate and the washings into the beaker which was used for the precipitation. The hydroxides are then reprecipitated. The precipitate is filtered off and washed 3–4 times with a solution of ammonium sulfate and 1–2 times with water. The precipitate is dissolved on the filter in 15 *N* sulfuric acid, and the filtrate collected in a tube and diluted to 10 ml with 15 *N* sulfuric acid. After cooling, 0.15–0.20 g of iron are added and the mixture is left to stand for 1 hour, with occasional stirring. The solution is transferred to another tube, 100 mg of iron are added and the solution is again left to stand for 30 minutes. A 5-ml aliquot is withdrawn and treated exactly as described in the preparation of the scale of standards.

Isolation of indium by extraction [4]

The method is based on the extraction of indium with butyl acetate from 5 N hydrobromic acid.

REAGENTS

Hydrochloric acid, $d = 1.19$ g/cm^3 and 6 N solution.
Hydrobromic acid, concentrated, 5 N and 2.5 N solutions.
Nitric acid, $d = 1.4$ g/cm^3.
Sulfuric acid, $d = 1.84$ g/cm^3.
Hydrogen peroxide, 30% solution.
Butyl acetate.

PROCEDURE

A 0.1-g sample of the ore is treated, at first in the cold, with 10 ml of nitric and 5 ml of hydrochloric acid. The solution is evaporated to dryness, 3–4 ml of concentrated hydrobromic acid are added to the residue and the solution is again evaporated to dryness. The residue is dissolved in 5 ml of 5 N hydrobromic acid, the solution is transferred to a separating funnel, washing the walls of the beaker with 5 N HBr. Ten ml of butyl acetate are added to the solution and the solution is extracted for 1 minute. After phase separation, the extract is washed twice, each time with 2–3 ml of 5 N hydrobromic acid, shaking 10–15 seconds each time. Indium is reextracted with 2 portions of 6 N hydrochloric acid (10 ml each time) with the addition of 1–2 drops of hydrogen peroxide. The reextract is evaporated to dryness, sulfuric acid added to the residue and the solution evaporated to the appearance of white fumes. The dry residue is treated with 2.5 N hydrobromic acid.

Indium is then determined photometrically, using Rhodamine C.

Separation of indium from tin by extraction [23]

Indium and tin are extracted together from sulfuric acid solution with a mixture of alkylphosphoric acids in octane with subsequent reextraction of indium with hydrobromic acid. Indium can be separated in this way from 1000-fold amounts of tin. The determination of indium is terminated photometrically by Rhodamine 6G. The method is suitable for the determination of 0.01–0.0001% indium in sulfide and oxide concentrates containing major amounts of tin (15–50%).

REAGENTS

Ethylhexylphosphoric acid. To a 400–500-ml beaker are added 150 ml of 2-ethylhexyl alcohol and 150 ml of octane. Phosphorus pentoxide (70–75 g) is added with stirring, not allowing the temperature to rise above 70–75°C. The mixture is left to stand for 1½–2 hours. The extractant is then diluted with octane or isooctane in the ratio of 1:1.5. The diluted extractant is washed free from phosphate ions with 1 *N* sulfuric acid at a phase volume ratio of 3:1.
Ammonium fluoride, 0.5 *N* solution.
Hydrobromic acid, dilute (5:1) solution.
Sulfuric acid, 1 *N* and 3–4 *N* solutions.
Ascorbic acid, crystalline.
Ammonium sulfate, 1–2% solution.
Sodium peroxide, crystalline.
Magnesium oxide, crystalline.

PROCEDURE

A 0.1–2-g sample of the material is fused in a corundum crucible with 6–7 times the amount of sodium peroxide in the presence of 0.1–0.2 g of magnesium oxide. The crucible is placed in a cold muffle furnace, the temperature is gradually raised to 650–700°C and the sample is fused for 6–8 minutes until a uniform melt is obtained. The cooled crucible is carefully transferred to a beaker containing 60–70 ml of water and the beaker is covered with a watch glass. After the vigorous reaction has ceased, the solution is filtered and the precipitate on the filter paper is washed with a solution of ammonium sulfate containing free ammonia. The precipitate on the filter is washed with 20–30 ml of 3–4 *N* sulfuric acid into the beaker which was used for the precipitation, the mixture is heated until the precipitate has fully dissolved, the solution is transferred into a 100-ml separating funnel and 0.1–0.2 g ascorbic acid is added. To 30–35 ml solution is added an approximately equal volume of the extractant and the mixture is shaken for 1–1½ minutes. If a stable emulsion is formed in the aqueous phase within 5–10 minutes, 8–10 ml of ammonium fluoride solution are added. After phase separation the extract is washed twice with 7–10 ml of 1 *N* sulfuric acid and with three 2-ml portions of hydrobromic acid. The solution in hydrobromic acid is used in the photometric determination of indium by Rhodamine 6G.

Amperometric determination of indium in concentrates [24, 25]

Indium is titrated at pH 1.0 against a solution of Complexone III. The equivalence point is established by the disappearance of the diffusion current of the reduction of In^{3+} ion on dropping mercury electrode at a potential between −0.7 and −0.8 volt relative to saturated calomel electrode.

Many elements encountered in indium-containing products such as zinc, manganese, cadmium, cobalt and aluminum do not interfere. Less than 10 mg ferrous iron do not interfere; ferric iron is reduced to the bivalent state. The effect of less than 5 mg of tin and less than 2 mg of antimony is removed by adding tartaric acid. The determination can be carried out in the presence of small amounts of copper (less than 0.5 mg), if this element is masked with thiourea; less than 2 mg lead and arsenic do not interfere. Larger amounts of these elements make the location of the equivalence point difficult, since copper, lead and arsenic give diffusion current. However, both lead and arsenic are readily separated from indium in the course of the analysis; arsenic and lead are separated during the decomposition of the mixture with hydrochloric and sulfuric acid, and evaporation of the solution to the appearance of sulfuric acid fumes; copper is removed during the precipitation of the hydroxides by ammonia in excess. Bismuth ions interfere.

REAGENTS

Hydrochloric acid, $d = 1.19$ g/cm^3.
Sulfuric acid, $d = 1.84$ g/cm^3.
Ammonia, solution, $d = 0.9$ g/cm^3.
Tartaric acid, crystalline.
Ascorbic acid, 4% solution.
Thiourea, 5% solution.
Tropeoline OO, indicator solution.
Buffer solution, pH 1.0: 50 ml of 0.2 *M* potassium chloride and 97 ml of 0.2 *N* hydrochloric acid are diluted with water to 200 ml.
Complexone III, 0.005 or 0.02 *M* solutions.

PROCEDURE

A 0.5-g sample is decomposed with a mixture of 5 ml each of hydrochloric and sulfuric acids and the solution is evaporated to the appearance of sulfuric acid fumes. The residue is diluted to 50 ml with water and the precipitate is filtered off. Hydroxides are precipitated from the filtrate by adding ammonia solution in excess. The precipitate is filtered off and dissolved in hydrochloric acid; the solution is transferred to a 50–100-ml volumetric flask and diluted to mark with water. To effect the titration, a 10–20-ml aliquot of the solution is withdrawn into the titration beaker. To the beaker are then added 0.5–1 g tartaric acid, 1–2 ml ascorbic acid solution and 0.2 ml thiourea solution. The solution is neutralized with ammonia solution to Tropeoline OO until the color changes from red to yellow. Buffer solution (15–20 ml) is then added and the solution is titrated against a solution of Complexone III added from a 5-ml semimicro buret. The equivalence point is found graphically from the variation of the diffusion current of indium as a function of the volume of Complexone III solution added.

Polarographic determination of indium in sulfuric acid solutions [26]

Prior to taking the polarogram, all the interfering elements are removed by zinc amalgam in the presence of not less than 20% of an alkali metal sulfate. This results in a vigorous reduction of As, Sb, Bi, Cu, Tl, Se, Sn, Ti, Fe(III), V(V), Cr(VI), ions to the elementary state or to lower valency compounds. Sodium chloride (10 wt % of the resultant solution) is added, and indium is determined polarographically.

The method is suitable for the determination of indium in industrial products and wastes. The half-wave potential of indium is −1.06 volt relative to saturated calomel half-cell.

REAGENTS

Nitric acid, d = 1.4 g/cm^3.
Sulfuric acid, d = 1.84 g/cm^3.
Zinc amalgam. Metallic zinc is dissolved in mercury, with heating, in the presence of dilute (1:5) sulfuric acid.
Sodium chloride, crystalline.

PROCEDURE

To a 1–3-g sample of the material is added a mixture of nitric and sulfuric acids and the mixture is heated until white fumes are evolved. To the cooled residue is added hot water and the precipitate is filtered off. An equal amount of sulfuric acid (or an equivalent amount of sulfate) is added to the solution so as to produce a concentration of not less than 20%. Zinc amalgam is then added to the solution, which is mechanically stirred at 350–400 rpm for 45 minutes. At the end of the cementation the solution is filtered, 10 weight % of sodium chloride are added and a polarogram is taken at −1.06 volt, the indium being determined by the method of additions.

Polarographic determination of indium in sulfide ores [27]

Indium is polarogrammed in 3 *N* hydrochloric acid medium between −0.4 and −0.8 volt relative to a saturated calomel electrode.

Copper, lead, cadmium and tin interfere. Indium is separated by precipitation as hydroxide, using ferric hydroxide as scavenger. The precipitation is conducted in hot, strongly alkaline solution in the presence of Complexone III. The precipitate is dissolved in hydrochloric acid and oxygen and ferric iron are reduced by metallic iron.

REAGENTS

Hydrochloric acid, $d = 1.19$ g/cm^3 and 3 *N* solution.
Perchloric acid, 60%.
Complexone III, 5% solution.
Sodium hydroxide, 10% and 5% solutions.
Iron, reduced by hydrogen.
Sodium chloride, 1% solution.
Standard solution of indium salt, 0.1 mg In/ml.

PROCEDURE

A 0.1–1.0-g sample is decomposed in the heat with 3 ml of concentrated hydrochloric acid and a few ml of perchloric acid. The solution is evaporated almost to dryness. Hot water (40–50 ml) is added to the residue and the mixture is heated to boiling. The undissolved residue is filtered off, washed with hot water and discarded. Complexone III solution (20–25 ml) is added to the filtrate and the mixture is neutralized with 10% sodium hydroxide until alkaline (adding 15–20ml in excess). The solution is diluted to 150–200 ml, slowly heated to boiling, boiled for 3–5 minutes and left to stand on a boiling water bath to a full coagulation of the precipitate (30–40 minutes). The precipitate is separated on a filter paper previously washed with hot 5% solution of NaOH. The precipitate is washed on the filter paper with hot sodium chloride solution and once or twice with hot water. The precipitate is dissolved in hot 3 *N* HCl, the solution is evaporated, if necessary, transferred to a 25-ml volumetric flask and diluted to the mark with 3 *N* HCl. The electrolyzer is filled with the solution, 0.2–0.3 g metallic iron is added on the tip of a spatula and the polarogram is taken after 40–50 minutes. The amount of indium is found by the method of additions.

BIBLIOGRAPHY

1. Ivanov-Emin, B.N. and E.A. Ostroumov.—*ZhOKh*, **14**, 777. 1944.
2. Bock, R., H. Kusche, and E. Bock.—*Z. anal. Chem.*, **138**, 167. 1953.
3. Poluektov, N.S., L.I. Kononenko, and R.S. Lauer.—*ZhAKh*, **13**, 396. 1958.
4. Blyum, I.A. and G.I. Dushina.—*Zav. Lab.*, **25**, 137. 1959.
5. Blyum, I.A., I.T. Solov'yan, and G.N. Shebalkova.—*Zav. Lab.*, **27**, 950. 1961.
6. Babko, A.K. and Z.I. Chalaya.—*ZhAKh*, **18**, 570. 1963.
7. Busev, A.I. *Analiticheskaya khimiya indiya (Analytical Chemistry of Indium)*.—Izd. AN SSSR. 1958.
8. Busev, A.I., T.N. Zholondkovskaya, and Z.M. Kuznetsova.—*ZhAKh*, **15**, 49. 1960.
9. Patrovsky, V.—*Chem. Listy*, **47**, 676, 1338. 1957; *Z. anal. Chem.*, **142**, 66. 1954; **143**, 50. 1954.
10. Busev, A.I. and V.M. Ivanov.—*Izvestiya Vuzov, Khimiya i Khimicheskaya Tekhnologiya*, **5**, 202. 1962.
11. Cheng, K.L.—*Anal. Chem.*, **27**, 1582. 1955.

12. FLASCHKA, H. and H. ABDINE.—*Chemist-Analyst*, **45**, 58. 1956.
13. BUSEV, A.I. and N.A. KANAEV.—*Nauchnye Doklady Vysshei Shkoly, Khimiya i Khimicheskaya Tekhnologya*, No. 2, 299. 1959.
14. BUSEV, A.I. and L.L. TALIPOVA.—*ZhAKh*, **17**, 447. 1962.
15. BUSEV, A.I. and L.L. TALIPOVA.—*Vestnik MGU*, chem. ser. II, No. 2, 63. 1962.
16. NAZARENKO, V.A., E.A. BIRYUK, and R.V. RAVITSKAYA.—*ZhAKh*, **13**, 445. 1958.
17. CHERKASHINA, T.V. and G.P. GORYANSKAYA.—*Trudy GIREDMET*, **13**, 322. 1964.
18. SAYUN, M.G. and S.P. TIKHANINA.—*Zav. Lab.*, **28**, 544. 1962.
19. CHENG, K.L. and R.A. BRAY.—*Anal. Chem.*, **27**, 782. 1955.
20. KISH, P.P. and S.T. ORLOVSKII.—*ZhAKh*, **17**, 1057. 1962.
21. TIÊNG WEI-CH'ÜN.—In: *Khimicheskie, fiziko-khimicheskie i spektral'nye metody issledovaniya rud redkikh i rasseyannykh elementov*, p. 47. Gosgeoltekhizdat. 1961.
22. VLADIMIROVA, V.M. and L.S. RAZUMOVA.—In: *Metody analiza khimicheskikh reaktivov i preparatov (IREA)*, No. 4, 82. 1962.
23. LEVIN, I.S. and T.G. AZARENKO.—*Zav. Lab.*, **28**, 1313. 1962.
24. TSYVINA, V.S. and V.M. VLADIMIROVA.—*Zav. Lab.*, **24**, 278. 1958.
25. VLADIMIROVA, V.M.—*Trudy GIREDMET*, **2**, 223. Metallurgizdat. 1959.
26. NIZHNIK, A.T. and I.S. CHAUS.—*ZhAKh*, **14**, 37. 1959.
27. *Metody khimicheskogo analiza mineral'nogo syr'ya (Chemical Analysis of Mineral Raw Materials)*, No. 2, 85.—Gosgeoltekhizdat. 1956.

Thallium Tl

Thallium displays valencies of + 1 and + 3. The standard electrode potentials relative to standard hydrogen electrode in aqueous medium at 25°C are the following:

$2Tl + S^{2-} \rightleftharpoons Tl_2S + 2e$	-0.93 V
$Tl + I^- \rightleftharpoons TlI + e$	-0.753 V
$Tl + Br^- \rightleftharpoons TlBr + e$	-0.658 V
$Tl + Cl^- \rightleftharpoons TlCl + e$	-0.557 V
$Tl \rightleftharpoons Tl^+ + e$	-0.336 V
$TlOH_{(S)} + 2OH^- \rightleftharpoons Tl(OH)_{3(S)} + 2e$	-0.05 V
$Tl + H_2O \rightleftharpoons TlOH_{(S)} + H^+ + e$	$+0.778$ V
$Tl \rightleftharpoons Tl^{3+} + 2e$	$+1.252$ V

Tl^{3+} ions are colorless and are readily hydrolyzed. The solutions contain the ions $TlOH^{2+}$, TlO^+ and other ions. The pK value for the reaction

$$Tl^{3+} + H_2O \rightleftharpoons TlOH^{2+} + H^+$$

is -0.8 for an ionic strength of 3.0 at 25°C.

In acid solutions thallium(III) ions behave as strong oxidizing agents and are readily reduced to the colorless Tl^+ ions (Figure 17).

The formation of the sparingly soluble TlCl enhances reduction. However,

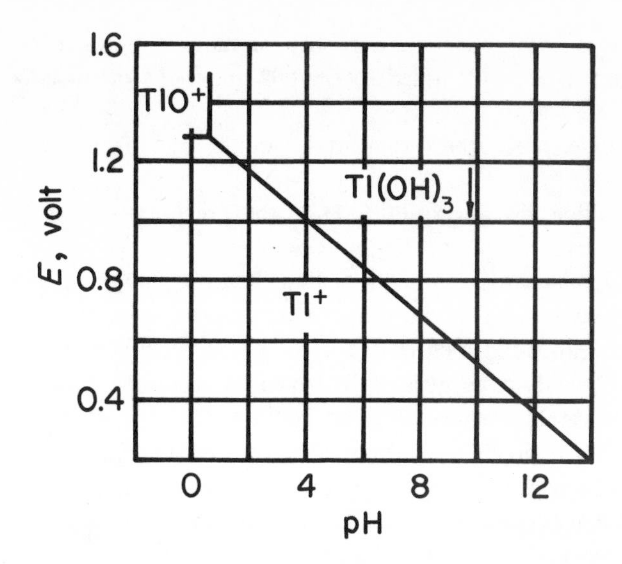

Figure 17
Approximate formal potentials of the system Tl(III)/Tl(I) as a function of pH. The concentration of thallium is 10^{-2} *M*.

chlorides, bromides and iodides decrease the oxidation potential of thallium(III) owing to complex formation.

Thallium (I) is oxidized only by strong oxidizing agents MnO_4^-, Cl_2^- in strongly acid medium. On the other hand, the thallous ion is a strong reudcing agent in alkaline medium.

As regards their chemical properties, thallous ions resemble the ions of alkali metals, silver and also lead(II) ions.

Thallous hydroxide TlOH is soluble in water and is a strong base.

Thallium(I) salts of strong acids have a neutral reaction; most of them are soluble in water except for the halides TlCl, TlBr, TlI (TlF alone is soluble), azide TlN_3, sulfide Tl_2S, chromate Tl_2CrO_4, chloroplatinate $Tl_2[PtCl_6]$ and cobaltinitrite $Tl_3[Co(NO_2)_6]$.

Thallium(I) does not form stable complexes. Thus, it does not form complexes with thiosulfate ions (unlike lead(II) and silver(I)) or with ammonia (unlike silver(I)).

Thallous thallium belongs to the analytical subgroup of silver, since its chloride is sparingly soluble in water.

Dilute hydrochloric acid precipitates the white thallous chloride TlCl from solutions containing thallous ions. The precipitate is sparingly soluble in water at room temperature and in dilute acids, but more so in hot water, $pL_{TlCl} = 3.7$.

Potassium iodide precipitates the sparingly soluble yellow TlI; $pL_{TlI} = 7.2$. The precipitate is insoluble in solutions of sodium thiosulfate, unlike silver and lead iodides. Complexone III does not prevent the precipitation of thallous iodide [1]. In this way the selectivity of the precipitation of the iodide may be greatly enhanced. The interfering elements may also be masked by tartrates or cyanides. Thallous thallium may be identified as crystals under the microscope in the form of iodides.

Hydrogen sulfide precipitates the black sulfide Tl_2S from neutral solution; $pL_{Tl_2S} = 22.2–23.9$.

Ammonium sulfide also precipitates Tl_2S; the precipitate is insoluble in excess reagent.

Potassium chromate precipitates the yellow Tl_2CrO_4, which is sparingly soluble in dilute acids and is used in the gravimetric determination of thallium (p. 283) in the cold. The orange-red bichromate $Tl_2Cr_2O_7$ is precipitated out of acid solutions. On boiling with concentrated hydrochloric acid the precipitate dissolves and beautiful red crystals separate out on cooling.

Chloroplatinic acid precipitates the orange-colored Tl_2PtCl_6 which is insoluble in cold water. Yellowish octahedral crystals separate out of hot acid solutions of thallous salts; these crystals will identify thallium under the microscope.

Potassium ferricyanide in alkaline medium oxidizes thallous ions with the separation of the brown precipitate of $Tl(OH)_3$:

$$2[Fe(CN)_6]^{3-} + Tl^+ + 3OH^- \rightleftharpoons Tl(OH)_3 + 2[Fe(CN)_6]^{4-}$$

Alkali metal iodobismuthates form a red precipitate with thallous ions [2]. The reaction is selective for thallous ions in the presence of sodium thiosulfate, which eliminates the effect of Fe(II), Ce(IV), Hg(II), Cu(II), Pt(IV), As(V), and Sb(V) and of antimony(V). Selenious and tellurous acids interfere (a black precipitate separates out).

Dipicrylamine is a selective reagent for thallous ions within the hydrochloric acid group Ag^+, Hg^{2+}, Pb^{2+}. A red precipitate separates out. K^+, NH_4^+, Rb^+ and Cs^+ ions interfere. The reagent is used [3] in the detection of thallous thallium crystals under the microscope.

Thiourea precipitates a white crystalline solid out of solutions acidified with perchloric acid. The reaction is highly selective [4].

Thallium(III) salts are soluble only in the presence of strong acids. These salts are readily hydrolyzed.

The hydroxide $Tl(OH)_3$ is a reddish-brown amorphous precipitate. It begins to precipitate out at pH ~ 0.3 (0.01 *M*). It does not display amphoteric properties.

The trioxide Tl_2O_3 is black.

Thallium(III) ions typically form various complex compounds. Thus, complexes of the type $[TlX_4]^-$ and $[TlX_6]^{3-}$ are readily formed, where X is a fluoride, chloride, bromide, or other ion. The complex compound $H[TlCl_4]$ is extracted with ether from hydrochloric acid solution.

Thallium(III) forms sulfate complexes which are not very stable. It forms complex compounds with organic hydroxy acids (tartaric, citric and other acids), oxalic acid and other dicarboxylic acids.

Thallium(III) is rapidly and quantitatively reduced to thallium(I) by hydrazine sulfate in dilute sulfuric acid solutions [5]. In this way thallium(III) ions can be potentiometrically titrated with the use of platinum-graphite electrodes.

The complex formation of thallium(III) in solutions with chlorides or bromides takes place in several steps [6]. The following instability constants were found for the chloride complexes: $pK_1 = 7.50$; $pK_2 = 4.50$; $pK_3 = 2.75$; $pK_4 = 2.25$: $pK_5 = 1.95$; $pK_6 = 1.75$ (for the ionic strength of 0.4 and a temperature of 20°C; concentration of $HClO_4$ 0.34 *M*). The instability constants of the bromide complexes of thallium(III) in solutions with the same ionic strength and perchloric acid concentration are: $pK_1 = 8.3$; $pK_2 = 6.3$; $pK_3 = 4.6$; $pK_4 = 3.1$; $pK_5 = 2.5$; $pK_6 = 1.7$ (20°C).

Chlorides and bromides interfere with complexometric titration of thallium(III) and with the color reactions of thallium with certain organic reagents.

Diantipyrylpropylmethane quantitatively precipitates thallium(III) ions from solutions containing chloride ions as the compound [7] $C_{26}H_{30}O_2N_4 \cdot HTlCl_4$, which is a suitable gravimetric form (p. 283).

The derivatives of pyrazolone—pyramidon, antipyrin, diantipyrylmethane, diantipyrylphenylmethane, and diantipyrylpropylmethane—precipitate thallium(III) ions in the presence of bromides in the form of crystalline compounds [7, 8]. Thallium can be gravimetrically determined as the compounds $C_{23}H_{24}O_2N_4 \cdot HTlBr_4$ and $C_{26}H_{30}O_2N_4 \cdot HTlBr_4$ in the presence of copper, zinc, cadmium, bismuth, iron and antimony ions [7, 8].

Thallium (III) ions react with basic dyes in the presence of chlorides and bromides to form sparingly soluble compounds. Thus, the bromide and chloride complexes of thallium form dark violet precipitates [9] with the basic dye Rhodamine B, which gives an intense red-colored extract with organic solvents; the pure reagent is not extracted. A similar color reaction in weakly acid solution is given by $TlCl_4^-$ and $TlBr_4^-$ anions with Methyl Violet [10–12] and other basic dyes of diaminotriphenylmethane series [12–15]. The benzene extract of the compound $HTlCl_4$ with Rhodamine B fluoresces strongly in the ultraviolet.

These reagents are used in the extraction-photometric determination of thallium and also in the detection of this element.

When a solution of potassium iodide is gradually added to a weakly acid dilute solution of $Tl_2(SO_4)_3$ the following reactions take place in succession [16]:

$$Tl^{3+} + 2I^- \rightleftharpoons Tl^+ + I_2 \quad \text{(I)}$$

$$I_2 + I^- \rightleftharpoons I_3^- \quad \text{(II)}$$

$$Tl^+ + I^- + I_3^- \rightleftharpoons TlI_4^- \quad \text{(III)}$$

Under certain conditions the oxidation of iodide ions to elementary iodine by trivalent thallium is quantitative. If the solution contains iodide ions in excess, the initially formed thallium (I) ions are oxidized by elementary iodine with the formation of stable complex iodide anions of trivalent thallium TlI_4^- (complex formation proceeds stepwise). Such a reaction is possible owing to the strong decrease of the oxidation potential of thallium (III) during complex formation. If the TlI_4^- ions which are formed are removed from the reaction zone by precipitation with organic bases, e.g., by pyrazolone derivatives, reaction III is fully shifted from left to right.

Thallium (III) ions form fairly stable tartrate complexes under certain conditions [17]. At pH 3–7 tartrate ions are practically incapable of reducing trivalent thallium to the monovalent state in several hours, whereas in an acid solution (pH 0.8–0.9) the reduction is 8–10% complete after 2 hours. The composition of the precipitate which separates out of solutions containing Tl^{3+} and $C_4H_4O_6^{2-}$ ions at pH 2 corresponds to the formula $Tl_2(C_4H_4O_6)_3$; the solubility of this compound in water at 20°C is $(1.0 \pm 0.1) \cdot 10^{-4}$ moles/liter and increases rapidly in the presence of tartrate at pH above 2, probably owing to the formation of $Tl(C_4H_4O_6)_2^-$ and other complexes. At pH 4 thallium forms part of the complex anion. At an ionic strength of 1 and at 20°C the successive stability constants of the tartrate complexes of thallium (III) are: $\beta_1 = 3.7 \cdot 10^{11}$, $\beta_2 = 6.5 \cdot 10^{12}$ and $\beta_3 = 2.2 \cdot 10^{13}$.

Complexone III reacts with thallium (III) ions to form a stable complex compound [18]. The instability constant of thallium (III) complexonate at 20°C and ionic strength of 0.4 is

$$K_{TlY} = \frac{[Tl^{3+}][Y^{4-}]}{[TlY^-]} = (3.3 \pm 3.5) \cdot 10^{-22}.$$

At 20°C and 0.1 ionic strength $pK = 22.5 \pm 0.5$.

Several variants of complexometric determination of thallium have been reported [19].

Different hydroxylated azo compounds form colored compounds with trivalent thallium.

1-(2-Pyridylazo)-2-naphthol reacts with thallium (III) in the molar ratio of 1:1 with formation of a red-violet compound which is sparingly soluble in water; the solubility of the compound increases in the presence of water-miscible organic solvents: methanol, ethanol, acetone. The absorption maximum of the solutions of the colored compound at pH 4.5 is at 560 mμ (maximum absorption of the solutions of the reagents under these conditions is at 470 mμ). The molar extinction coefficient is 22,000 (solvent: 50% methanol, pH 2.2). The equilibrium constant of the formation reaction of the complex compound is $1.9 \cdot 10^2$. The reagent is used as complexometric indicator [20] in the titration of thallium (III).

1-(2-Pyridylazo)-resorcinol also reacts with thallium (III) ions to give a red-colored compound which is readily soluble in water, butanol and isoamyl alcohol, but is insoluble in ether, benzene, carbon tetrachloride and chloroform. The absorption maximum of the aqueous solution of this compound is at 550 mμ at pH 2.2; the absorption maximum of the solution of the reagent is at 415 mμ. The molar extinction coefficient of the thallium compound is $1.9 \cdot 10^4$. The constant complex formation is $0.80 \cdot 10^4$ [20]. The reagent is used as complexometric indicator in the titration of thallium (III).

7-(1-Naphthylazo)- and 7-(5,7-disulfo-2-naphthylazo)-8-hydroxyquinoline-5-sulfonic acids react with thallium (III) ions at pH 2 and above to form soluble yellow compounds. If Complexone III is added to such solutions, the complexonate of Ti(III) is formed, free azo compounds separate out and the solution turns violet. The color change is sharp. The reagents are employed as complexometric indicators in the titration of thallium (III) [21].

Xylenol Orange reacts with thallium (III) at pH 2.50 in the molar ratio of 1:1; the absorption maximum of the solution of the compound is at 580 mμ. The molar extinction coefficient is $2 \cdot 10^4$. The constant of the complex formation reaction is $7 \cdot 10^4$.

The reagent is used as complexometric indicator during the titration of thallium (III).

Thallium compounds color the flame emerald-green. The wide thallium line in the visible part of the spectrum at 535 mμ is found near the barium line, but the thallium line is more intense. Thallium can be detected and determined by spectroscopy and by flame coloration.

The reader is referred to [22] for a survey of quantitative analytical methods for the determination of thallium and to [23] for qualitative detection methods.

Gravimetric determination of thallium (I) as the thionalidate [24]

Thallous thallium [24] is precipitated as the sparingly soluble thionalidate $C_{10}H_7NHCOCH_2STl$.

Cations of heavy metals also react with thionalide, giving sparingly soluble sulfides. Thionalide precipitates Cu^{2+}, Cd^{2+}, Hg^{2+} and Au(III) as well as thallium, from solutions containing tartrates and sodium hydroxide; it precipitates Pb^{2+}, Bi^{3+}, Sb(III), Sn(II), Au(III) ions, in addition to thallium, from solutions containing tartrates and potassium cyanide. If the solution contains tartrates, potassium cyanide and sodium hydroxide, thionalide will precipitate only thallous and auric ions.

REAGENTS

Thionalide, 5% solution in acetone. 0.4–0.5 g of thionalide in 8–10 ml of acetone should be taken for each 0.1 g of Tl.

Sodium tartrate, crystalline.

Potassium cyanide, crystalline.

Sodium hydroxide, 2 *N* solution.

PROCEDURE

About 10–20 ml of the sample solution, which should contain 0.1–0.025 g of Tl, is neutralized to phenolphthalein with sodium hydroxide solution, 2 g of sodium tartrate, 3–5 g of KCN and an equal volume of sodium hydroxide solution are added, and the mixture is diluted to 100 ml with water. A 4–5-fold excess of thionalide solution is added. The mixture is stirred and heated to boiling, when the amorphous precipitate becomes crystalline. After cooling, the lemon-yellow precipitate is filtered through a sintered glass crucible, washed with cold water to remove KCN and with acetone to remove the excess reagents. The precipitate is dried at 100°C and weighed. The conversion factor to Tl is 0.4859.

Gravimetric determination of thallium (I) as chromate [25, 26]

Thallium (I) is precipitated from alkaline solution by potassium chromate as the sparingly soluble Tl_2CrO_4 (solubility in water at 20°C is $8.15 \cdot 10^{-5}$ *M*, and in 96% ethanol $1.14 \cdot 10^{-5}$ *M*).

The interfering ions are those which reduce chromates (e.g., sulfites), and also the ions which are precipitated from ammoniacal solutions of chromates as hydroxides (Al, Fe, Cr) or as chromates (Ag, Pb, Bi, Ba, Sr). The effect of the foreign ions is eliminated by introducing Complexone III, which forms stable complex compounds with the interfering ions in alkaline medium.

REAGENTS

Potassium chromate, 10% and 1% solutions.
Ammonia, 25% solution.
Ethanol, 50% solution.

PROCEDURE

The neutral sample solution, which should contain about 1 mg Tl/ml, is heated to 70–80°C, and 4 ml ammonia solution and 10 ml of 10% potassium chromate are added per 100 ml of solution. The mixture is cooled and is left to stand for not less than 12 hours. The precipitate of thallium chromate is separated by filtering on No. 3 sintered glass crucible, with the aid of 1% solution of potassium chromate. The precipitate is washed 2–3 times with small portions of ethanol until the filtrate is no longer yellow, dried to constant weight at 120–130°C and weighed. The conversion factor to Tl is 0.7789.

Gravimetric determination of thallium (III) by diantipyrylpropylmethane [7]

Thallium (III) is precipitated from hydrochloric acid solution by diantipyrylpropylmethane as $C_{26}H_{30}O_2N_4 \cdot HTlCl_4$. The solubility of the precipitate in water at 20°C is $5.7 \cdot 10^{-5}$ *M*. The results of the determination are not affected by variations in acidity between 1 *N* and 6 *N*, provided the concentration of chloride ions is greater than 0.1 *M*.

In 3 *N* HCl In^{3+}, Bi^{3+}, Zn^{2+}, Cd^{2+}, Fe^{3+}, Mg^{2+}, PO_4^{3-}, AsO_4^{3-} do not interfere.

REAGENTS

Diantipyrylpropylmethane, 1% solution in 1:1 acetic acid.
Washing solution, 0.05% solution of diantipyrylpropylmethane in 1 *M* HCl.
Hydrochloric acid, 1 *M* solution.
Sodium chloride, crystalline.

PROCEDURE

Sodium chloride is added to the acid solution of thallium (III) salt until the chloride concentration is not less than 0.1 *M*, and then the solution of diantipyrylpropylmethane is introduced with stirring until the formation of the precipitate is complete. The precipitate is allowed to coagulate and is separated on No. 3 sintered glass filter, washing several times with small portions of washing solution and then with hydrochloric acid; the precipitate is then dried at 110–120°C to constant weight and weighed. The conversion factor to Tl is 0.2628.

Bromatometric determination of thallium (I) [27]

Thallous thallium is oxidized by potassium bromate in a hydrochloric acid medium (5–8% HCl):

$$3Tl^+ + BrO_3^- + 6H^+ \rightarrow 3Tl^{3+} + Br^- + 3H_2O.$$

The equivalence point is established potentiometrically by the change in the potential of the platinum electrode. The measurements are carried out with a compensation potentiometer P-4, using a saturated calomel half-cell as the comparison electrode.

As, Sb, Te, Cu interfere. In the presence of ferric ions the titration is effected after the addition of ammonium phosphate.

REAGENTS

Potassium bromate, 0.02 N solution.
Hydrochloric acid, $d = 1.19$ g/cm^3.

PROCEDURE

The sample solution, which should contain 0.05–0.01 g Tl, is diluted to mark with water in a 50-ml volumetric flask. A 20-ml aliquot is withdrawn into the titration beaker. Hydrochloric acid (15 ml) is added and the mixture is titrated against a solution of potassium bromate until a sharp potential jump is noted. The first (orientation) titration is carried out adding the solution of bromate in 1-ml portions; the second, accurate titration is carried out by adding the bromate solution in 0.1-ml portions near the end point. One ml of 0.0200 N solution of potassium bromate is equivalent to 2.044 mg of thallium.

Complexometric determination of thallium (III) in alloys [19]

Thallium (III) is titrated at about pH 2 against standard solution of Complexone III. The equivalence point is established with the aid of 1-(2-pyridylazo)-2-naphthol or 1-(2-pyridylazo)-resorcinol, which react with trivalent thallium to form red-violet and red complexes respectively; these are converted into thallium complexonate when acted upon by Complexone III. This is accompanied by a sharp color change at the equivalence point to yellow, which is the color of the indicator itself.

Thallium may be determined in the presence of major amounts of zinc, lead and cadmium and also in the presence of metals which form stable complexonates in neutral or weakly alkaline solutions only (calcium, magnesium, aluminum).

The ions Fe^{3+}, Bi^{3+}, In^{3+}, Ga^{3+}, Cu^{2+}, MnO_4^-, and also chlorides, bromides and iodides, which decompose the complex of thallium with the dye, interfere.

Thallium may be determined in the presence of bismuth (III) ions by titrating the sum of bismuth (III) and thallium (III) ions at pH 4–5; thallium (III) is then reduced to thallium (I) and the liberated Complexone III is titrated against a standard solution of a copper salt.

The method is suitable for the analysis of magnesium and other alloys containing not less than 0.5% Tl.

The error in the determination of thallium does not exceed 0.5 rel. %. The duration of one determination is 20–25 minutes.

REAGENTS

Sulfuric acid, dilute (1:2) solution.
Ammonium persulfate, crystalline.

Sodium fluoride, crystalline.
Ammonia, 2 *N* solution.
Monochloroacetic acid, 1 *M* solution.
Potassium bromide, 2 *M* solution.
Sulfosalicylic acid, 20% solution.
Complexone III, 0.01 *M* solution.
Ammonium acetate, crystalline.
1-(2-*Pyridylazo*)-2-*naphthol*, 0.1% ethanolic solution.
1-(2-*Pyridylazo*)-*resorcinol*, 0.1% aqueous solution.

PROCEDURE

Determination of thallium in magnesium alloys

A 0.2–0.5-g sample is dissolved in 10 ml sulfuric acid, the solution is diluted to 100 ml with water (final acidity of the solution 1–2 *N*), 0.5 g ammonium persulfate is added and the mixture is boiled for several minutes to remove the excess. Potassium fluoride (1–2 g) is added to the solution (if the alloy contains zirconium or major amounts of aluminium), followed by ammonia solution until the solution becomes light yellow (incipient separation of $Tl(OH)_3$), after which monochloroacetic acid is added, the pH is adjusted to 2 (Universal Indicator paper), and a few drops of 1-(2-pyridylazo)-2-naphthol solution are added. The mixture is titrated against a solution of Complexone III to a color change from red-violet to yellow.

Determination of thallium in the presence of iron

Between 5 and 10 ml of potassium bromide solution are added to 50–70 ml solution containing ferric ions and trivalent thallium to mask thallium (III), and the pH is adjusted to 2 by adding ammonia solution drop by drop. The solution is heated to 40–60°C, 3–5 drops of sulfosalicylic acid solution are added and ferric iron is titrated against the solution of Complexone III until the red-brown coloration disappears. Ammonium acetate is added to pH 4–5, then a few drops of the 1-(2-pyridylazo)-resorcinol solution, until the solution is colored distinctly red. Thallium (III) is titrated against Complexone III until the solution turns orange-yellow.

If the contents of thallium and iron are approximately equal, the titration may be conducted with the same solution of Complexone III. If more iron is present than thallium, the solution of Complexone III used in the first titration should be correspondingly more concentrated.

Spectrophotometric determination of thallium (III) [28]

The method is based on the formation of the iodide complex of thallium (III) with diantipyrylmethane or diantipyrylpropylmethane ($C_{23}H_{24}O_2N_4 \cdot HTlI_4$ or $C_{26}H_{30}O_2N_4 \cdot HTlI_4$) with subsequent extraction of the complex with benzene. The molar extinction coefficient of the benzene extract at 400–405 mμ is 12,000 and is independent of the nature of the base.

In the presence of Complexone III at pH 2–3, Cu^{2+}, Fe^{3+}, Bi^{3+}, Cd^{2+}, Pb^{2+}, Zn^{2+}, Ga^{3+} and In^{3+} ions do not interfere.

REAGENTS

Diantipyrylmethane, or *diantipyrylpropylmethane*, 0.05% solution in 1:10 acetic acid.
Potassium iodide, 0.1% solution.
Sulfuric acid, 1 *N* solution.
Bromine water.
Benzene.
Standard solution of thallium (I) *salt*, 40 mμ Tl/ml.

CONSTRUCTION OF CALIBRATION CURVE

Into 5 small beakers are placed 1,2,3,4, and 5 ml of the standard solution of thallium (I) salt and 2–3 ml bromine water are added to each beaker; the solutions are heated to boiling and are boiled to expel excess bromine. The cooled solutions are transferred to 25–50-ml separating funnels, the beakers are rinsed with 2–3 ml of sulfuric acid, and 2–3 ml of potassium iodide solution and 1–2 ml of diantipyrylmethane or diantipyrylpropylmethane are added to each funnel. The solutions are extracted with 10 ml of benzene. The layers are allowed to separate, and the lower layer is decanted, while the upper layer is filtered through a dry filter paper and the optical density of the filtrate is measured at 400 mμ in a 1-cm cell of an SF-4 spectrophotometer relative to benzene. The calibration curve is then plotted.

PROCEDURE

All the above reagents are added to the sample solution as described in the construction of the calibration curve, the optical density of the solution is measured and the thallium content is found from the calibration curve.

Extraction-photometric determination of small amounts of thallium (III) by Methyl Violet [11, 12, 29, 30]

The colored compound of the anion $TlCl_4^-$ with the cation of the dye Methyl Violet is extracted with toluene or benzene.

The absorption maximum of the extract is at 570 mμ. The molar extinction coefficient is about 50,000. Variation of acidity between 0.2 and 0.8 *N* HCl has no effect on the results of the determination.

The effect of antimony is eliminated by oxidizing thallous ions to thallium (III) in the presence of chloride concentrations not exceeding 0.1–0.2 g-eq/liter; under these conditions the ions $SbCl_6^-$ are not formed. The interfering gold (III) and mercury (II) are eliminated by cementation on metallic copper.

REAGENTS

Methyl Violet, 0.2% aqueous solution.
Phosphoric acid, $d = 1.16$ g/cm^3.
Hydrogen peroxide, 30% solution.
Ferric chloride, 20% solution.
Copper wire spiral.
Standard solution of thallous salt, 10 μg Tl/ml.
Hydrochloric acid, $d = 1.19$ g/cm^3.
Nitric acid, $d = 1.4$ g/cm^3.
Toluene.

CONSTRUCTION OF CALIBRATION CURVE

Into five 100-ml beakers are placed portions of the standard solution of thallous salt containing 1, 5, 10, 15 and 20 μg of Tl. 30 ml of phosphoric acid, 1 ml of ferric chloride solution and 7–8 drops of hydrogen peroxide solution are added to each beaker, and the mixtures are left to stand for 30–40 minutes or overnight. The solutions are transferred to separating funnels and diluted to 40 ml with water; 25 ml toluene and 1.0 ml Methyl Violet solution are added and the mixtures are shaken for 1 minute. The layers are allowed to separate and the extract is decanted into a dry 50-ml beaker. After 20 minutes the optical density of the extract is measured in an electrophotocolorimeter with a green filter and the calibration curve is plotted.

PROCEDURE

A 1-g sample of the ore is placed in a 150-ml beaker and 20–40 ml of 3:1 mixture of hydrochloric and nitric acids are added. When the vigorous reaction has ceased, the solution is heated and evaporated to a residual volume of 3–5 ml. The beaker is transferred onto a water bath and the solution is evaporated to dryness. The residue is treated with 5–10 ml of hydrochloric acid and the solution is again evaporated to dryness on a water bath. During the analysis of materials in which thallium is bound with a silicate base, silicon is distilled off as SiF_4. The residue is treated with concentrated hydrochloric acid, and the excess of the latter is removed by evaporation on a water bath.

Thirty ml of phosphoric acid are added to the resulting solution, which is then heated to boiling. The copper spiral, which has been washed with dilute nitric acid and with water, is immersed in the solution. Ten minutes later the beaker with the solution is removed from the plate and the spiral is taken out and washed down with a little water. The insoluble residue is filtered off and the filter is washed once or twice with water. The volume of the filtrate and of the washings should not exceed 30–35 ml. The subsequent procedure is as described for the construction of the calibration curve.

Separation of thallium (III) from ores and ore dressing products by extraction [11]

Thallium (III) is separated by ether extraction from 1 *N* solution of hydrobromic acid. Mercury (II), antimony (V) and gold (III) are extracted together with thallium (III).

REAGENTS

Nitric acid, $d = 1.4$ g/cm³.
Hydrochloric acid, dilute (1:1) and 10% solutions.
Sulfuric acid, dilute (1:1) solution.
Hydrobromic acid, 40% solution, saturated with bromine.
Hydrobromic acid, 1 *N* solution saturated with bromine.
Hydrobromic acid, 1 *N* solution, saturated with ether.
Ether, saturated with 1 *N* hydrobromic acid.
Bromine.

PROCEDURE

A 0.2–0.5 g sample of the material is dissolved with heating on a sand bath in 20–50 ml of hydrochloric acid (1:1), 2–3 ml nitric acid are added to the solution and the mixture is evaporated to dryness. The residue is treated twice with 2–3 ml of 1:1 hydrochloric acid, the solution being evaporated to dryness each time. The last residue is dissolved in 20–30 ml of 10% hydrochloric acid, the solution is heated and allowed to cool. The undecomposed residue and $PbCl_2$ are filtered off, washing 2–3 times with 10% hydrochloric acid. The filtrate is evaporated to dryness, the residue is treated with 5–7 ml of 40% hydrobromic acid saturated with bromine and the solution is evaporated almost to dryness; this operation is repeated three times.

The dry residue is dissolved in 8–10 ml of 1 *N* hydrobromic acid saturated with bromine and the solution is transferred to a separating funnel, washing the beaker 1–2 times with 2–3 ml of this acid. An equal volume of ether is added to the solution and the mixture is shaken for 1 minute. The layers are allowed to separate and the aqueous layer is decanted into a second separating funnel, washed with 1 *N* hydrobromic acid saturated with ether and the extraction is repeated. The ether extracts are washed twice with 5 ml of the acid, shaking for 15–20 seconds each time. The ether is expelled by heating on a water bath and the solution is evaporated to dryness. Sulfuric acid (0.5 ml) is added and the solution is evaporated on a sand bath to the appearance of sulfuric acid fumes, after the addition of a few drops of nitric acid to oxidize the organic components. Water (2–3 ml) is added and the solution is again evaporated to the evolution of sulfuric acid fumes. This operation is repeated 2–3 times. The last evaporation should be continued almost to dryness. Thallium is determined in the residue by the photometric or polarographic method.

Polarographic determination of thallium (I) in cadmium [31]

Thallous ions are reversibly reduced on dropping mercury electrode at about − 0.50 volt relative to saturated calomel electrode. The half-wave potential is independent of the composition of the electrolyte. The concentration of thallium in the solution is proportional to the wave height if ammonia or hydrochloric acid are used as supporting electrolytes. Thallous thallium in ammoniacal medium, in the absence of copper, is readily determined in cadmium and in cadmium salts. If hydrochloric acid is used as supporting electrolyte, the half-wave potentials of thallous thallium merge with those of tin and lead. In order to separate the tin wave from the thallium wave, tartrate ions are introduced which suppress the tin wave, while Complexone III, which shifts the half-wave potential of lead in a weakly acid medium towards more negative values (− 1.1 volt), is added to separate the lead from the thallium wave. This is the method employed in the determination of thallium in lead. Under these conditions copper is reduced at − 0.3 volt. The polarographic determination is usually conducted after thallium has already been separated by the extraction method.

REAGENTS

Sulfuric acid, dilute (1:1) solution.
Ammonia, 25% solution.
Sodium sulfite, crystalline.
Carpenter's glue, 0.25% solution.
Hydrogen peroxide, 30% solution.

PROCEDURE

The ether extracts from thallium-containing solutions in 1 *M* hydrobromic acid are evaporated on a water bath to expell the ether. Three ml of 1:1 sulfuric acid are added to the residue and the solution is evaporated to the appearance of sulfuric acid fumes. Hydrogen peroxide (3–5 drops) is added to the hot solution to decolorize it. After cooling, the walls of the flask are rinsed with 5 ml of water and the solution is again heated to the evolution of sulfuric acid fumes.

Water (10–15 ml) and sodium sulfite (0.2–0.3 g) are added to the residue which is heated for 5 minutes until the odor of sulfur dioxide disappears. The solution is transferred to a 25-ml volumetric flask, neutralized with ammonia solution to Congo Red and 2.5 ml of ammonia in excess are added. The solution is cooled, made up to mark with water and thoroughly mixed. Fifteen ml of this solution are transferred into a beaker, about 1 g of sodium sulfite and 5 drops of carpenter's glue solution are added, the mixture is stirred and the polarogram is taken.

BIBLIOGRAPHY

1. PŘIBIL, R. *Komplexony v chemické analyse.*—Prague, Natl. Českosl. Akad. Věd. 1957.
2. WENGER, P. and Y. RUSCONI.—*Helv. chim. acta,* **26**, 2263. 1943.
3. SHEINTSKIS, O.G.—*Zav. Lab.,* **4**, 1047. 1935.
4. MAHR, C.—*Z. anal. Chem.,* **115**, 254. 1938.
5. BERKA, A. and A.I. BUSEV.—*Anal. chim. acta,* **27**, 493. 1962.
6. BUSEV, A.I., V.G. TIPTSOVA, and T.A. SOKOLOVA.—*Vestnik MGU,* chem. ser. II, No. 6, 42. 1960.
7. BUSEV, A.I. and V.G. TIPTSOVA.—*ZhAKh,* **15**, 291. 1960.
8. BUSEV, A.I. and V.G. TIPTSOVA.—*ZhAKh,* **14**, 28. 1959.
9. FEIGL, F., V. GENTIL, and D. GOLDSTEIN.—*Anal. chim. acta,* **9**, 393. 1953.
10. KUZNETSOV, V.I.—*ZhAKh,* **16**, 249. 1947.
11. GUR'EV, S.D.—*Sbornik trudov Gintsvetmet,* No. 10, 371. Metallurgizdat. 1955.
12. SCHEMELEVA, G.G. and V.I. PETRASHEN'.—*Trudy Novocherkasskogo Politekhnicheskogo Instituta,* **31**, 87. 1955.
13. KOVAŘIK, M. and M. MOUČKA.—*Anal. chim. acta,* **16**, 249. 1957.
14. LAPIN, L.N. and V.O. GEIN.—*Trudy komissii po Analiticheskoi Khimii,* **7**, 217. 1956.
15. VOSKRESENSKAYA, N.T.—*ZhAKh,* **11**, 585. 1956.
16. BUSEV, A.I. and V.G. TIPTSOVA.—*Nauchnye Doklady Vysshei Shkoly, Khimiya i Khimicheskaya Tekhnologiya,* No. 3, 486. 1958.
17. BUSEV, A.I., V.G. TIPTSOVA, and L.M. SOROKINA.—*ZhNKh,* **7**, 2122. 1962.
18. BUSEV, A.I., V.G. TIPTSOVA, and T.A. SOKOLOVA.—*ZhNKh,* **5**, 2749. 1960.
19. BUSEV, A.I. and V.G. TIPTSOVA.—*ZhAKh,* **13**, 180. 1958; **15**, 573. 1960; **16**, 275. 1961.
20. BUSEV, A.I. and V.G. TIPTSOVA.—*ZhAKh,* **15**, 573. 1960.
21. BUSEV, A.I. and L.L. TALIPOVA.—*Uzbekskii Khimicheskii Zhurnal,* No. 3, 24. 1962.
22. BUSEV, A.I. and V.G. TIPTSOVA.—In: *Metody opredeleniya i analiza redkikh elementov,* p. 182. Izd. AN SSSR. 1961.
23. HECHT, H.—In: *Handbuch der analytischen Chemie,* R. FRESENIUS and G. JANDER, Editors, **III**, part 2, p. 70. Berlin, Springer-Verlag. 1944.
24. BERG, R. and E.S. FAHRENKAMP.—*Z. anal. Chem.,* **109**, 305. 1937.
25. BASHILOVA, N.I.—In: *Khimiya redkikh elementov,* No. 3, 105. 1957.
26. EFREMOV, G.V.—*Uchenye Zapisky Leningradskogo Gosudarstvennogo Pedagogicheskogo Instituta im. Pokrovskogo,* **5**, No. 2, 97. 1949.
27. ZINTL, E. and G. REINACKER.—*Z. anorg. allgem. Chem.,* **153**, 278. 1926.
28. BUSEV, A.I. and V.G. TIPTSOVA.—*ZhAKh,* **14**, 550. 1959.
29. BLYUM, I.A. and I.A. YL'YANOVA.—*Zav. Lab.,* **23**, 283. 1957.
30. BLYUM, I.A. and I.T. SOLOV'YAN.—*Zav. Lab.,* **27**, 950. 1961.
31. AREF'EVA, T.V., A.A. POZDNYAKOVA, and R.G. PATS.—In: *Obogashchenie i metallurgiya tsvetnykh metallov,* No. 8, 129. Metallurgizdat. 1953.

Germanium Ge

Germanium displays valencies of + 4, + 2 and −4. In analytical chemistry the compounds of Ge^{4+} (GeO_2,$GeCl_4$, numerous complex compounds etc.) are the most important. The compounds of germanium (II) are readily oxidized to compounds of Ge^{4+}. The compounds of Ge^{4-} include the gaseous hydride GeH_4, which decomposes into elementary germanium and elementary hydrogen at 360°C. It is more stable than SbH_3. The hydride GeH_4 is formed by the reduction of germanium compounds by metallic zinc or sodium amalgam. Silver germanide Ag_4Ge is known; the compound is black.

Standard electrode potentials relative to the standard hydrogen electrode in aqueous solution at 25°C are the following:

$$Ge^{2+} + 3H_2O \rightleftharpoons H_2GeO_3 + 4H^+ + 2e \qquad -0.363\ V$$

$$Ge + 2H_2O \rightleftharpoons GeO_2 + 4H^+ + 4e \qquad -0.246\ V$$

$$Ge + 3H_2O \rightleftharpoons H_2GeO_3 + 4H^+ + 4e \qquad -0.182\ V$$

$$Ge \rightleftharpoons Ge^{2+} + 2e \qquad 0.000\ V$$

Germanium (II) forms a tartrate complex; germanium cannot be precipitated by ammonia or by hydrogen sulfide from solutions of this complex.

Germanium sulfide GeS is reddish-orange, soluble in concentrated hydrochloric acid and in solutions of ammonium sulfide. Concentrated hydrochloric acid solutions contain anions $GeCl_3^-$ and other anions. Sulfo-salts exist in solutions of ammonium sulfide.

The orange-red germanium hydroxide $Ge(OH)_2$ is soluble in hydrochloric acid and in alkalis, but very sparingly soluble in solutions of ammonia.

Germanium (II) compounds are readily oxidized in air. Germanous oxide GeO, which is dark grey or black, is known.

Germanium (II) ions are formed by reduction of compounds of germanium (IV) by hypophosphorous acid in hydrochloric acid solutions. Under these conditions arsenic is reduced to the elementary state and precipitates out. This fact is used to advantage in separating germanium from arsenic [1].

Germanium (IV) in aqueous solutions may be present as different anions and cations (Ge^{4+}, $Ge(OH)_3^+$ etc.).

Aqueous solutions of germanium dioxide contain ions of metagermanic (H_2GeO_3) and pentagermanic ($H_2Ge_5O_{11}$) acids, the values of the equilibrium constants between these acids depending on the pH and the total concentration of germanium (IV) ions. The dissociation constants of metagermanic acid are:

$$K_1 = 1.86 \cdot 10^{-9} \text{ and } K_2 = 1.9 \cdot 10^{-13}.$$

In hydrochloric acid solutions of germanium dioxide equilibrium is established between germanic acid or germanium dioxide and $GeCl_5^-$ or $GeCl_6^{2-}$ anions, and it seems probable that one or two intermediate acido-complexes $[Ge(OH)_xCl_{5-x}]^-$ or $[Ge(OH)_xCl_{6-x}]^{2-}$, where $x = 3$ or 4, participate in the equilibrium.

Solutions of germanium dioxide contain both cations and anions [2,3]. Germanium cations are absent in an alkaline medium, but appear at pH below 7. Ions of germanium (IV), germanyl GeO^{2+} or $Ge(OH)_2^{2+}$ ions are formed in solutions. Despite the fact that their concentration in weakly and moderately acid solutions is small, many reactions of germanium with organic reagents take place entirely because of the presence of cations. Thus, Complexone III reacts with germanium cations in acid solution.

Hydrogen sulfide precipitates from strongly acid solutions the white germanium disulfide GeS_2, which is sparingly soluble in 3 *N* hydrochloric acid. It is readily hydrolyzed and it is soluble in solutions of alkali sulfides, where it forms GeS_3^{2-} and other ions.

Germanium (IV) typically forms complex compounds with chlorides, fluorides, oxalates, polyhydric alcohols (glycerol, mannitol), and with phenols which contain at least two ortho-hydroxy groups.

The fluoride (GeF_6^{2-}) and oxalate complexes of germanium (IV) are very weakly dissociated. Germanium may be masked by potassium fluoride or potassium oxalate. In the presence of KF germanium is not precipitated by hydrogen sulfide from acid solutions. In this way arsenic (III) ions can be separated from germanium, since arsenic sulfide As_2S_3 is precipitated in the presence of KF. Hexafluorogermanates are fully decomposed by alkalis.

Ammonium molybdate reacts with germanic acid in nitric acid, sulfuric acid or hydrochloric acid medium, similarly to silicic acid, with formation of the yellow germanomolybdic acid $H_4[Ge(Mo_3O_{10})_4]$. Germanium in this compound may be determined by photometry, either directly or after reduction to the blue compound (see p. 299).

Complexone III reacts with germanium (IV) ions in acid solutions (0.02–0.05 *N*) in the molar ratio of 1:1. The reaction is terminated after 10 minutes of boiling provided the solution contains not less than a 2.5-fold excess of Complexone III. The reaction proceeds slowly in the cold. Germanium cannot be displaced from the resulting complex by zinc ions in alkaline medium, by bismuth ions in weakly acid medium or by zirconium ions in strongly acid medium. In this way germanium can be determined by titrating the excess of Complexone III against solutions of a zinc salt, bismuth salt or zirconium salt, using suitable indicators [4]. The method has been employed in the analysis of technologically important materials.

Polyhydric alcohols or monosaccharides (glycerol, mannitol, fructose, glucose, galactose, etc.), taken in excess, react with solutions of germanium dioxide in the molar ratio of 2:1 with formation of monobasic complex acids. The thermodynamic ionization and instability constants of these acids at 25°C are given below:

	Ionization constants [5]	*Instability constants* [6]
Difructosogermanic acid	$(1.04 \pm 0.01) \cdot 10^{-4}$	$4.24 \cdot 10^{-5}$
Digalactosogermanic acid	$(2.39 \pm 0.01) \cdot 10^{-5}$	$7.64 \cdot 10^{-3}$
Dimannitolgermanic acid	$(1.21 \pm 0.01) \cdot 10^{-5}$	$4.04 \cdot 10^{-4}$
Diglucosogermanic acid	$(8.35 \pm 0.03) \cdot 10^{-6}$	$3.54 \cdot 10^{-2}$
Diglycerolgermanic acid	$(5.05 \pm 0.02) \cdot 10^{-6}$	$7.85 \cdot 10^{-2}$

The relatively large increase in the acidity of the solutions noted as a result of the interaction between the weak germanic acid with mannitol or fructose makes it possible to determine germanium by the titrimetric method, using a suitable indicator.

Germanium dioxide in aqueous solutions reacts with simple diphenols, forming two series of complex compounds: the sparingly soluble complexes $GePh_2 \cdot 2H_2O$— the so-called germaniumdiphenols—and the soluble complexes H_2GePh_3—the so-called phenolgermanic acids.

Tripyrocatecholgermanic and tripyrogallolgermanic acids in 0.01 *M* and in more dilute solutions are fully ionized in the first stage. The second ionization constants are respectively $1.85 \cdot 10^{-2}$ and $2.21 \cdot 10^{-2}$ at 25°C. The dissociation constants of the complexes according to the equation

$$HGePh_3^- + 3H_2O \rightleftharpoons HGeO_3^- + 3H_2Ph$$

are $2.16 \cdot 10^{-9}$ and $0.88 \cdot 10^{-9}$ at 25°C.

When diphenol is added to an aqueous solution of GeO_2, only the soluble complex acid is formed [7].

Phenolgermanic acids give salts with pyridine, piperidine, quinoline, triphenylmethane dyes, and cations of o-phenanthroline complexes of Co, Ni, Cu, Zn, and Cd.

The formation of tripyrocatecholgermanic acid forms the basis of the alkalimetric method for the determination of germanium (the end point of the titration is found potentiometrically or with the aid of Bromocresol Purple). The method gives highly reproducible and accurate results [8] (p. 296). This reaction also forms the basis of the gravimetric method for the determination of germanium (p. 295).

o-Dihydroxyazo-compounds react with tetravalent germanium with formation of colored products [9].

The various derivatives of 2,3,7-trihydroxy-6-fluorone react with germanium (IV) ions with formation of colored compounds [10–12]. Phenylfluorone (p. 300) is the most suitable for the detection and photometric determination of microgram amounts of germanium. The reagent is used in photometric determination of 0.0001–0.2% germanium in silicates, sulfide and other ores and in coals. Germanium is first isolated as $GeCl_4$ by extraction with carbon tetrachloride from 9 *N* hydrochloric acid or by distillation from solutions in hydrochloric and phosphoric acids.

o-Dihydroxychromophenols form colored compounds with germanium (IV) ions [13]. The most sensitive reagent is 6,7-dihydroxy-2,4-diphenylbenzopyrylium chloride. The reagent is used in photometric determination of germanium. The method is used in the analysis of different materials. Complexone III masks Zr, Th, Fe(III), Sn(IV), Bi(III) ions; phosphoric acid masks Ti(III) ions, while hydrogen peroxide minimizes the effect of the following ions: Mo(VI), W(VI), Sb(III) and V(V).

Purpurogallin interacts with germanium (IV) in the molar ratio of 2:1 in a wide pH range [14], with formation of pale yellow or pale pink soluble compounds; a pink precipitate separates out on standing. In the presence of ethanol and gelatin the solutions remain clear. The light absorption curves of the purpurogallin complex of germanium have a maximum at 340 mμ. Purpurogallin is suitable for the spectrophotometric determination of germanium. In 3 *N* hydrochloric acid the molar extinction coefficient is 34,000 at 340 mμ. The optical density is proportional to the concentration between 0.1 and 3.4 μg Ge/ml.

In its detection and quantitative determination germanium is separated by distilling off as germanium tetrachloride $GeCl_4$ (b. p. 86°C) from hydrochloric acid solution in a stream of chlorine in the presence of an oxidizing agent; under these conditions arsenic (III) is transformed into the nonvolatile arsenic (V). It is more convenient to separate germanium from many accompanying elements by extracting germanium tetrachloride with carbon tetrachloride from solutions in 8–10 *N* hydrochloric acid.

Small amounts of germanium are isolated by coprecipitation with ferric hydroxide and also by precipitation with tannin and by some other methods.

Relatively large amounts of germanium are determined by alkalimetric titration of the monobasic mannitolgermanic acid or the dibasic tripyrocatecholgermanic acid. These methods give satisfactory results. Gravimetric methods for the determination of germanium (weighing as GeO_2, Mg_2GeO_4, germanomolybdates of organic bases, etc.) are used relatively rarely. A method based on weighing cadmiumphenanthroline tripyrocatecholgermanate $[CdPhen_2]$ $[GeIO_2C_6H_4)_3]$ is noteworthy.

Small amounts of germanium are determined by photometric methods, as germanomolybdic complex of 9-phenyl-2,3,7-trihydroxyfluorone, and also by other hydroxylated organic reagents. Polarographic (p. 304) and spectroscopic methods are of practical importance in the determination of germanium.

The reader is referred to a survey of the analytical methods for the determination of germanium [15] and a survey of the chemistry of germanium [16] (complex formation and valencies of germanium in solutions).

Gravimetric determination of germanium as tripyrocatecholgermanate [17]

Tripyrocatecholgermanic acid is precipitated at pH 3.5–4.5 by the doubly charged cationic complex of cadmium with o-phenanthroline. The resulting compound $[Cd\ Phen_2]$ $[Ge\ (O_2C_6H_4)_3]$ is weighed. In the determination of 0.5–20 mg Ge the error is less than 2%. The method is applicable after the separation of germanium by distillation or by extraction.

REAGENTS

Complex of cadmium with o-phenanthroline, 0.01 *M* solution. 0.21 g of anhydrous cadmium sulfate is dissolved in 50 ml of water; 0.364 g of o-phenanthroline is dissolved in 50 ml of water and the two solutions are mixed.

Acetate buffer solution, pH 4.

Pyrocatechol, 10% solution, freshly prepared.

Sodium sulfate, 5% solution.

Ethanol, 96%.

PROCEDURES

Fifteen ml of sample solution, which should contain 0.5–20 mg Ge, are mixed with 5 ml of pyrocatechol solution, 2 ml of sodium sulfate solution (to produce a better coagulation of the precipitate) and 8 ml of acetate buffer solution; 25 ml of the solution of cadmium–phenanthroline complex is then introduced with stirring. After 15–30 minutes the precipitate is separated on No. 3 or No. 4 sintered glass filter, washed with two 5-ml portions of water, once with 3 ml of ethanol, dried at 100–110°C to constant weight and weighed. The conversion factor to Ge is 0.08347.

Potentiometric titration of tripyrocatecholgermanic acid [8, 18]

Tripyrocatecholgermanic acid is titrated against a solution of sodium hydroxide, the end point being located with the aid of a glass electrode. The titration curve is represented in Figure 18. The end point may also be found with the aid of Bromocresol Purple. B, Sb(III), Sn(IV), Fe(III) interfere. Germanium is precipitated as $GeCl_4$ by extraction or distillation.

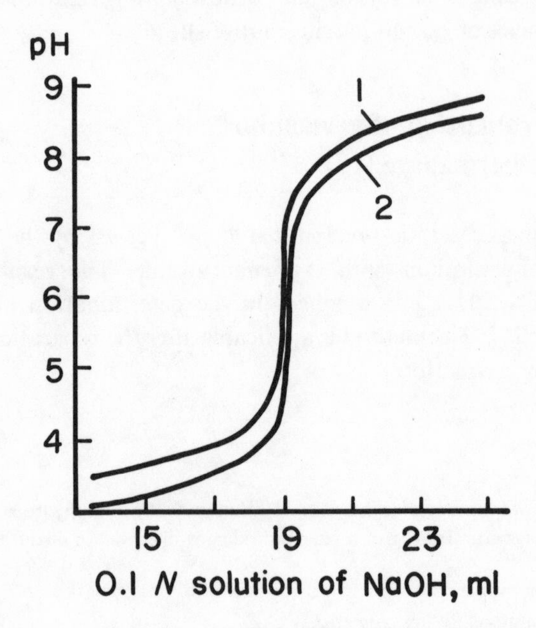

Figure 18

Potentiometric titration curves of tripyrocatecholgermanic acid (100 mg GeO_2 in 200 ml solution): 1) in the absence of NaCl; 2) in the presence of NaCl (1mole/liter).

REAGENTS

Pyrocatechol, crystalline.

Sodium hydroxide, 1 *N* and 0.02 *N* solutions, free from Na_2CO_3. The 0.02 *N* solution of NaOH is standardized against pure GeO_2. To do this a weighed amount of GeO_2 is dissolved in a solution of sodium hydroxide free from sodium carbonate and the preparation is continued as described in the procedure.

Hydrochloric acid, 1 *N*.

PROCEDURE

About 250 ml of sample solution containing 20–100 mg of Ge are adjusted to pH 5 with 1 *N* sodium hydroxide or HCl (pH verified with glass electrode). Pyrocatechol

(3 g) is added and the solution is titrated with 0.02 *N* NaOH to pH 5 at the rate of 1 drop per second.

Complexometric determination of germanium (IV) [4]

The acid solution of germanium dioxide is heated with excess Complexone III, when germanium (IV) complexonate is quantitatively formed. The excess of Complexone III is titrated against a solution of zirconium or zinc salt in the presence of a suitable indicator.

Magnesium, alkaline earth metals, zinc, lead, aluminium and many trivalent and bivalent metals do not interfere with the titration of excess Complexone III. One hundred mg of arsenic, less than 50 mg of antimony and less than 15 mg of molybdenum do not interfere.

When germanium is determined by titration against a solution of zinc salt, magnesium, alkaline earth metals and all metals which form complexonates in ammoniacal medium interfere. Arsenic (100 mg), antimony (up to 10 mg) and molybdenum (up to 15 mg) do not interfere.

Iron and tin interfere during the titration of excess Complexone III both in alkaline and acid medium. For this reason germanium in multicomponent specimens can only be determined following its separation from many other elements. This can be done by distilling germanium off as $GeCl_4$ or by extracting germanium tetrachloride with carbon tetrachloride.

The method will determine between 0.05% and a few tenths of one per cent of germanium in technological materials.

The method is used in the analysis of cinders, germanium-enriched coals and germanium dioxide concentrates containing 20–60% Fe_2O_3, 10–20% SiO_2, a few per cent of As_2O_3, Al_2O_3, CaO, and oxides of other elements.

In the analysis of cinders germanium is isolated by extraction with carbon tetrachloride, whereas during the analysis of the concentrates it is distilled off from 6 *N* hydrochloric acid. There is no need to separate germanium from arsenic during the extraction or distillation, since arsenic does not interfere with subsequent complexometric determination of germanium.

REAGENTS

Nitric acid, $d = 1.4$ g/cm^3.
Hydrofluoric acid, 40%.
Orthophosphoric acid, $d = 1.8$ g/cm^3.
Hydrochloric acid, $d = 1.19$ g/cm^3, and 1 *N* and 6 *N* solutions.
Carbon tetrachloride, pure.
Hydroxylamine hydrochloride, 2% solution in 9 *N* hydrochloric acid.
Phenolphthalein, 0.1% ethanolic solution.

Sodium hydroxide, 2 *N* and 18 *N* solutions.
Complexone III, 0.005 *M*, 0.05 *M* and 0.1 *M* solutions.
Xylenol Orange, 0.1% solution.
Zirconium oxychloride, 0.005 *M*, 0.05 *M* and 0.1 *M* solutions in 1 *N* hydrochloric acid.
Ammonia, 25% solution.
Chromogene Black ET-OO, 0.1% solution.
Zinc sulfate, 0.005, 0.05 and 0.1 *M* solutions.

PROCEDURE

Determination of germanium in cinders

Cinders (0.5–1 g) in a platinum dish are mixed with 2 ml of nitric acid and 5 ml hydrofluoric acid, and the result is evaporated on a water bath to a residual volume of 1–1.5 ml; 2 ml of nitric acid, 5 ml of hydrofluoric acid and 3 ml of orthophosphoric acid are added and the mixture is evaporated first on a water bath and then on a sand bath to a syrupy consistency; it is then cooled, the walls of the dish are rinsed with water and the mixture is again evaporated to a thick, syrupy consistency. It is then rinsed into a separating funnel with 25 ml of water; 75 ml of concentrated hydrochloric acid are added and the $GeCl_4$ is extracted with three 10-ml portions of carbon tetrachloride. The combined extracts are washed with three 10-ml portions of hydroxylamine hydrochloride, and germanium tetrachloride is reextracted with three 10-ml portions of water in a second separating funnel.

One drop of phenolphthalein is added to the reextract in a conical flask and then 18 *N* solution of sodium hydroxide is added drop by drop to the appearance of a pink coloration which is then removed by adding 2–3 drops of 1 *N* HCl. Hydrochloric acid (2.5 ml, 1 *N*) is added, the solution is diluted to 50 ml with water and 25 ml of 0.005 *M* Complexone III solution are added. The mixture is heated to boiling and gently boiled for 15 minutes, making up the evaporated water from time to time.

The excess Complexone III is titrated against a solution of zirconium oxychloride. To do this, 3 ml of 1 *N* HCl and 3 drops of Xylenol Orange solution are added, the mixture is cooled to 55°C and titrated against 0.005 *M* solution of zirconium oxychloride in 1 *N* hydrochloric acid until the yellow solution turns bright pink.

The excess of Complexone III may be titrated against a solution of zinc sulfate. To do this the solution is cooled to room temperature, 2 ml of ammonia solution, and 5 drops of Chromogene Black ET-OO are added and the solution is titrated against 0.005 *M* solution of zinc sulfate until the color changes from blue-green to pink without a blue tinge.

One ml of 0.005 *M* solution of Complexone III is equivalent to 0.363 mg of Ge.

Determination of germanium in concentrates

The sample size of the concentrate should be between 0.25 and 1.0 g, depending on the germanium content. Germanium tetrachloride is distilled from 15 ml of 6 *N* HCl.

The distillation of $GeCl_4$ is conducted in a special apparatus. A round-bottomed, 100-ml flask, with a 9-cm long neck is connected to a fractionating column by a black rubber stopper; the height of the fractionating column should be 7–8 cm from the widening point to the outlet tube. The column is filled with glass fragments and is closed with a black rubber stopper. The outlet tube of the column is connected end-to-end by a black rubber tube with the tube of vertical condenser bent at an obtuse angle. The length of the internal tube in the condenser should be between 20 and 25 cm. Its lower end protrudes into a 100-ml conical flask containing 20 ml of water.

During the distillation the receiver is immersed in cold water, while a strong current of cold water is passed through the condenser.

The concentrate sample is placed in the flask of the apparatus and 15 ml of 6 *N* HCl are added. The flask is connected to the fractionating column by the stopper, the apparatus is assembled, the flask is heated over a gas flame and the distillation is continued for 15 minutes until the residual volume of the liquid in the flask is 2–3 ml. After cooling, 15 ml of 6 *N* hydrochloric acid are added to the distillation flask through the column, and the distillation is repeated at the previous rate until 2–3 ml of liquid remain in the flask. This operation is then repeated once more.

The resulting distillates are transferred into a 500-ml conical flask, 2–3 drops of phenolphthalein solution are added and 18 *N* sodium hydroxide is introduced drop by drop to a pink coloration, which is removed by adding 3–5 drops of 1 *N* hydrochloric acid. Hydrochloric acid (1 *N*, 4.5 ml), 50 ml of 0.1 *M* or 0.05 *M* Complexone III solution are added and then water to a total volume of 150 ml. The mixture is heated to boiling and gently boiled for 15 minutes, making up the evaporated water. The excess of Complexone III is titrated against 0.1 *M* or 0.05 *M* solution of zirconium oxychloride. To do this the solution is diluted with water to 300 ml, the temperature is brought to 55°C, 20 ml of 1 *N* hydrochloric acid and 0.5 ml of 0.1% Xylenol Orange solution are added and the mixture is titrated against 0.1 *M* or against 0.05 *M* solution of zirconium oxychloride.

The excess Complexone III may also be titrated against a solution of zinc sulfate. The sample solution is cooled to room temperature, water is added to a volume of 300 ml, 6 ml of ammonia solution, 1 ml of Chromogene Black ET-OO is added and the mixture is titrated against 0.1 *M* or 0.05 *M* solution of zinc sulfate.

One milliliter of a 0.1 *M* solution of Complexone III corresponds to 7.26 mg of germanium.

Photometric determination of germanium as germanomolybdic hetropoly acid [19]

Germanium is separated by carbon tetrachloride extraction from 9 *N* hydrochloric acid (distribution coefficient 500). Arsenic (III) and osmic acid are extracted at the same time. Germanium is reextracted with water and the determination is terminated photometrically, using the yellow coloration of the germanomolybdic acid.

Thousandfold amounts of silicon, ferric iron and arsenic (V) do not interfere.

REAGENTS

Standard solution of germanium, 0.2 mg GeO_2/ml.
Carbon tetrachloride, pure.
Hydrochloric acid, $d = 1.19$ g/cm^3 and 9 *N* solution.
Ammonium molybdate, 5% solution.
Sulfuric acid, 2 *N* solution.

CONSTRUCTION OF CALIBRATION CURVE

Into five separating funnels are placed 1, 2, 3, 4 and 5 ml of the standard solution of germanium; 20 ml of hydrochloric acid ($d = 1.19$ g/cm^3) and, if necessary, water to a total volume of 25 ml are added to each flask; the solutions are cooled, 10 ml of carbon tetrachloride are added to each funnel and the mixtures are shaken for 2 minutes. The extraction is repeated twice more with 5-ml portions of carbon tetrachloride. The extracts are collected into other separating funnels, washed with 5 ml of 9 *N* hydrochloric acid, the phases are separated and germanium is extracted twice, each time with 6 ml of water for 2 minutes. To the aqueous solutions are added 10 ml of ammonium molybdate solution and 5 ml of sulfuric acid, the solutions are made up to the mark with water in a 25-ml volumetric flask and their optical densities are measured after 10 minutes in a 5-cm cell in an electrophotocolorimeter with a blue filter. The data thus obtained are used to plot a calibration curve.

PROCEDURE

All the reagents as above are added to the sample solution and the procedure is continued as above. The optical density is measured and the content of germanium is read off the calibration curve.

Photometric determination of germanium by phenylfluorone [10, 12, 20]

Germanium (IV) ions react with phenylfluorone (9-phenyl-2,3,7-trihydroxy-6-fluorone) in acid solution with formation of a red precipitate in which the molar ratio is 1:2. In the presence of small amounts of germanium a suspension appears which can be stabilized by the addition of a protective colloid. Variations in the acidity of the solution in a relatively wide range (between 0.3 and 1.5 *N*) is without effect on the formation of the phenylfluorone compound of germanium (Figure 19). At acidities below 0.3 *N* the reagent precipitates out.

In 0.5–1.5 *N* solutions of hydrochloric acid the color develops within 15–20 minutes.

The absorption maximum of colloidal solutions is at 500 mμ (Figure 20), but the

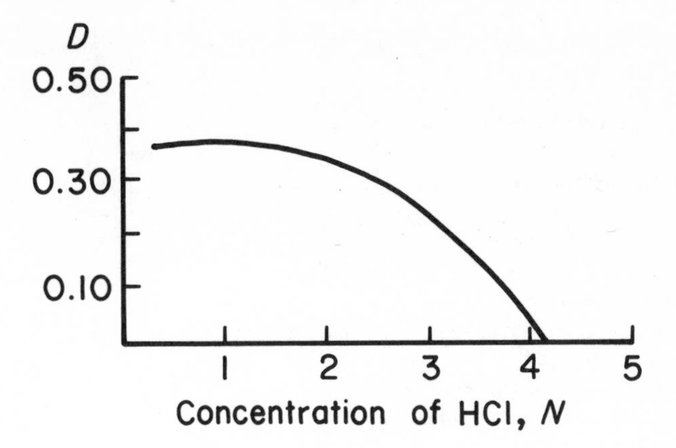

Figure 19

Effect of hydrochloric acid concentration on the color intensity of solutions of the compound of germanium with phenylfluorone

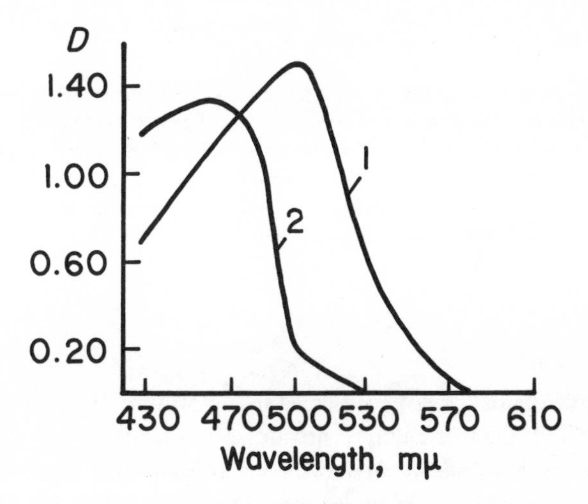

Figure 20

Absorption of light by the solution of the complex of germanium with phenylfluorone:
1) solution of complex; 2) solution of phenylfluorone.

optical density should be measured using a filter with maximum transmission at 530 mμ. The molar extinction coefficient is 38,500 $\pm$ 800.

Ti, Zr, Hf, Sn(IV), Nb, Ta, Sb(III), Te, Mo(VI), W(VI), V(V), Cr(VI), Mn(VII), and Ce(IV) interfere with the determination. In acid solutions Ga and As do not react with phenylfluorone. F^- in amounts of less than 1 mg in 10 ml and Fe^{3+} in amounts of less than 100 μg in 10 ml do not interfere with the determination.

In the analysis of different specimens germanium is separated from the interfering ions by extraction with carbon tetrachloride from 9 N solution of hydrochloric acid or by distilling off the $GeCl_4$ in the presence of an oxidizing agent.

REAGENTS

Calcium oxide, solid.
Calcium nitrate, saturated aqueous solution.
Nitric acid, $d = 1.4$ g/cm^3.
Hydrofluoric acid, concentrated (38–40%).
Phosphoric acid, $d = 1.7$ g/cm^3.
Sulfuric acid, dilute (1:1) solution.
Sodium peroxide, solid.
Ammonium nitrate, 2% solution.
Hydroxylamine hydrochloride, 2% solution in 9 *N* hydrochloric acid.
Carbon tetrachloride, pure.
Standard solution of germanium, 5 μg of Ge/ml.
Hydrochloric acid, 6 *N*, 9 *N* and 12 *N* solutions.
Gelatin, 1% solution.
Phenylfluorone, 0.05% solution. 50 mg of the reagent are dissolved with gentle heating in 100 ml of 96% ethanol and 0.5 ml of 6 *N* HCl is added.

CONSTRUCTION OF CALIBRATION CURVE

Into five 25-ml volumetric flasks are placed 1, 2, 3, 4 and 5 ml of the standard solution of germanium; 4 ml of 6 *N* hydrochloric acid, water to a volume of 20 ml and 1 ml of gelatin solution are added to each flask. After mixing, 1.5 ml of phenylfluorone solution are added to each flask, the solutions are again mixed and made up to the mark with water. After 30 minutes the optical densities of the solutions are measured in an electrophotocolorimeter or a photometer using a filter with maximum transmission at 530 mμ relative to a blank solution. The calibration curve is then plotted.

PROCEDURE

Solubilization of samples

A one-gram sample of coal or coke is placed in a platinum dish, 0.5 g CaO is added, the product is mixed, 6 ml of calcium nitrate solution are added and the mixture is evaporated almost to dryness, first on a water bath and then on a sand bath. It is then heated in a muffle furnace to 400–450°C at a slow rate (combustion of most of the carbonaceous matter) and the temperature is increased to 700–800°C. Too rapid heating at the beginning of the combustion leads to spattering and loss of material. Some coals must be ashed without the addition of calcium nitrate in order to avoid such spattering. When most of the coal has burnt up, the residue is mixed with a spatula 2 or 3 times to accelerate the ashing (the residue should be white or greyish). The residue is cooled, 5 ml of nitric acid are added drop by drop with stirring, and the mixture is evaporated to dryness on a water bath. To the residue are added 5 ml of hydrofluoric acid and the mixture is again evaporated to dryness. 5 ml of hydrofluoric and 10 ml of phosphoric acid are added and the mixture is evaporated, first on a water bath and then on a sand bath to remove hydrofluoric acid. The syrupy residue is washed into a beaker with the aid of 25 ml of water and is heated until the lumps have disintegrated.

In the analysis of ferric ores germanium may be extracted by decomposing the specimen with hydrofluoric, nitric and phosphoric acids. A 0.5–1.0-g sample of the ore is placed in a platinum dish, 5 ml of nitric acid are added drop by drop from a pipet, taking care to avoid a vigorous reaction, and the mixture is evaporated to dryness on a water bath. The treatment with nitric acid and the evaporation are then repeated once or twice more. To the residue are added 5 ml of hydrofluoric acid and 5 ml of phosphoric acid and the mixture is evaporated to remove the hydrofluoric acid. If tiny drops of sulfur are noted on the surface of the liquid mass, 3–4 ml of nitric acid are again added and the mixture is evaporated until a syrupy residue is obtained. The syrupy residue is washed into a beaker with the aid of 25 ml of water and heated until dissolved.

Silicate samples are decomposed by a mixture of hydrofluoric, sulfuric and nitric acids (or with a mixture of hydrofluoric, phosphoric and nitric acids). A 0.5-g sample is placed in a platinum dish, and treated with 3 ml of sulfuric acid (1:16), 0.5 ml of concentrated nitric acid and 5 ml of hydrofluoric acid. The mixture is evaporated until white fumes appear. To the residue 5 ml of water are added and the mixture is heated until dissolved.

In the analysis of sulfide ores the sample (0.5–1 g) in a crucible may be heated in a muffle furnace at 400–500°C and the residue treated with the acid mixture as above or heated with calcium oxide and calcium nitrate following the method used in the case of coals.

Separation of germanium by extraction

The resulting solution, which should be 25 ml in volume, is cooled to room temperature, 75 ml of concentrated hydrochloric acid are added (any turbidity is ignored), the solution is cooled, transferred to a 200-ml separating funnel, 20 ml of carbon tetrachloride are added and the solution is extracted twice for 2 minutes each time. The extract is washed three times with 2% solution of hydroxylamine hydrochloride in 9 *N* hydrochloric acid (using 10 ml portions), with shaking. Germanium is then re-extracted twice with 6 ml of water, shaking for 1 minute each time.

The aqueous solutions are transferred to a 25-ml volumetric flask and the determination is continued according to the description for the construction of the calibration curve.

Photometric determination of germanium by 6,7-dihydroxy-2,4-diphenylbenzopyranol [13]

This method is based on the formation of an orange-red complex of germanium with 6,7-dihydroxy-2,4-diphenylbenzopyranol hydrochloride; the absorption maximum of the solution of this complex is at 470 mμ. The molar extinction coefficient of the complex is 25,000. The optimum acidity is between 0.1 and 0.6 *N* in HCl.

Zr, Hf, Ti, Th, Mo, W, V, Ta, Nb and Sn interfere; for this reason germanium is separated by extraction.

The method will determine between 0.1 and 25 μg of Ge/10 ml.

REAGENTS

6,7-Dihydroxy-2,4-diphenylbenzopyranol hydrochloride, 0.2% ethanolic solution.
Gelatin, 1% solution.
Hydrochloric acid, 5 *N* solution.
Standard solution, 5 μg of Ge/ml.

CONSTRUCTION OF CALIBRATION CURVE

Into five 25-ml volumetric flasks are introduced 1, 2, 3, 4 and 5 ml standard solution of germanium; 1 ml of hydrochloric acid, 1 ml of gelatin solution, and 1.25 ml of reagent solution are then added to each flask and the solutions are diluted with water to 25 ml. The optical density is measured after 3 minutes in an electrophotocolorimeter with a green filter against a blank solution. The calibration curve is then plotted.

PROCEDURE

All the above reagents are added to the sample solution, the optical density is measured and the content of germanium is read off the calibration curve.

Polarographic determination of germanium in sulfide ores [21]

Germanium tetrachloride is separated by extraction with carbon tetrachloride and its polarogram is taken in carbonate–complexonate solution at pH 8.0–9.0. The half-wave potential of germanium is −1.5 V, relative to saturated calomel electrode.

The method will determine 0.0005–0.01% Ge in a 1-gram sample or 0.001–0.02% Ge in a 0.5-g sample.

REAGENTS

Nitric acid, $d = 1.4$ g/cm^3.
Hydrochloric acid, $d = 1.19$ g/cm^3 and 9 *N* solution.
Sulfuric acid, dilute (1:100) solution.
Phosphoric acid, $d = 1.7$ g/cm^3.
Hydrofluoric acid, 40% solution.
Carbonate mixture: 2 g of anhydrous sodium carbonate are ground in a mortar with 20 g of sodium bicarbonate.
Sodium sulfite, crystalline.
Sodium hydroxide, 10 *M* solution.

Carbon tetrachloride, pure.
Complexone III, 0.1 *M* solution.
Phenolphthalein, 0.1% ethanolic solution.
Standard solution of germanium, 25 μg of Ge/ml. This solution is diluted 10 times prior to the construction of the calibration curve.

CONSTRUCTION OF CALIBRATION CURVE

Solutions corresponding to 0.0, 0.2, 0.4, 1.0, 2.0 and 4.0 μg Ge are introduced into a number of 100-ml beakers, all the reagents listed under *Procedure* are added, germanium tetrachloride is extracted, then reextracted, polarogrammed, and the calibration curve is plotted.

PROCEDURE

A 0.5–1.0 g finely ground sulfide ore is solubilized with the aid of nitric acid, hydrofluoric acid and phosphoric acid. Germanium tetrachloride is separated by extraction with carbon tetrachloride. The procedure is described in detail on p. 298.

To perform the polarographic determination on the acid solution (after the reextraction of $GeCl_4$), 2.5 ml of sodium hydroxide solution are added, the excess alkali is neutralized with hydrochloric acid until colorless to phenolphthalein and 1–2 drops of the alkali are then added until the solution becomes pink. The weakly alkaline solution is boiled for 3–5 minutes, evaporated to a residual volume of 4–7 ml and, depending on the amount of germanium to be expected, transferred to a 25-ml volumetric flask or to a 10-ml tube. Each 10 ml of the final solution is mixed with 0.4 g of the dry sodium carbonate–sodium bicarbonate mixture, then 1 ml Complexone III solution and 0.2 g sodium sulfite are added and the solution is diluted to mark with water. A part of the solution is transferred to the electrolyzer, a stream of purified hydrogen or nitrogen is passed through and the polarogram is taken between −1.2 and −1.8 V. The height of the polarographic wave is measured.

The concentration of germanium is found from the calibration curve constructed from standard solution.

BIBLIOGRAPHY

1. IVANOV-EMIN, B.N.—*Zav. Lab.*, **13**, 161. 1947.
2. NAZARENKO, V.A., G.V. FLYANTIKOVA, and N.V. LEBEDEVA.—*Ukrainskii Khimicheskii Zhurnal*, **28**, 266. 1962.
3. DRANITSKAYA, R.M., A.I. GAVRIL'CHENKO, and A.A. MOROZOV.—*Ukrainskii Khimicheskii Zhurnal*, **28**, 866. 1962.
4. NAZARENKO, V.A., N.V. LEBEDEVA, and L.I. VINAROVA.—*ZhAKh*, **19**, 87. 1964; **21**, 192. 1966.
5. NAZARENKO, V.A. and G.V. FLYANTIKOVA.—*ZhNKh*, **8**, 1370. 1963.
6. NAZARENKO, V.A. and G.V. FLYANTIKOVA.—*ZhNKh*, **8**, 2271. 1963.
7. ANDRIANOV, A.M. and V.A. NAZARENKO.—*ZhNKh*, **8**, 2276, 2281. 1963.

8. NAZARENKO, V.A. and A.M. ANDRIANOV.—*Ukrainskii Khimicheskii Zhurnal,* **30**, 620. 1964.
9. NAZARENKO, V.A. and G.V. FLYANTIKOVA.—*ZhAKh,* **18**, 172. 1963.
10. GILLIS, J., J. HOSTE, and A. CLAEYS.—*Anal. chim. acta,* **1**, 302. 1947.
11. NAZARENKO, V.A. and N.V. LEBEDEVA.—*Trudy GIREDMET,* **2**, 63. Metallurgizdat. 1959.
12. NAZARENKO, V.A., N.V. LEBEDEVA, E.A. BIRYUK, and M.B. SHUSTOVA.—*ZhNKh,* **7**, 2731. 1962.
13. KONONENKO, L.I. and N.S. POLUEKTOV.—*ZhAKh,* **15**, 61. 1960.
14. NAZARENKO, V.A. and E.N. POLUEKTOVA.—*ZhAKh,* **19**, 1459. 1964.
15. NAZARENKO, V.A.—In: *Metody opredeleniya i analiza redkikh elementov,* p. 400. Izd. AN SSSR. 1961.
16. NAZARENKO, V.A. and A.M. ANDRIANOV.—*Uspekhi Khimii,* **34**, 1313. 1965.
17. NAZARENKO, V.A. and A.M. ANDRIANOV.—*Zav. Lab.,* **29**, 795. 1963.
18. WUNDERLICH, E. and E. GÖHRING.—*Z. anal. Chem.,* **169**, 346. 1959.
19. SHAKHOVA, Z.F., R.K. MOTORKINA, and N.N. MAL'TSEVA.—*ZhAKh,* **12**, 95. 1957.
20. NAZARENKO, V.A., N.V. LEBEDEVA, and R.V. RAVITSKAYA.—*Zav. Lab.,* **24**, 9. 1958.
21. SOCHEVANOV, V.G.—*Metallurgicheskaya i khimicheskaya promyshlennost' Kazakhstana. Nauchno-tekhnicheskii sbornik* No. 3, 7. 83. Alma-Ata, Kazakhskoe Pravlenie NTO Tsvetnoi Metallurgii. 1960.

Bismuth Bi

Bismuth displays valencies of +3 and +5. Compounds of bismuth (IV) may also exist. In the hydride BiH_3 bismuth has a valency of −3. The compounds of bismuth (III) are the most important ones in analytical chemistry.

The standard electrode potentials in aqueous medium at 25°C have the following values (relative to standard hydrogen electrode):

$Bi + H_2O + Cl^- \rightleftharpoons BiOCl + 2H^+ + 3e$	+ 0.16 V
$Bi + 4Cl^- \rightleftharpoons BiCl_4^- + 3e$	+ 0.16 V
$Bi \rightleftharpoons Bi^{3+} + 3e$	+ 0.215 V
$Bi + H_2O \rightleftharpoons BiOH^{2+} + H^+ + 3e$	+ 0.254 V
$Bi + H_2O \rightleftharpoons BiO^+ + 2H^+ + 3e$	+ 0.320 V
$2Bi + 3H_2O \rightleftharpoons Bi_2O_3 + 6H^+ + 6e$	+ 0.371 V
$2Bi_2O_3 + H_2O \rightleftharpoons Bi_4O_7 + 2H^+ + 2e$	+ 1.338 V
$Bi_4O_7 + H_2O \rightleftharpoons 2Bi_2O_4 + 2H^+ + 2e$	+ 1.541 V
$Bi_2O_4 + H_2O \rightleftharpoons Bi_2O_5 + 2H^+ + 2e$	+ 1.607 V
$2Bi^{3+} + 5H_2O \rightleftharpoons Bi_2O_5 + 10H^+ + 4e$	+ 1.759 V

The compounds of bismuth (V)—alkali metal bismuthates—are known only in the solid state. The ions of bismuth (V) do not exist in solution. Bismuthates are strong oxidizing agents; in acid solutions they oxidize the ions of manganese (II) to permanganate, and the ions of cerium (III) to cerium (IV).

The Bi^{3+} cations predominate in bismuth perchlorate at pH 0 or higher acidity. In less acidic solutions the ions $BiOH^{2+}$, BiO^{+}, Bi_2O^{4+}, $Bi_3O_2^{5+}$ and other ions also exist. All these ions are colorless.

The yellow oxide of bismuth (III) Bi_2O_3 is practically insoluble in water but is soluble in acids. It is a suitable gravimetric form produced during the ignition of bismuth hydroxide, basic bismuth carbonate and a number of compounds of bismuth with organic reagents.

Bismuth hydroxide $Bi(OH)_3$, unlike antimony (III) hydroxide, is distinctly basic and does not display amphoteric properties; it is soluble in acids and practically insoluble in solutions of alkalis. When boiled with concentrated (50%) solutions of alkali hydroxides, small amounts of $[Bi(OH)_4]^-$ are formed.

Bismuth (III) salts are hydrolyzed in aqueous solutions at pH 2 with the formation of precipitates of basic salts. At higher alkalinity the basic salts are converted to the amorphous hydroxide $Bi(OH)_3$; when this is heated, the yellow BiO(OH) is formed.

Bismuth salts are hydrolyzed in solution on the addition of solutions of ammonia alkali metal hydroxides, alkali metal carbonates, sodium formate, sodium acetate, different organic bases, and of solid oxides of zinc or mercury.

Of bismuth (III) salts, the chloride is most easily hydrolyzed followed by the nitrate and the perchlorate. The capacity of $BiCl_3$ to be hydrolyzed in solutions decreases with increasing concentration of hydrogen ions and chloride ions (addition of alkali metal chloride) owing to the formation of the chloride complexes of bismuth. At lower acidities of $BiCl_3$ and $Bi(NO_3)_3$ solutions, white precipitates of basic chlorides and nitrates separate out. Their composition approximately corresponds to the formulas BiOCl and $BiONO_3$. Precipitates of the basic salts are insoluble in 2 *N* NaOH, unlike the products of hydrolysis of the salts of Sb (III), Sn (II) and Sn (IV).

Hydrogen sulfide precipitates the dark-brown Bi_2S_3 which is insoluble in cold dilute acids and in solutions of ammonium sulfide and ammonium polysulfide (in contrast to Sb_2S_3, SnS and SnS_2. The sulfide Bi_2S_3 is soluble in hot dilute nitric and concentrated hydrochloric acids.

Sodium hydrogen phosphate precipitates the white $BiPO_4$ out of solutions of bismuth (III) salts; the precipitate is practically insoluble in 0.2 *N* nitric acid, unlike phosphates of other metals. The solubility product is $pL_{BiPO_4} = 19–20$. The solubility of $BiPO_4$ in HNO_3 rapidly increases in the presence of chlorides, owing to complex formation. In 0.2 *N* nitric acid Zr (IV), Th (IV) and Ti (IV) alone give sparingly soluble phosphates. The phosphate $BiPO_4$ is relatively readily soluble in dilute hydrochloric acid solutions.

Bismuth (III) forms different complex ions with chlorides $BiCl_4^-$, $BiCl_5^{2-}$, $BiCl_6^{3-}$, $Bi_2Cl_9^{3-}$ etc., and with mannitol, SCN^-, $S_2O_3^{2-}$ etc. The orange-colored complexes BiI_4^-, BiI_5^{2-}, BiI_6^{3-}, etc., are somewhat more stable.

The tartrate complexes of bismuth are less stable than the corresponding complexes of antimony (III). Citrate and oxalate complexes are known.

Bismuth gives no complexes with ammonia, unlike copper (II), silver (I), cadmium (II), etc.

The black-brown iodide BiI_3, which precipitates out on the addition of potassium iodide to a solution of a bismuth salt, and the red-colored oxyiodide BiOI (product of hydrolysis of the triiodide) is soluble in excess potassium iodide solution; this is accompanied by the formation of the yellow-colored solutions of iodobismuthite.

The ions of iodobismuthite are precipitated by various organic bases (quinine, cinchonine and many others) as $BH[BiI_4]$ and in the form of other yellow or red-colored compounds, where B is the molecule of the base; the precipitates are soluble in polar organic liquids.

Complexone III reacts [1] with bismuth ions in the molar ratio of 1:1. The reagent is used in titrimetric determination of bismuth in different materials (p. 310). The end point of the titration is determined amperometrically [1] or with the aid of chemical indicators such as Pyrocatechol Violet [1], Xylenol Orange [1], 1-(2-pyridylazo)-2-naphthol [2], etc. As regards its selectivity, rapidity, accuracy and reliability, the complexometric method is superior to all other titrimetric methods of determination of bismuth.

8-Hydroxyquinoline quantitatively precipitates bismuth ions from dilute acetic acid solution in the form of the compound $Bi(C_9H_6ON)_3$ which has a constant composition [3]. The formation of this compound forms the principle of gravimetric, titrimetric and also photometric methods of determination of bismuth. A convenient procedure is to titrate the 8-hydroxyquinoline bromatometrically, using the apparatus employed for non-compensated potentiometric titration [4], after dissolving the filtered and washed precipitate of bismuth 8-hydroxyquinolate.

Thiourea $CS(NH_2)_2$ reacts with bismuth ions in nitric acid or sulfuric acid solutions to form an intense yellow coloration [3]. Thiourea is a highly selective reagent for the detection and photometric determination of small amounts of bismuth in different materials (p. 311).

The o,o-diethyl ester of dithiophosphoric acid $(C_2H_5O)_2PSSH$ quantitatively precipitates bismuth ions as the compound $[(C_2H_5O)_2PSS]_3Bi$ from acid solutions. The resulting greenish-yellow precipitate, m.p. 55°C, has a constant composition, is practically insoluble in water and in acids, but is readily soluble in organic solvents. The solutions of $[(C_2H_5O)_2PSS]_3Bi$ in organic solvents are greenish-yellow and absorb in the ultraviolet and in the visible. Solutions of $[(C_2H_5O)_2PSS]_3Bi$ in carbon tetrachloride are fairly stable and obey Beer's law. The molar extinction coefficient at 330 mμ is $1.6 \cdot 10^4$, while at 400 mμ it is $4.5 \cdot 10^3$. Diethyldithiophosphoric acid in the form of the nickel salt is used in the gravimetric, titrimetric and extraction-photometric determination of bismuth in the presence of a number of other elements [5].

Bismuthol I and Bismuthol II form a red precipitate with Bi^{3+} ions in acid medium;

ions of other metals give differently colored precipitates. The reagents are employed in the identification of bismuth after separation from other elements [6].

1,5-Diphenylthiocarbazone (dithizone), dinaphthylthiocarbazone and their numerous derivatives form intensely colored compounds with bismuth ions; these compounds are extractable with chloroform and with other organic solvents [3,7]. The extraction equilibria of bismuth compounds of the above reagents, with numerous organic solvents in the presence of different masking compounds have been studied [7].

Dithizone, dinaphthylcarbazone and some of their derivatives are valuable reagents in the extraction-photometric determination of small amounts of bismuth.

Zinc reduces trivalent ions of bismuth and antimony in hydrochloric acid solutions to the black-colored metal which, unlike tin, is insoluble in hydrochloric acid after the removal of zinc. Silver ions and mercury ions are masked by adding KCN in excess [9,10].

The solution of sodium stannite Na_2SnO_2 reduces bismuth salts in alkaline medium separating the dark-brown precipitate of elementary bismuth.

Chromium (II) ions (solutions of the salts $CrCl_2$, $CrSO_4$ quantitatively reduce the ions of bismuth (III) to the elementary state [11–13]. The reaction is strictly stoichiometric and proceeds in accordance with the equation:

$$Bi^{3+} + 3Cr^{2+} \rightarrow Bi\downarrow + 3Cr^{3+}.$$

At the end of the reduction a sharp jump in the potential of the platinum or tungsten indicator electrode is noted. In this way bismuth can be potentiometrically titrated in the presence of lead, cadmium and other elements.

Potassium hexacyanomanganate $K_4[Mn(CN)_6]$, a solution of which may be obtained by reacting a manganous salt with excess potassium cyanide, yields a black precipitate on addition to an acidified solution of a bismuth salt; the precipitate probably consists of a mixture of the oxide BiO with metallic bismuth. The reagent becomes oxidized to $[Mn(CN)_6]^{3-}$ at the same time. The ions Ag^+, Hg_2^{2+} and Hg^{2+} are masked with cyanide [14].

The presence of traces of bismuth induce the reduction of Pb(II) by formaldehyde in the presence of excess alkali [15]. This fact is used to advantage in detecting small amounts of bismuth in the presence of lead ions.

For a survey of methods for the determination of bismuth the reader is referred to [3].

Complexometric determination of bismuth in the presence of 1-(2-pyridylazo)-2-naphthol [2]

Bismuth ions react with Complexone III to form a compound which is stable at pH 1–2; under these conditions similar complexes of many other ions are fully or almost fully dissociated. A suitable complexometric indicator is 1-(2-pyridylazo)-2-naphthol which forms intensely pink-colored complexes with bismuth ions and also with cupric and ferric ions at a pH of about 1. Above pH 1 the selectivity of 1-(2-pyridylazo)-2-naphthol as reagent for bismuth decreases.

Large amounts of Pb, Cd, Zn, Sr, Ba, Ca, Mg, Al, In, Ga, Ag and Mn do not interfere; traces of Cu, Co, Ni and Cr ions, which interfere with the location of the end point at high concentrations owing to their own coloration, do not interfere; traces of mercury ions, which form a stable complex with Complexone III at pH 1, also do not interfere.

REAGENTS

Complexone III, 0.01 *M* solution.
Ammonia, 1 *N* solution.
1-(2-*Pyridylazo*)-2-*naphthol*, 0.1% solution in methanol.

PROCEDURE

To the solution containing 2–20 mg bismuth as nitrate, ammonia solution is added to pH 1 (universal indicator paper), the solution is heated to 70–80°C and 2–3 drops of 1-(2-pyridylazo)-2-naphthol are introduced. The pink-colored solution is titrated against a solution of Complexone III until yellow.

One ml of 0.01 *M* Complexone III solution is equivalent to 2.09 mg of bismuth.

Photometric determination of bismuth by thiourea [16–18]

Bismuth reacts with thiourea to form compounds, to which different compositions have been assigned:

$$Bi(CSN_2H_4)_3Cl_3;\ Bi(CSN_2H_4)(NO_3)_2OH;\ Bi(CSN_2H_4)Cl_3$$

The determination is conducted in nitric acid medium, optimum concentration 0.4–1.2 *N*, concentration of thiourea 1%. The optical density of the solution of the thiourea complexes is proportional to the concentration of bismuth between 2 and 16 μg/liter. The color of the solution is not changed [1] during approximately 1.5 hours. The method is suitable for the determination of bismuth in lead and aluminum alloys and in many other materials.

Lead, copper and many other elements do not interfere. Large amounts of antimony (III), which forms a yellow compound with thiourea, interfere; gold (III) (brown coloration) and large amounts of mercury (I) and selenium (IV) also interfere.

DETERMINATION OF BISMUTH IN ALUMINUM ALLOYS [17]

The alloy is dissolved in hydrochloric acid. Bismuth and most of the copper and lead remain in the precipitate. The precipitate is filtered off, dissolved in nitric acid, and bismuth is determined photometrically in the solution thus obtained.

The method is suitable for the determination of bismuth in aluminum alloys with 3.5–6% copper and 0.2–0.6% lead.

REAGENTS

Standard solution of bismuth nitrate, 0.1 mg Bi/ml. Bismuth oxide (0.1115 g) is dissolved in 20 ml of 1:1 nitric acid and the solution is cautiously diluted with water to one liter in a volumetric flask.

Hydrochloric acid, dilute (1:1).

Nitric acid, dilute (1:1 and 1:10).

Thiourea, saturated aqueous solution.

CONSTRUCTION OF CALIBRATION CURVE

Into a number of 50-ml volumetric flasks are introduced standard portions of solution of bismuth nitrate corresponding to contents of 0.05–0.25 mg bismuth and 7 ml of 1:1 nitric acid are added to each flask (in the determination of bismuth and aluminum alloys); 1:4 acid is added in the determination of bismuth in lead. Ten ml of thiourea solution are then added to each flask, the solutions are diluted to mark with water and mixed. The optical density of the solutions is measured in an electrophotocolorimeter with a blue filter and the calibration curve is plotted.

PROCEDURE

A 0.1-g sample of aluminum alloy is treated in the cold with 5 ml of hydrochloric acid in a 100–150 ml beaker. As a result, aluminum, magnesium and other elements pass into solution, whereas all of the bismuth and most of the lead and copper remain in the precipitate. When the reaction has ended, 5 ml of distilled water are immediately added and the undissolved residue is separated on a small filter paper which is then rapidly washed with two small portions of hot water. The filtering and the washing of the residue should be conducted as rapidly as possible or the results of determination of bismuth will be low. The washed precipitate is dissolved on the filter in 5–10 ml of hot 1:1 nitric acid, collecting the liquid into a 50-ml volumetric flask. The filter paper is washed with small portions of 1:10 nitric acid and then with water. The washings are collected in the same flasks. Ten ml of thiourea solution are introduced into the flask and the solution is diluted to 50 ml with water. The optical density of the solution is measured on an electrophotocolorimeter.

The content of bismuth is found from the calibration curve.

DETERMINATION OF BISMUTH IN LEAD [18]

Bismuth is separated from the bulk of the lead by precipitation as basic bismuth carbonate. Bismuth ions are quantitatively precipitated from dilute nitric acid solution at pH 4.1 ± 0.2, and lead is precipitated as the basic salt from a dilute solution of lead nitrate at pH 6. The ions Zn, Cd, Co, Ni, Cu, As (≦ 1 %), Sn(<0.1 %) and Fe(III)(<0.05 %) which are contained in the lead do not interfere; the same applies to small amounts of chlorides. Ions which form colored compounds with thiourea interfere. The effect of antimony is eliminated by introducing tartaric acid into the solution.

REAGENTS

Standard solution of bismuth nitrate, see p. 312.
Nitric acid, dilute (1:4 and 1:1) solutions.
Sodium carbonate, 2 *N* solution.
Tartaric acid, crystalline.
Thiourea, saturated aqueous solution.

CONSTRUCTION OF CALIBRATION CURVE

See page 312.

PROCEDURE

The sample of lead, which should be 1 g for bismuth contents between 0.005 and 0.05 % and 10 g for bismuth contents between 0.0005 and 0.005 %, is dissolved in nitric acid with heating. The excess acid is eliminated by evaporation, the residue is dissolved in a small amount of distilled water and neutralized with a solution of sodium carbonate until turbidity appears (separation of basic salts), after which another 4–5 ml of sodium carbonate solution are added and the mixture is boiled for 1–2 minutes. The resulting precipitate contains all the bismuth and some lead.

The liquid is decanted from the precipitate as completely as possible through a small filter paper and the filter paper is transferred without washing, together with the precipitate, into the beaker which was used for the solubilization of the sample.

The precipitate is dissolved in 5–10 ml of nitric acid. If the lead sample is suspected to contain major amounts of antimony, 1–2 g tartaric acid are added. In a 50-ml volumetric flask 10 ml of thiourea solution are added to the resulting solution which is then diluted to the mark with water and its optical density measured on an electrophotocolorimeter.

The amount of bismuth is read off the calibration curve.

BIBLIOGRAPHY

1. PŘIBIL, R. *Komplexony v chemické analyse.*—Prague, Nakl. Českosl. Akad. Věd. 1957.
2. BUSEV, A.I.—*ZhAKh,* **12**, 386. 1957.
3. BUSEV, A.I. *Analiticheskaya khimiya vismuta (Analytical Chemistry of Bismuth).*—Izd. AN SSSR. 1953.
4. BUSEV, A.I.—*ZhAKh,* **8**, 299. 1953.
5. IVANYUTIN, M.I. and A.I. BUSEV.—*Nauchnye Doklady Vysshei Shkoly. Khimiya i Khimicheskaya Tekhnologiya,* No. 1, 73. 1958.
6. MAJUMDAR, A.K. and M.M. CHAKRABARITY.—*Z. anal. Chem.*, **154**, 262. 1957.
7. BUSEV, A.I. and L.A. BAZHANOVA.—*ZhNKh,* **6**, 2210, 2805. 1961.
8. BUSEV, A.I. and L.A. BAZHANOVA.—*ZhAKh,* **16**, 399. 1961.
9. KARAOGLANOV, Z.—*Z. anal. Chem.*, **114**, 81. 1938.
10. TANANAEV, N.A. and A.V. TANANAEVA.—*ZhPKh,* **8**, 1457. 1935.
11. BUSEV, A.I.—*Doklady AN SSSR,* **74**, 55. 1950.
12. BUSEV, A.I.—*ZhAKh,* **6**, 178. 1951.
13. BUSEV, A.I. *Primenenie soedinenii dvukhvalentnogo khroma v analiticheskoi khimii (Utilization of Divalent Chromium Compounds in Analytical Chemistry),* p. 73.—Izd. VNIGI. 1960.
14. TANANAEV, N.A.—*Z. anal. Chem.*, **105**, 419. 1936.
15. BUSEV, A.I.—*ZhAKh,* **5**, 255. 1950.
16. LUR'E, YU.YU. and L.B. GINZBURG.—*Zav. Lab.*, **15**, 21. 1949.
17. BUSEV, A.I.—*Zav. Lab.*, **16**, 103. 1950.
18. BUSEV, A.I. and N.P. KORETS.—*Zav. Lab.*, **15**, 30. 1949.

Selenium Se and Tellurium Te

Selenium and tellurium display valencies of +2, +4, +6, and −2. Selenium and tellurium compounds most important to analytical chemistry are those in which the elements display a valency of +4.

Many compounds of selenium and tellurium in the same valency states greatly resemble each other and resemble corresponding compounds of sulfur.

Elementary selenium and tellurium exist in several allotropic modifications. When selenites are reduced or when hydrogen selenide H_2Se is oxidized, the red-colored amorphous selenium separates out. This modification of selenium is sparingly (0.05%) soluble in carbon disulfide; it is soluble in solutions of ammonium sulfide. Selenium separates out of carbon disulfide solutions as red-brown monoclinic prisms or needles. When heated to 90–100°C, the red modification of selenium undergoes a transition into the gray modification (m.p.217°C, b.p.690°C). The gray modification of selenium is insoluble in CS_2 and conducts electricity, the electric conductivity depending on the temperature and on the intensity of illumination. A vitreous modification of amorphous selenium is also known; at 90°C this modification becomes crystalline.

In colloidal solutions selenium is present in the amorphous form. Colloidal solutions of selenium are brick-red in color and are used in the photometric determination of this element.

Tellurium is known in a number of brown or black modifications. The reduction of telliurium compounds yields amorphous tellurium which is gray-black. It readily forms colloidal solutions which are used in photometric determinations and is soluble in solutions of ammonium sulfide $(NH_4)_2S$.

The standard electrode potentials with reference to the standard hydrogen electrode in aqueous solutions at 25°C, are the following:

For selenium:

$$Se^{2-} \rightleftharpoons Se + 2e \qquad -0.92\ V$$

$$Se + 3H_2O \rightleftharpoons H_2SeO_3 + 4H^+ + 4e \qquad +0.741\ V$$

$$2Se + 2Cl^- \rightleftharpoons Se_2Cl_2 + 2e \qquad +1.1\ V$$

For tellurium:

$$Te_2^{2-} \rightleftharpoons 2Te + 2e \qquad -0.84\ V$$

The most important valency states of selenium and their corresponding formal redox potentials are shown in Figure 21.

Selenium and tellurium dissolve in nitric acid forming selenious and tellurous acids.

Tellurium is soluble in a mixture of nitric and hydrochloric acids forming tellurous and telluric acids. Hydrochloric acid alone is without effect on tellurium. Tellurium is soluble in concentrated sulfuric acid, and black tellurium precipitates out of the intense red solution on dilution.

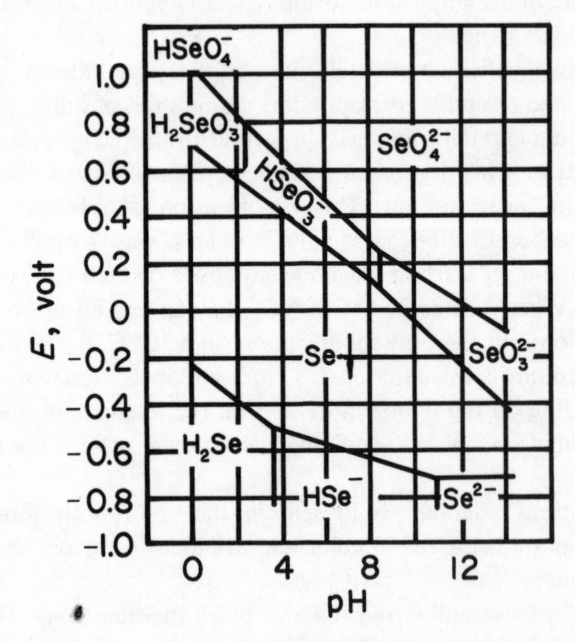

Figure 21

Approximate values of formal redox potentials for selenium as a function of pH. Selenium concentration, 10^{-2} mole/liter.

Selenious acid can be reduced to the red modification of selenium even by weak reducing agents (ferrous salts, sulfur dioxide, iodides, organic acids, thiourea, etc.). The reducing agents employed include hydrazine, hydroxylamine, ascorbic acid, ammonium thiocyanate, thiourea, chlorides of stannous tin, titanium (III) and chromium (II), metallic zinc, magnesium, iron, copper and many other reducing agents. Very strong reducing agents partly reduce selenium to hydrogen selenide H_2Se.

Tellurous acid is reduced to tellurium by stronger reducing agents (stannous chloride, zinc, etc.) than is selenious acid. Sulfurous acid reduces tellurous acid only in weakly hydrochloric acid solutions but not in strongly acid solutions. Hydroxylamine reduces tellurous acid in ammoniacal solutions and does not reduce it in acid solutions. Elementary tellurium separates out as a brownish-black powder.

It is possible to produce experimental conditions favorable for the reduction of selenious acid, e.g., by reducing with hydroxylamine hydrochloride or potassium iodide to elementary selenium, while the tellurous acid remains in solution.

When hydrogen sulfide is passed through a solution of selenious acid acidified with hydrochloric acid, a yellow precipitate separates out in the cold and an orange-colored precipitate separates out in the heat (mixture of sulfur and selenium), which is soluble in solutions of ammonium sulfide.

Sulfur dioxide reduces selenious acid to the red modification of selenium, even in the cold, in dilute and concentrated hydrochloric acids (p. 322). Tellurous acid is reduced in 10% hydrochloric acid in the heat, but not in the cold. Selenious acid is also reduced in 50% sulfuric acid, while tellurous acid is not reduced under these conditions.

Selenious acid is sometimes reduced by a solution of sulfur dioxide in acetone.

The reduction of selenious acid

$$SeO_3^{2-} + 4I^- + 6H^+ \rightleftharpoons Se + 2I_2 + 3H_2O$$

forms the basis of the various methods for the detection and the photometric and titrimetric determination of selenium.

Ascorbic acid quantitatively reduces selenious acid in 0.1 *N* hydrochloric acid to the elementary state (with the formation of dehydroascorbic acid). One mole of selenious acid consumes 2 moles of ascorbic acid. This reaction forms the basis of the microtitrimetric method [1] of determination of selenium in the commercial product. The excess ascorbic acid is titrated against a solution of iodine in the presence of starch.

Ammonium thiocyanate NH_4SCN reduces selenious acid in hydrochloric acid solutions (approximately 0.5 *N* HCl) at boiling point. Tellurous and telluric acids are not reduced under these conditions.

Thiourea reduces selenious acid in acidified solutions to elementary selenium. Tellurous acid reacts with thiourea to form yellow soluble complex compounds which are used in the photometric determination of tellurium [2]. The solutions contain the complex ions of bivalent tellurium $[Te(CSN_2H_4)_4]^{2+}$. When made alkaline, a black precipitate of elementary tellurium separates out.

Selenium separates out together with tellurium on the addition of stannous chloride, hydrazine salts or a solution of sulfur dioxide and a hydrazine salt.

For the oxidation of selenious acid in solution to selenic acid stronger oxidizing agents are required than for the oxidation of sulfurous acid.

Selenic acid, unlike selenious acid, is difficult to reduce and is reduced only by strong reducing agents ($TiCl_3$, $CrCl_2$, hypophosphorous acid, etc.). Hydrogen sulfide, sulfur dioxide and other mild reducing agents do not reduce selenious acid.

Selenic acid H_2SeO_4 reacts with concentrated (12 *N*) hydrochloric acid as follows:

$$H_2SeO_4 + 2HCl \rightleftharpoons Cl_2 + H_2SeO_3 + H_2O$$

The reaction is reversible. In alkaline medium chlorine oxidizes selenious acid to selenic acid.

Telluric acid, like selenic acid, is reduced only by strong reducing agents. When reduced by tin, stannous chloride or sodium hypophosphite in an acid medium, the black precipitate of elementary tellurium precipitates out.

Stannite ions in a strongly alkaline medium reduce tellurates and tellurites to elementary tellurium:

$$TeO_3^{2-} + 3SnO_2^{2-} \rightleftharpoons Te + 3SnO_3^{2-} + 2e$$

$$TeO_4^{2-} + 4SnO_2^{2-} \rightleftharpoons 4SnO_3^{2-} + 2e.$$

Selenium and tellurium form the gaseous hydrogen selenide H_2Se and hydrogen telluride H_2Te. In aqueous solutions hydrogen selenide behaves as a somewhat stronger acid than hydrogen sulfide (for H_2Se at 18°C $pK_1 = 3.77$ and $pK_2 = 11.0$, whereas for H_2S at 25°C $pK_1 = 7.2$ and $pK_2 = 14.0$). For hydrogen telluride H_2Te $pK = 3.0$ at 25°C.

The tendency to become oxidized increases in the sequence H_2S, H_2Se, H_2Te. These compounds are oxidized by atmospheric oxygen.

Hydrogen telluride H_2Te spontaneously decomposes into elementary hydrogen and tellurium. Aqueous solutions of hydrogen telluride are not very stable and they are readily oxidized by atmospheric oxygen with resulting separation of black tellurium.

Selenides and tellurides of different metals resemble sulfides in their properties. Solutions of alkali metal selenides and tellurides become oxidized by atmospheric oxygen on standing and become turbid owing to the separation of elementary selenium and tellurium.

Alkali metal and alkaline earth metal selenides and tellurides are water-soluble. Selenides and tellurides of other metals are sparingly soluble. Silver telluride is less soluble than silver sulfide.

Polyselenides and polytellurides are red.

Selenium dioxide SeO_2 is acidic. On being dissolved in water it forms the weak selenious acid H_2SeO_3 ($pK_1 = 2.46$; $pK_2 = 7.3$).

Solutions of sodium and potassium selenites have an alkaline reaction.

Selenites of most metals are sparingly soluble in water, but are soluble in solutions of strong acids.

Barium chloride precipitates the white $BaSeO_3$ out of solutions of alkali selenites; the precipitate is soluble in nitric and hydrochloric acids ($pL_{BaSeO_3} = 7.0$).

Lead acetate precipitates the white $PbSeO_3$, which is sparingly soluble in hydrochloric and nitric acids ($pL_{PbSeO_3} = 7.0$).

Silver nitrate yields the white precipitate Ag_2SeO_3.

Mercurous nitrate $Hg_2(NO_3)_2$ in dilute nitric acid forms the crystalline precipitate Hg_2SeO_3 which is suitable for identifying microcrystals of selenium under the microscope.

The white tellurium dioxide TeO_2 is amphoteric, soluble in hydrochloric acid and in alkali hydroxide solutions, and sparingly soluble in water. When the solution of this dioxide in an acid is diluted with water, TeO_2 precipitates out.

Tellurous acid H_2TeO_3 is a weak acid ($pK_1 = 2.5$; $pK_2 = 7.7$). The solubility of tellurous acid is small. It is precipitated out of acetate buffer solution and is relatively sparingly soluble in strong acids, the ions formed being mainly $TeOOH^+$ and also Te^{4+}. In 1 *N* hydrochloric acid and in 1 *N* perchloric acid, the solubility of tellurous acid is 0.01 mole per liter. Alkaline solutions contain TeO_3^{2-} ions.

The following equilibria of the compounds of tetravalent tellurium have been studied:

	pK
$TeO_2 + H_2O \rightleftharpoons HTeO_2^+ + OH^-$	16.1
$H_2TeO_3 \rightleftharpoons HTeO_2^+ + OH^-$	11
$3H_2O + TeO_3^{2-} \rightleftharpoons Te^{4+} + 6OH^-$	46
$Te^{4+} + 2H_2O \rightleftharpoons HTeO_2 + 3H^+$	0.7

Most of the salts of tellurous acid are sparingly soluble in water. Tellurites are soluble in strong acids.

Barium chloride reacts with tellurite solutions to form a white precipitate which is soluble in dilute nitric acid.

Selenic acid H_2SeO_4 is almost as strong as sulfuric acid. Selenates of most metals are soluble in water. The compounds $BaSeO_4$, $SrSeO_4$, $PbSeO_4$($pL_{BaSeO_4} = 7.0$; $pL_{SrSeO_4} = 5.0$) are sparingly soluble.

Telluric acid $H_2TeO_4 \cdot 2H_2O$ is a colorless crystalline substance which is readily soluble in water. It is a weak acid ($K_1 = 2 \cdot 10^{-8}$; $K_2 = 5 \cdot 10^{-11}$). The salts Ag_6TeO_6 and Hg_3TeO_6 are known; these correspond to the hexabasic (orthotelluric) acid H_6TeO_6. Tellurates of most metals are sparingly soluble in water. Potassium tellurate

$K_2H_4TeO_6 \cdot 3H_2O$ is readily soluble in water, whereas $Na_2H_4TeO_6$ is sparingly soluble. On heating, $H_2TeO_4 \cdot 2H_2O$ is converted to H_2TeO_4.

Hexavalent tellurium does not form cations.

A number of organic reagents of different types are used for the detection and quantitative determination of selenium and tellurium.

3,3′-Diaminobenzidine reacts with selenious acid at pH 2–3 with the formation of the yellow diphenylpiazselenol [3, 4]. The reagent is used in the extraction-photometric determination of selenium (p. 327).

3,3′-Diaminobenzidine makes it possible to determine [4, 5] microgram amounts of selenium in the presence of gram amounts of arsenic, without separation. The reagent will determine selenium in gallium arsenide without separation [5].

Major amounts of indium interfere with the determination of selenium [5]. In this case microgram amounts of selenium are separated by reducing with a mixture of $SnCl_2$ and hypophosphite in hydrochloric acid solution (1:7) in the presence of a collector (arsenic). This technique was used in the analysis of indium, indium antimonide and indium arsenide.

In the photometric determination of selenium 3,3′-diaminobenzidine may be replaced by o-phenylene diamine [6], which reacts with selenious acid at pH 1.5–2.5 with the formation of a practically colorless crystalline compound which can be extracted with toluene.

The optical density of the toluene extract is measured at 335 mμ. Many ions do not interfere with the determination. Ferric, stannic and iodide ions interfere. Ferric ions are masked with Complexone III.

Some of the sulfur-containing organic reagents which contain the sulfhydryl group —SH, react with selenious acid with the formation of colored compounds.

These reagents include 2-mercaptobenzimidazole. On the addition of a 1% ethanolic solution of the reagent to a dilute solution of selenious acid in 1–3 *M* hydrochloric acid a yellow coloration appears [7]. The resulting compound is readily extracted with a mixture of butanol and chloroform (1:5) and the color of the extract is very stable. At high concentrations of the reagent and of the selenious acid a yellow crystalline precipitate separates out immediately; the precipitate is not very stable. The absorption maximum of the extracts of the resulting compound of selenium is at 330 mμ, and the molar extinction coefficient is 10,500. Selenious acid reacts with the reagent in the molar ratio of 1:4. The reagent is used in the photometric determination of selenium; the optical density is not measured at the maximum absorption of the resulting compound, since the reagent itself also absorbs at this wavelength, but at 420 mμ. Mercury, silver, gold, cadmium and lead ions and other ions do not interfere; tellurium, bismuth and copper ions interfere.

If a solution of selenious acid is added to an acid solution of N-(mercaptoacetyl)-p-anisidine or N-(mercaptoacetyl)-p-toluidine, a yellow color, a yellow-brown turbidity or a precipitate appears immediately, depending on the concentration of selenious acid. The resulting compounds are readily extracted with chloroform, carbon tetrachloride, benzene, toluene, isoamyl and isobutyl alcohols. The extracts are yellow and are very stable. The extraction is best performed with a mixture of butanol with chloroform (1:5). The maximum optical density of the extracts is noted in 0.05–0.5 *M* hydrochloric acid. Elementary selenium separates out of solutions in 6 *M* hydrochloric acid. The maximum absorption of the extracts is at 315–320 mμ. The molar extinction coefficient is 1200. The reagents react [8] with selenious acid in the molar ratio of 2:1; they are used for the photometric determination of selenium in different technical materials. Te(IV), Cu(II), Bi(III) and Mo(VI) ions interfere with the determination.

The salt of selenium (II) $[(C_2H_5O)_2PSS]_2Se$ is a crystalline, yellow, very stable compound which is soluble in organic solvents [9].

Sodium diethyldithiocarbamate and other derivatives of dithiocarbamic acid are valuable reagents for the extraction-photometric determination of tellurium (pp. 334, 335). The diethyldithiocarbamate method makes it possible to determine microgram amounts of tellurium in the presence of up to 0.5 g of arsenic without separation [10]. In order to increase the sensitivity, tellurium diethyldithiocarbamate in carbon tetrachloride is converted into copper diethyldithiocarbamate. Large amounts of indium and gallium interfere with the determination of tellurium. Accordingly, in determining traces of tellurium in indium and in indium and gallium arsenides, tellurium is separated by reduction with a mixture of stannous chloride with hypophosphite in hydrochloric acid solution (1:7) using a collector (arsenic).

Bismuthiol II reacts with tellurous acid at pH 3.3–4.4 forming a yellow compound [11, 12].

Selenious acid also forms a yellow compound with Bismuthiol II; this compound, however, is extracted very slowly.

Tellurium (IV) in hydrochloric acid, hydrobromic acid and hydriodic acid solutions is present in the form of halogenated complex ions (pp. 333, 338). Different compounds containing anions (TeX_6^{2-} (X = Cl, Br, I) and cations of alkali metals or organic cations have been isolated in the crystalline state from such solutions.

Diantipyrylmethane, diantipyrylmethylmethane, diantipyrylpropylmethane and diantipyrylphenylmethane precipitate tellurium (IV) compounds from hydrobromic acid solutions.

The extraction-photometric method for the determination of tellurium in metallic lead and other materials [13] (see pp. 341 and 342) is based on the formation of some of these compounds.

The chloride complexes of tellurium (IV) are not precipitated by the derivatives of pyrazolone but are quantitatively extracted in their presence with dichloroethane.

The ions of selenium (IV) are not precipitated from hydrochloric acid solutions and are not extracted in the presence of pyrazolone derivatives. In this way tellurium and selenium can be readily separated. Chlorotellurous acid is extracted in the presence of Rhodamine B from solutions in 5–7% hydrochloric acid with a 2:1 mixture of benzene with ether. The content of tellurium is found from the intensity of fluorescence of the extract in the ultraviolet [14]. Rhodamine 6G is less suitable for the determination of tellurium than Rhodamine B.

Butylrhodamine B is a highly sensitive reagent for the extraction-photometric determination of tellurium [15]. More than $10^{-4}\%$ Te can be determined. The compound of butylrhodamine B with an anionic bromide complex of tellurium is extracted with a 3:1 mixture of benzene and ether. In, Sb, Au, Hg and large amounts of ferric and cuprous ions interfere. When tellurium is determined in a mineral sample, it is isolated together with selenium by hypophosphite from 6 *N* hydrochloric acid, using arsenic as collector or else with titanium trichloride from 5 *N* sulfuric acid and 2.5 *N* hydrochloric acid.

Selenium and tellurium may also be determined by the polarographic method (p. 345).

Gravimetric determination of selenium in steels [16]

Selenium is reduced to elementary selenium by sulfur dioxide and is weighed after drying at 105°C.

REAGENTS

Hydrochloric acid, $d = 1.19$ g/cm^3 and dilute (1:1).
Nitric acid, $d = 1.4$ g/cm^3.
Sulfur dioxide, obtained by the action of concentrated sulfuric acid on sodium sulfite.
Ethanol, 96%.
Ether.

PROCEDURE

A 5-g sample of steel containing 0.1–0.5% Se or a 2-g sample of steel containing 0.5–1% Se is dissolved in 60 ml of a hydrochloric–nitric acid mixture. The solution is evaporated to dryness on a water bath while avoiding calcination, 20 ml of hydrochloric acid (1:1) and 20 ml of water are added and the salts are dissolved by gentle heating. Silicic acid is filtered off and discarded. Sixty ml of concentrated hydrochloric acid are added to the filtrate and the solution is diluted to 100 ml with water. Gaseous sulfur dioxide is passed through the hydrochloric acid solution, which is stirred to accelerate the coagulation of the selenium precipitate.

The solution with the precipitate is left to stand overnight, filtered through a No. 3 or No. 4 sintered glass filter and the precipitate is washed with 1:1 hydrochloric acid, then with warm water until no longer acid, and finally three times with ethanol and once with ether and dried to constant weight at 105°C.

Determination of selenium and tellurium in ores and ore dressing products [17]

Selenium and tellurium are separated with hydrazine in 9.5 *N* hydrochloric acid solution. The method is based on the reaction between selenious acid and the excess of standard thiosulfate solution:

$$4Na_2S_2O_3 + H_2SeO_3 + 4HCl \rightleftharpoons Na_2SeS_4O_6 + Na_2S_4O_6 + 4NaCl + 3H_2O.$$

The unreacted thiosulfate is titrated against a solution of iodine.

Tellurium is determined by the reaction between tellurous acid and excess potassium iodide:

$$H_2TeO_3 + 4I^- + 4H^+ \rightleftharpoons Te + 2I_2 + 3H_2O$$

Small amounts of selenium and tellurium are isolated in the presence of a collector Hg_2Cl_2.

REAGENTS

Nitric acid, $d = 1.42$ g/cm^3.
Mixture of bromine and carbon tetrachloride (2:3).
Sulfuric acid, concentrated.
Hydrochloric acid, $d = 1.19$ g/cm^3, 5% solution and dilute (2:100) solution.
Mercurous chloride, solution with a concentration of 1 mg Hg/ml.
Stannous chloride, crystalline $SnCl_2 \cdot 2H_2O$ and 20% solution in 20% hydrochloric acid.
Hydrazine, 10% solution.
Urea.
Sodium thiosulfate, 0.02 *N* solution.
Iodine, 0.02 *N* solution.
Potassium iodide, 30% solution.

PROCEDURE

Solubilization of sample material

Between 1.0 and 10 g of finely ground sample material are placed in a 300–800-ml, heat-resistant beaker. The bromine–carbon tetrachloride mixture (8–30 ml) is added and the result is left to stand for 10–20 min. Nitric acid (10–75 ml) is introduced in small portions with stirring, the solution is left to stand in the cold, then cautiously heated and evaporated to a small residual volume. After complete decomposition of the sample 10–60 ml of sulfuric acid are added and the solution is evaporated to the appearance of sulfuric acid fumes. The residue is cooled, some cold water added, and the solution is again evaporated until the evolution of sulfuric acid fumes begins.

This operation is repeated as many times as necessary for complete elimination of nitrous oxides and organic matter.

The residue is dissolved in 60–200 ml of water and 10–50 ml of hydrochloric acid by boiling in the beaker under a watch glass for 5–10 minutes. The insoluble residue is filtered off and washed with hot water. Depending on the sample size, the volume of the solution is made up to 100–300 ml by adding hydrochloric acid to a final concentration of 15% by volume, and the solution is heated to boiling. Five ml of mercurous chloride solution are introduced and ferric ions are reduced by adding a 20% solution of stannous chloride in small portions. After the solution becomes clear a 3–10-ml excess of the reducing agent is added, then a little macerated filter paper, the mixture is boiled for 5 minutes and left to stand for 10 minutes at a temperature close to the boiling point. The precipitate of selenium and tellurium is filtered through a macerated filter paper plug and washed three to four times with hot 2:100 hydrochloric acid.

Separation of selenium and tellurium

The precipitate of selenium and tellurium is placed in the beaker in which the precipitation was performed, and 20 ml of concentrated hydrochloric acid and 4–5 drops of nitric acid are added; the filter is loosened and heated on a water bath until the precipitate is dissolved (the filter paper whitens). Forty ml of concentrated hydrochloric acid are added, then 50 ml of water, the solution is heated almost to boiling and 15–20 ml of hydrazine solution are added; the mixture is stirred and left to stand for four hours in a warm place. The selenium precipitate is filtered through a double filter paper and washed 7–8 times with hot 5% hydrochloric acid.

To separate tellurium, the filtrate is diluted to 250 ml with water, a little macerated filter paper is added, and the mixture is heated to boiling. Then 0.3 g of stannous chloride is added and the mixture is boiled until the solution is quite clear. The separated precipitate of tellurium is filtered on double filter paper and washed 7–8 times with hot 5% hydrochloric acid.

Determination of selenium

The filter paper with the precipitate of selenium is transferred to a beaker, 10 ml of concentrated hydrochloric acid are added, the filter is loosened with the aid of a glass rod, 3–4 drops of nitric acid are added and the mixture is heated on a water bath until the precipitate is fully dissolved. Water (150 ml) is added to the solution, then 4 g of urea and the mixture is boiled for 2 or 3 minutes. It is then cooled to room temperature, 10–20 ml of sodium thiosulfate solution are added and the excess sodium thiosulfate is titrated against iodine in the presence of starch.

Determination of tellurium

The filter paper with the precipitate of tellurium is transferred to a flask, 10 ml of concentrated hydrochloric acid and 3–4 drops of nitric acid are added, and the mixture is heated on a water bath until the precipitate is fully dissolved. Water (100 ml) is added to the solution, 4 g of urea are added and the mixture is boiled for 2–3 minutes. The solution is then cooled, 10 ml of potassium iodide solution are added and the iodine is titrated against sodium thiosulfate in the presence of starch until the blue or dark cherry-red color (depending on the tellurium content) turns light yellow.

Photometric determination of selenium as sol [18]

Selenium is conveniently reduced by hydrazine. Selenium sols absorb light in a very wide range of wavelengths; beginning with 600 mμ the absorption increases gradually, attaining a maximum at 250–260 mμ (Figure 22). Unlike tellurium sols, the optical properties of selenium sols do not vary to a significant extent with the concentration of the reducing agent. Only in high concentrations of hydrazine or alkalis does the sol consist of particles of a different form with a greyish tinge and has a relatively low absorption maximum at 360 mμ. Since hydrazine displays a considerable absorption of light at wavelengths below 260 mμ, it is desirable to work at lower concentrations and to measure the optical density at 260 mμ.

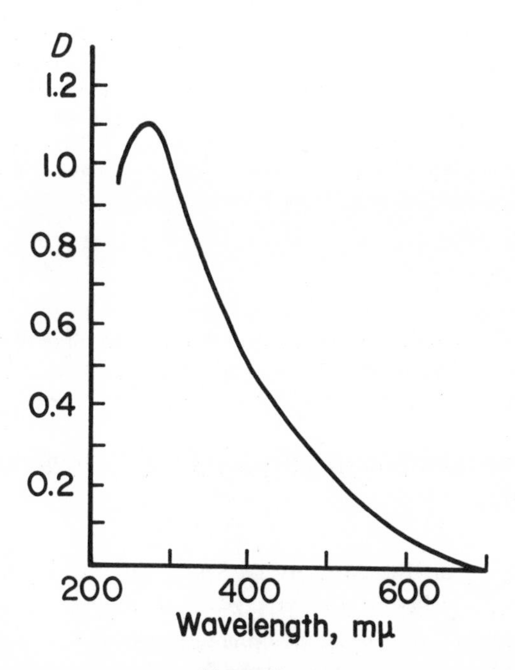

Figure 22

Light absorption by the sol of the red modification of amorphous selenium (reduction). Selenious acid (20 μg Se/ml) reduced by hydrazine.

The reduction of selenium(IV) ions by hydrazine is catalyzed by hydrogen ions. The reaction proceeds very slowly at pH 11 or above; at pH 9 the duration of the reduction is one minute and at lower pH values only a few seconds. The maximum homogeneity of the sol particles is observed at pH 8–11. Selenium sols are relatively stable in weakly alkaline medium, even in the absence of protective colloids. Hydrogen ions produce coagulation of selenium sols and for this reason a protective colloid must be used in acid media. The stability of the sols also depends on the salt concentration.

The optimum concentration of hydrazine is 0.9–1.5 moles per liter. At lower concentrations the rate of formation of the sol is reduced and the homogeneity of the particles decreases.

REAGENTS

Hydrazine hydrate, 85% (17 *M*) solution.

Standard solution of selenious acid, 0.2 mg Se/ml. The solution is prepared by dissolving a weighed amount of SeO_2 in water.

CONSTRUCTION OF CALIBRATION CURVE

Into five 50-ml volumetric flasks are placed 1, 2, 3, 4 and 5 ml of standard solution of selenious acid and the solutions are diluted to 35 ml with water. The solutions are heated to boiling and 2 ml of hydrazine hydrate solution is rapidly added to each flask (in the presence of coagulants 3 ml each of 4% gum arabic solution are introduced prior to the introduction of the reducing agent to ensure stabilization), and the heating is prolonged for 2–5 minutes. The solutions are cooled to room temperature and diluted to the mark with water. The optical density is measured at 260 mμ or at longer wavelengths. In the latter case the method becomes much less sensitive but remains almost as accurate.

PROCEDURE

All the reagents are added to the sample solution, the optical density is measured and the selenium content is read off the calibration curve.

Photometric determination of selenium by 3,3′-diaminobenzidine without extraction [3]

The method is based on the capacity of SeO_3^{2-} ions to react with 3,3′-diaminobenzidine forming the yellow diphenylpiazselenol which is readily soluble in water, in neutral, alkaline and acid media. Diphenylpiazselenol is stable in acid, neutral and alkaline solutions. It is readily extracted with toluene and other organic solvents at pH 6–7. At pH below 5 the extraction is not as satisfactory.

The light absorption of dilute aqueous solutions of diphenylpiazselenol obeys

Beer's law at concentrations between 0.25 and 2.5 μg Se/ml. The molar extinction coefficient is 10,200 at 347–349 mμ at an acidity of 0.1 *N*. The color of the solution develops within 50 minutes under these conditions and remains constant for 5 hours. Al, Ba, Ca, Cd, Co, K, Cr(III), Mg, Mn, Mo(VI), Ni, Ti(IV), Na, NH_4^+, Zn, Br^-, Cl^-, NO_3^-, PO_4^{3-}, SO_4^{2-}, citrate, oxalate and tartrate ions do not interfere. Ferric ions are masked by fluoride or oxalate.

REAGENTS

3,3'-Diaminobenzidine, 0.1% solution of tetrahydrochloride in dilute hydrochloric acid.
Hydrochloric acid, 5 *N* solution.
Standard solution of selenious acid, 20 μg Se/ml, is prepared by dissolving a known amount of pure SeO_2 in water.

CONSTRUCTION OF CALIBRATION CURVE

Into six 50-ml volumetric flasks are placed 0.5, 1, 2, 3, 4 and 5 ml of standard solution of selenious acid, 25 ml of reagent solution are added and the final acidity is adjusted to 0.1 *N* by adding 5 ml of hydrochloric acid. The solutions are diluted to the mark with water and the optical density is measured after 1 hour on an electrophotocolorimeter with a blue filter.

PROCEDURE

All the reagents are added as above to the sample solution, the optical density is measured as above and the selenium content is found from the calibration curve.

Extraction-photometric determination of selenium by 3,3'-diaminobenzidine [3–5, 15]

The toluene extract of diphenylpiazselenol has two absorption maxima at 340 and 420 mμ. 3,3'-Diaminobenzidine in toluene has its absorption maximum at 340 mμ. For this reason the optical density of solutions of diphenylpiazselenol in toluene is measured at 420 mμ. V(V), Fe(III), Cu(II) and other oxidizing agents, which form colored products with the reagent, interfere. Complexone III is used to mask all interfering ions except V(V). In the extraction all colored ions remain behind in aqueous solution. Substances which reduce selenium(IV) ions interfere as well.

The method will determine selenium in tellurium, arsenic, sulfur, indium, copper, stainless steel, certain semiconductor compounds and ores.

This method will determine $1 \cdot 10^{-4}\%$ selenium in a 1-g sample.

The sensitivity of the determination of selenium may be increased somewhat if instead of the light absorption, the fluorescence of the toluene extract of diphenylpiazselenol is measured. Diphenylpiazselenol fluoresces strongly at 580 mμ, whereas 3,3′-diaminobenzidine fluoresces at 420 mμ. To carry out the determination, the toluene extract (10 ml) is irradiated at 420 mμ and the resulting fluorescence is measured at 550–600 mμ. The amount of selenium is calculated relative to a standard containing 0.5 μg of Se in 10 ml.

Note. The naphthyl analog of 3,3′-diaminobenzidine is more sensitive to selenium both in the spectrophotometric and in the fluorimetric determinations.

REAGENTS

Standard solution of sodium selenite, 2 μg Se/ml.
Hydrochloric acid, $d = 1.19$ g/cm^3 and dilute (1:7) solutions.
Nitric acid, $d = 1.5$ g/cm^3.
Ammonia, dilute (1:1) solution.
Complexone III, 2.5% solution and 0.1 *M* solution.
Formic acid, dilute (1:9) solution.
Stannous chloride. Ten grams of the preparation are dissolved with heating in 4 ml of hydrochloric acid, $d = 1.19$ g/cm^3, and 16 ml water are added; a freshly prepared solution is employed.
Sodium hypophosphite, 10% solution in 1:7 hydrochloric acid.
Urea, 5% solution.
Citric acid, 25% solution and 40% solution.
Sodium arsenate solution, 50 mg As/ml.
Cresol Red, 0.1 solution in 20% ethanol.
3,3′-*Diaminobenzidine*, 0.5% aqueous solution. The solution should be stored in a cool, dark place.
Toluene or *benzene*.

CONSTRUCTION OF CALIBRATION CURVE

Portions of the standard solution containing 1.0, 3.0, 5.0 and 7.0 μg selenium are placed into four 50-ml beakers. Forty ml of water are added to each beaker, then 1 ml of 2.5% solution of Complexone III and 2 ml of formic acid. The solutions are neutralized to Cresol Red with ammonia to pH 2–3 (yellow color of indicator), 2 ml each of 0.5% freshly prepared solution of 3,3′-diaminobenzidine are introduced and the solutions are left to stand for 30 minutes. Ammonia solution is added to pH 8 (violet color of indicator), the solutions are transferred into 75–100-ml separating funnels and 11 ml each of toluene or benzene are added. The colored compound is extracted for 1 minute. The extracts are filtered through dry filter paper into 3-cm thick cells and their optical density is determined in a FEKN-54 electrophotocolorimeter at 415–420 mμ relative to a blank solution prepared in a similar manner. A calibration curve is plotted from the data obtained.

DETERMINATION OF SELENIUM IN TELLURIUM

A 1-g sample of finely ground tellurium is dissolved with heating in a mixture of 15 ml concentrated hydrochloric acid and 2 ml nitric acid, the solution is evaporated to a very small residual volume (about 1 ml) and 20 ml of water are added. The solution is cooled, transferred to a 50-ml volumetric flask and diluted to the mark with water.

A 25-ml portion of the solution is withdrawn with a pipet into a 100-ml beaker, and diluted to 50 ml with water. Three ml of 0.1 *M* solution of Complexone III, 2 ml of formic acid and 10 ml of 40% citric acid solution are added, and ammonia is introduced drop by drop until the pH is 2.5–3.0 (universal indicator paper). Two ml of 3,3′-diaminobenzidine solution are added, the solution brought to the boil, left to stand at room temperature for 10 minutes, then cooled under a stream of tap water. The pH is adjusted to 7.0–8.0 by adding ammonia solution (the initially formed precipitate of tellurous acid should redissolve in the course of this operation), the solution is transferred to a 100-ml separating funnel and the diphenylpiazselenol is extracted for 1 minute with 11 ml of benzene or toluene. The subsequent procedure is as described for the construction of the calibration curve.

DETERMINATION OF SELENIUM IN ARSENIC

One gram of arsenic is dissolved in 6 ml of a mixture of equal volumes of concentrated nitric and hydrochloric acids, heating gently to prevent the reaction from becoming too vigorous. The solution is evaporated almost to dryness, 15–20 ml of water are added and the mixture heated until a clear solution is obtained. After cooling the solution is diluted with water to 40 ml. Complexone III and formic acid are introduced and the procedure is continued as described for the construction of the calibration curve.

DETERMINATION OF SELENIUM IN INDIUM

Indium (0.5 g) is dissolved with gentle heating in 3 ml of nitric acid, and the solution is evaporated almost to dryness. The residue is dissolved in 20 ml of hydrochloric acid (1:7), the solution of sodium arsenate is added, then 1 ml of stannous chloride solution, 10 ml of sodium hypophosphite solution and a little macerated filter paper (the separated arsenic serves as a collector). This solution is heated to boiling, boiled for 5–7 minutes until a precipitate appears, diluted with 1:7 hydrochloric acid to 50 ml and left to stand overnight.

On the following day the precipitate is filtered through a plug of macerated filter paper and washed 8–10 times with 1:7 hydrochloric acid. The washed precipitate together with the plug is transferred to the beaker which was used for the precipitation, and 6 ml of concentrated hydrochloric acid and 3–5 drops of nitric acid are added. The beaker is immersed in a hot water bath and held at 70–80°C until the precipitate is dissolved. Hot water (10–15 ml) and 2 ml of urea solution are added to the solution to decompose the excess oxidizing agent. The solution is thoroughly stirred and filtered through White Ribbon paper; the filter paper is washed 2–3 times with warm water. Four ml of citric acid solution are added to the filtrate, the solution is cooled and the procedure is continued as described for the construction of the calibration curve.

DETERMINATION OF SELENIUM IN INDIUM ARSENIDE AND INDIUM ANTIMONIDE

The arsenide or the antimonide of indium (0.5 g) is dissolved with gentle heating in 6 ml of a mixture of equal volumes of hydrochloric and nitric acid. The solution is evaporated almost to dryness.

During the analysis of indium arsenide the residue is dissolved with heating in 20 ml of hydrochloric acid (1:7), 1 ml of stannous chloride solution is added, then 10 ml of sodium hypophosphite solution and a little macerated filter paper; the blank solution should contain all the above and also 1 ml of sodium arsenate solution (50 mg As).

In the analysis of indium antimonide the residue is dissolved with heating in acid, and 20 ml of 1:7 hydrochloric acid, 0.5 ml of sodium arsenate solution, 1 ml of stannous chloride solution, 10 ml of sodium hypophosphite solution and a little macerated filter paper are added.

In both cases the solution is brought to the boil, boiled for 5–10 minutes to the appearance of a precipitate, diluted to 50 ml with 1:7 hydrochloric acid and left to stand overnight. Further procedure is as described for the determination of selenium in metallic indium, but in the case of indium antimonide 4 ml rather than 2 ml of 3,3′-diaminobenzidine solution are introduced.

Determination of small amounts of selenium in ores

Selenium is separated by reduction with hypophosphite using arsenic as collector.

REAGENTS

Nitric acid, $d = 1.4$ g/cm^3.
Hydrochloric acid, $d = 1.19$ g/cm^3 and 3% solution.
Sulfuric acid, dilute (1:1) solution.
Solution of As_2O_3, 1 mg As/ml.
Copper sulfate, $CuSO_4 \cdot 5H_2O$, crystalline.
Sodium or *potassium hypophosphite*, crystalline.
Other reagents—see p. 328.

CONSTRUCTION OF CALIBRATION CURVE

See p. 328.

PROCEDURE

The sample material (2 g) is treated with 40 ml of nitric acid (in the analysis of sulfide ores 0.05–0.1 g of KI or 0.1–0.2 g of $KClO_3$ are added) and the mixture is left to stand for 24 hours. The solution is evaporated to a volume of 5–10 ml, 10 ml of sulfuric acid are added and the solution is evaporated to the appearance of white fumes. The residue is dissolved in 5–10 ml of water and the evaporation is repeated, after which 25–30 ml of water are added. The solution is then heated to boiling, the insoluble residue filtered off, and the filtrate collected in a 150-ml beaker. The precipitate on the filter is washed 2 or 3 times with small portions of hot water. Fifty ml of filtrate are mixed with an equal volume of hydrochloric acid, then with 1 ml of As_2O_3 solution and 0.1 g copper sulfate. A little macerated filter paper is introduced, the mixture is heated to 80–90°C, and enough sodium or potassium hypophosphite is added to the hot solution in small portions, while stirring, to reduce all ferric ions and then 1–2 g in excess. The solution is boiled for 10 minutes and left to stand for 2–3 hours or overnight. The precipitate is filtered off, washed 4–5 times with hot 3% hydrochloric acid and 3–4 times with water. The solution is dissolved on the filter in 5–8 ml of acid mixture (5 ml of concentrated nitric acid and 2–3 drops of concentrated hydrochloric acid) and the filter paper is washed with water. Two ml of sulfuric acid are added and the solution is evaporated to white fumes. The residue is washed into a test tube 10 ml in volume. To determine selenium 3 ml of the solution are withdrawn and the analysis is continued as shown on page 328.

Photometric determination of tellurium as sol [19, 20]

One of the most frequently used and most reliable techniques of photometric determination of tellurium is based on the determination of the optical density of elementary tellurium sol. Sols obtained by reduction with stannous chloride probably contain major amounts of co-precipitated tin oxides, and are very different in their properties, mostly as regards stability, from the sols obtained by reduction with hypophosphorous acids. The color of tellurium sols depends on the concentration of hypophosphorous acid.

Concentration of H_3PO_2, M	Color of sol	Absorption maximum, mμ
0.06	Blue	580
0.12	Purple	500
0.25	Red	400
0.4	Yellow-orange	350

The concentration of hydrogen ions may vary between 0.01 and 0.03 *N*. If the concentration is lower, the rate of reduction slows down and the particles coagulate in more acid medium.

Gum arabic or gelatin solutions are used to stabilize the sols.

The rate of reduction strongly varies with the temperature. In order to reduce the temperature effect to the minimum, the reduction is performed at the boiling point. Light absorption obeys Beer's law across the entire visible spectrum.

As the concentration of tellurium increases under otherwise similar conditions, the absorption spectrum of the sols changes (Figure 23): a) the absorption bands become narrower, b) the bands shift towards shorter wavelengths, c) the absorption coefficient at the maximum increases. For this reason, in plotting the calibration curve the optical densities of the series of standard solutions are measured at three or four wavelengths near the maxima to establish the optimum wavelengths.

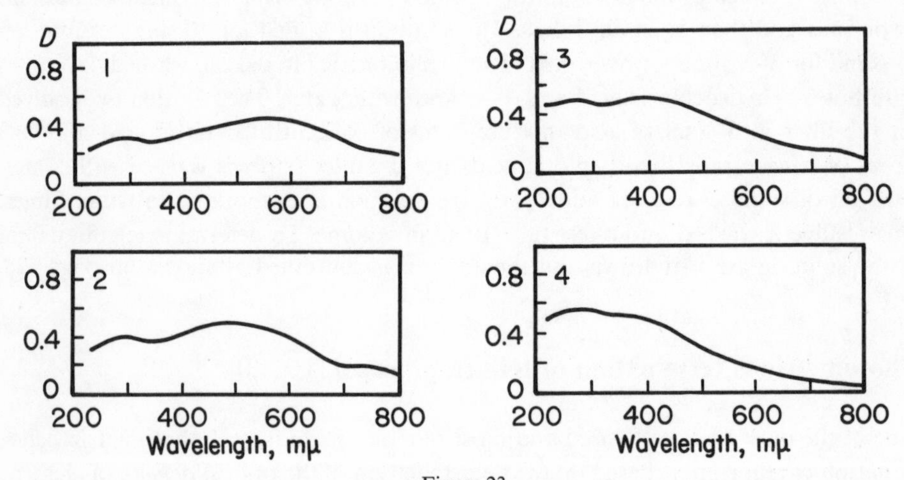

Figure 23

Light absorption of tellurium sols:

1) blue sol; 2) purple sol; 3) red sol; 4) yellow-orange sol.

REAGENTS

Hypophosphorous acid, 3 *M* solution: 32 ml of purified 50% (9.5 *M*) hypophosphorous acid are diluted with water to 100 ml. If purified 50% acid is not available, 50 ml of 30% (5 *M*) hypophosphorous acid are employed.

Gum arabic, 4% solution, freshly prepared.

Standard solution of tellurous acid, 0.1 mg Te/ml. Metallic tellurium (0.25 g) is dissolved in 5 ml of concentrated hydrochloric acid, while adding nitric acid by small portions. When the dissolution is complete, the solution is evaporated almost to dryness. The dry residue is wetted with concentrated hydrochloric acid and the solution is again evaporated almost to dryness. Twenty-five ml of 2 *N* hydrochloric acid are added to the residue, the solution is transferred to a 250-ml volumetric flask, made up to the mark with water and thoroughly mixed. A 25-ml portion of the resulting solution is withdrawn with a pipet into a 250-ml volumetric flask, diluted to the mark with water and well mixed.

CONSTRUCTION OF CALIBRATION CURVE

Into five 125-ml conical flasks are placed 1, 2, 3, 4 and 5 ml portions of a standard solution of tellurous acid, 1–8 mg-equiv. hydrochloric acid, and 3 ml of gum arabic solution are added to each flask and the solutions are made up to 35 ml with water. The solutions are heated to boiling. Five ml of hypophosphorous acid are rapidly added with a pipet to each flask while swirling. The temperature is maintained near the boiling point for 15 minutes, and the solutions are then cooled under the tap for 15 minutes. The sols are transferred to 50-ml volumetric flasks, diluted to the mark with water and the optical density of the red sol is measured at 240–290 mμ or using an electrophotocolorimeter with a blue filter.

Sols of other colors are obtained by varying the concentration of hypophosphorous acid.

PROCEDURE

All the above reagents are added to the solution of the sample, the optical density is measured and the content of tellurium is read off the calibration curve.

Photometric determination of tellurium in steel [21]

The optical densities of the solutions (which are yellow owing to the presence of the iodide complex TeI_6^{2-}) are measured; the solutions have two absorption maxima, at 285 and 333 mμ. The ions Fe(III), Cu(II), Bi(III), Se(IV) interfere. Tellurium may be separated from the three first named elements by reduction with stannous chloride. Selenium is also precipitated in the process. For this reason, if the steel contains selenium, this element must be removed. However, steels contain as a rule either selenium only or tellurium only.

The following are the optimum conditions for the determination of tellurium: concentration of sulfuric acid 1 *N*, concentration of potassium iodide 0.2 *M*, duration of standing not less than 10 minutes.

REAGENTS

Solution of tellurous acid, 0.1 mg Te/ml (p. 332).

Stannous chloride, 10% solution. The reagent (100 g) is dissolved with heating in 250 ml of concentrated hydrochloric acid and this solution is diluted to 1 liter with water.

Urea, crystalline.

Hydrochloric acid, $d = 1.19$ g/cm^3 and dilute (1:20) solution.

Nitric acid, $d = 1.4$ g/cm^3.

Potassium iodide, 2 *M* solution, freshly prepared.

Sulfuric acid, dilute (1:5).

CONSTRUCTION OF CALIBRATION CURVE

Into five 100-ml volumetric flasks are introduced 2, 4, 6, 8 and 10 ml of standard solution of tellurous acid. Twenty-five ml of water, 10 ml of concentrated hydrochloric acid, 10 ml of potassium iodide and

3 g of urea are introduced into each flask, the solutions are thoroughly stirred and left to stand in a dark place for 20 minutes. The optical density of the solutions is measured in an electrophotocolorimeter with a blue filter. The calibration curve is constructed.

PROCEDURE

In the analysis of steel, a 0.5–1.0-g sample (depending on the concentration of tellurium) is dissolved with moderate heating in 50–100 ml of sulfuric acid. When the shavings have fully dissolved, the insoluble residue is filtered off and washed several times with hot water. The filter with the precipitate is placed in the beaker which was used for the solubilization, 10 ml of concentrated hydrochloric acid and 5–6 drops of nitric acid are added, and the mixture is heated for 2–3 minutes until the insoluble residue is fully decomposed (sometimes another addition of nitric acid is required). To the solution are added 25–30 ml of hot water and the filter paper is taken out and washed a few times with hydrochloric acid (1:20). The filtrate is heated to boiling, 3 g of urea are introduced and then enough stannous chloride solution to decolorize the solution, then 5 ml in excess, and the solution is boiled for 10 minutes until the precipitate of tellurium coagulates. A little macerated filter paper is added, the solution is heated to boiling and left to stand in a dark place for 10–15 minutes, after which the precipitate is separated on a White Ribbon filter paper and washed with hot 1:20 hydrochloric acid. The filter with the precipitate is placed in the beaker which was used for the precipitation of tellurium, and 10 ml of hydrochloric acid ($d = 1.19$ g/cm^3) and 5–6 drops of nitric acid are added. The solution is heated, the flask being vigorously shaken from time to time. After the tellurium has completely dissolved, the solution is diluted to 30–40 ml, filtered into a 100-ml volumetric flask through White Ribbon filter paper and the paper is thoroughly washed with hot water.

Further procedure is exactly as described for the construction of the calibration curve.

The blank solution in the measurement of the optical density is prepared as follows: 25–30 ml distilled water are introduced into a 100-ml volumetric flask; 10 ml of concentrated hydrochloric acid and 2 drops of nitric acid, 3 g of urea and 10 ml of potassium iodide solution are added and the solution is diluted to the mark with water.

Spectrophotometric determination of tellurium in selenium [22]

The method is based on the formation of the yellow compound of tellurium with 3,5-diphenylpyrazoline-1-sodium dithiocarbamate, which is soluble in chloroform and carbon tetrachloride. The compound is extracted in a wide range of acidities, from 12 *N* HCl to pH 9. The absorption maximum of chloroform extracts is at

415 mμ. The apparent molar extinction coefficient is 20,400. The optical density of the extracts remains unchanged for a long time.

At approximately pH 8 in the presence of tartrate ions microgram amounts of tellurium may be determined directly if the ratio of tellurium to selenium is 1:100.

Ions of all elements which react with dithiocarbamates under these conditions interfere. Selenium (more than 1 g) and As, Bi, Cd, Ni, Fe(III), Cu, Mn(II), Pd(II), Sb, Sn, Zn (more than 1 mg each) and other elements may be masked in the extraction of more than 2 μg of Te. Thallic and thallous ions also interfere. The latter may be removed prior to the determination by extraction of their pyrazolinedithiocarbamates in strongly alkaline medium (pH 14 and above).

The method is suitable for separating and determining tellurium in selenious acid in semiconductor, metallic selenium and also in sulfur and phosphorus. The ions of selenium and of the accompanying elements are masked with cyanide, Complexone III, and tartrate in alkaline medium at pH 8–9.

REAGENTS

Standard solution of tellurium compound, 1 μg Te/ml.

Selenious acid. Selenious acid is purified as follows. The sample is dissolved in water, the pH is adjusted to 8.7–8.9 with 10% sodium hydroxide solution and 2 ml of 0.1% solution of 3,5-diphenylpyrazoline-1-sodium dithiocarbamate are added. The resulting turbidity is extracted several times with 10-ml portions of chloroform, shaking for 2–3 minutes and adding 0.5–1.0 ml of the reagent prior to each extraction. The extraction of the impurities is continued until the solution no longer becomes turbid when the reagent is added. The extracts are discarded, and the solution is used in the preparation of standard solutions.

3,5-*Diphenylpyrazoline*-1-*sodium dithiocarbamate*, 1% solution.

Chloroform.

Masking mixture. Ten g sodium cyanide, 1 g of Complexone III and 5 g of sodium tartrate are mixed and the volume of the solution is made up to 100 ml with water.

Nitric acid, $d = 1.4$ g/cm^3.

CONSTRUCTION OF CALIBRATION CURVE

A solution of selenious acid previously purified from tellurium (about 1 g Se) is placed in each of five separating funnels, and 10–15 ml masking solution is added to each funnel. The pH is adjusted to about 8.6 and 0.5, 1.0, 1.5, 2.0 and 2.5 ml of standard solution of tellurium compound is added, then 1–2 ml of 1% solution of the reagent. The solution is extracted with 10-, 10- and 5-ml portions of chloroform, and the extracts are filtered into 25-ml volumetric flasks; the extracts are made up to the mark with chloroform, mixed, and their optical density is determined at 415 mμ.

PROCEDURE

In the determination of tellurium in elementary selenium the sample of selenium (0.5–1.0 g) is dissolved in nitric acid ($d = 1.4$ g/cm^3) with very gentle heating on a water bath, adding a few drops of concentrated hydrochloric acid. The acid solution is transferred to a 500-ml separating funnel, the solution is neutralized with sodium hydroxide, the masking mixture is added, the pH adjusted to 8.5 and the determination is continued exactly as in the construction of the calibration curve.

Photometric determination of tellurium in technical grade indium, arsenic and their semiconductor compounds [10]

Tellurium forms a colored diethyldithiocarbamate compound, which is extractable with carbon tetrachloride. In the presence of potassium cyanide and Complexone III bismuth, thallium, and antimony are the only elements which are extracted together with tellurium as diethyldithiocarbamates. The extract of tellurium diethyldithiocarbamate is then shaken with a solution of a copper salt, when the intensely colored copper diethyldithiocarbamate in carbon tetrachloride is formed. The optical density of the solution containing 1 μg of copper as diethyldithiocarbamate is equal to the optical density of the solution containing 7 μg of tellurium as the same compound. In this way the sensitivity of the photometric determination is increased.

The method will determine $5 \cdot 10^{-5}$% tellurium in a 0.5 g of sample of the material.

REAGENTS

Standard solution of sodium tellurite, 1 μg Te/ml.
Hydrochloric acid, *d* = 1.19 g/cm^3, and 1:7 and 1 *N* solutions.
Nitric acid, *d* = 1.4 g/cm^3.
Sulfuric acid, dilute (1:1) solution.
Potassium hydroxide, 20% solution.
Potassium cyanide, 10% solution.
Buffer solution, pH 8.4–8.5. Five g of boric acid, 10 g of Complexone III, and 10 g of KH_2PO_4 are dissolved in 500–600 ml of hot water, neutralized to pH 8.5 with 20% solution of potassium hydroxide (potentiometric control) and the solution is diluted to 1 liter with water.
Sodium diethyl dithiocarbamate, 0.5% aqueous solution.
Copper sulfate, 0.02% solution. Anhydrous copper sulfate (0.1 g) is dissolved in 0.5 liter of water containing 7 ml of sulfuric acid (1:1); prior to use, anhydrous sodium carbonate is added to a part of the solution, until a pale blue color appears; the solution should be transparent.
Sodium arsenate solution, 50 mg As/ml.
Sodium hypophosphite, 10% solution in 1:7 hydrochloric acid.
Citric acid, 25% solution.
Phenolphthalein, 1% solution in 60% ethanol.
Carbon tetrachloride.

CONSTRUCTION OF CALIBRATION CURVE

Tellurium solutions in volumes corresponding to 0, 1.0, 2.0, 3.0 and 5.0 μg of tellurium are introduced into five 50-ml beakers. Twenty ml water, a drop of phenolphthalein solution, 4 drops of potassium cyanide solution and 5 ml of buffer solution are introduced into each beaker. The pH is then adjusted to 8.4–8.5 and the procedure is continued as described in the determination of tellurium in arsenic.

The calibration curve is constructed.

DETERMINATION OF TELLURIUM IN ARSENIC

Ground arsenic (0.5 g) is dissolved in 6 ml of a mixture of equal volumes of concentrated hydrochloric and nitric acids. The bulk of the solution is evaporated, 1 ml of sulfuric acid (1:1) is added and the evaporation is continued to the appearance of sulfuric acid fumes. The solution is cooled, 15 ml of water and 1 drop of phenolphthalein are added and the solution is neutralized with 20% potassium hydroxide until pale pink. Potassium cyanide solution (8–10 drops) is added to the solution, then 1 *N* hydrochloric acid until the solution is decolorized and 5 ml of buffer solution (if the solution becomes pink, the color is removed by introducing 1 *N* hydrochloric acid). The solution is transferred to a 75–100-ml separating funnel, 2 ml solution of sodium diethyldithiocarbamate are added and the mixture is left to stand for 10 minutes in a dark place. Twelve ml of carbon tetrachloride are added and the mixture is vigorously shaken for 2 minutes. The extract is filtered through a dry filter paper into another separating funnel which contains 10 ml copper sulfate solution, and the mixture is vigorously shaken for 1 minute. The organic layer is placed in a 3-cm cell and its optical density is measured against chloroform using a FEKN-54 electrophotocolorimeter with a No. 2 filter (maximum transmission at 415 mμ).

The tellurium content in the sample is found from the calibration curve.

DETERMINATION OF TELLURIUM IN METALLIC INDIUM

Indium (0.5 g) is dissolved with gentle heating in 3 ml of nitric acid. The solution is evaporated almost to dryness, the residue is dissolved in 20 ml of hydrochloric acid (1:7), 1 ml of sodium arsenate solution is introduced, then 1 ml of stannous chloride solution, 10 ml solution of sodium hypophosphite in hydrochloric acid (1:7) and a little macerated filter paper. The solution is boiled for 5–10 minutes until a precipitate appears, diluted to 50 ml with 1:7 hydrochloric acid and left to stand overnight. The precipitate is filtered through a filter paper plug and washed 8–10 times with hot 1:7 hydrochloric acid, dissolved on the filter paper in 6–8 ml of hot nitric acid containing 2–3 drops of concentrated hydrochloric acid and the filter paper is washed 2 or 3 times with hot water. One ml of sulfuric acid (1:1) is added to the filtrate and the mixture is evaporated until sulfuric acid fumes appear. The solution is cooled, 2 ml of citric acid solution added, neutralized, and the determination is continued as for the determination of tellurium in arsenic.

DETERMINATION OF TELLURIUM IN SEMICONDUCTOR INDIUM ARSENIDE

The determination proceeds as described for the determination of tellurium in metallic indium; arsenic serves as collector in the precipitation of micro amounts of tellurium while sodium arsenate solution must also be introduced into the blank solution. Otherwise the determination is performed as described for the determination of tellurium in metallic indium.

Extraction-photometric determination of tellurium in technical grade selenium [23, 24]

The chloride complex of tellurium(IV) with diantipyrylpropylmethane is quantitatively extracted with dichloroethane from hydrochloric acid solutions. Selenium(IV) remains behind in the aqueous phase at all acid concentrations. In this way any amount of tellurium can be separated from selenium.

In the determination of tellurium the chloride complex is converted to the bromide complex with diantipyrylpropylmethane and the optical density (Figure 24) of the resulting solution is measured.

The method is very simple, fairly rapid, and gives reproducible results.

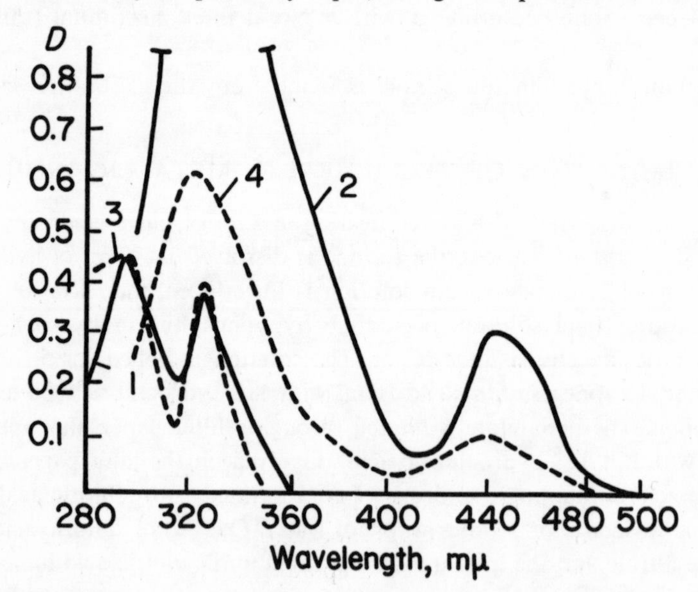

Figure 24

Light absorption by chloride extracts (curves 1 and 2) and bromide extracts (curves 3 and 4) of complexes of Te(IV) with diantipyrylpropylmethane:
1, 3) relative to dichloroethane; 2, 4) relative to water.

Other methods of separation of small amounts of tellurium from selenium in the analysis of materials such as red selenium and stamp selenium, are very complicated, time-consuming (more than 2 days), involve the use of the highly toxic potassium cyanide and require a large number of operations, which reduce the accuracy of the results.

The results obtained by the extraction-photometric method with the use of diantipyrylpropylmethane can be illustrated by the comparison of the results obtained by this method and by the USSR State Standard method (GOST) which is given in Table 6.

Table 6

DETERMINATION OF TELLURIUM IN COMMERCIAL SAMPLES OF SELENIUM

Sample	Found, % Te	
	by extraction-photometric method	by the GOST method
Technical grade selenium	2.2; 2.0	2.0
Commercial grade selenium		
No. 1	0.078; 0.080	0.076
No. 2	0.064; 0.065	0.065
No. 3	0.020; 0.019	0.024

If the content of tellurium is higher than 0.5%, the determination of tellurium in selenium may be completed by the titrimetric method.

REAGENTS

Nitric acid, $d = 1.4\ g/cm^3$.

Sulfuric acid, dilute (1:1).

Hydrochloric acid, $d = 1.19\ g/cm^3$, dilute (1:1).

Hydrobromic acid, 40%, redistilled.

Diantipyrylpropylmethane, 5% and 1% solutions in 1:1 acetic acid. To prepare diantipyrylpropylmethane 5 g of antipyrin are dissolved in a small amount of water, 2 ml of hydrochloric acid are added, then 2 ml of redistilled butyraldehyde. The mixture is heated for 30 minutes on a water bath. After cooling water is added to about 200 ml of the total volume and then 25% solution of ammonia until a faint odor is perceived. The oily liquid rapidly crystallizes out. The crystalline substance is filtered using a Buchner funnel, washed with water and recrystallized from ethanol or from methanol. M.p. 156°C.

1,2-*Dichloroethane.*

Standard solution of tellurous acid, 0.001 g Te/ml. One gram of C.P. tellurium is dissolved in nitric acid ($d = 1.4\ g/cm^3$), adding small amounts of hydrochloric acid if the dissolution of tellurium is slow. The solution is transferred to a one-liter volumetric flask and made up to the mark with 5% hydrochloric acid. One ml of the solution contains 0.001 g Te (solution A). One ml of solution A is placed in a 100-ml beaker and evaporated twice with 5 ml of 1% sulfuric acid. The residue is transferred to a 100-ml volumetric flask with the aid of 1:8 sulfuric acid and the solution is made up to the mark. One ml of this solution contains 0.00001 g of Te (solution B).

Complexone III, 0.02 *M* solution.

Ascorbic acid, freshly prepared solution containing 1 g in 20 ml.

CONSTRUCTION OF CALIBRATION CURVE

Into ten 60-ml separating funnels are introduced 1, 2, 3, 4, 5, 6, 7, 8, 9 and 10 ml of solution of tellurous acid (solution B), the solutions are diluted to a volume of 10 ml, 5 ml of redistilled hydrobromic acid are added to each funnel, and the solutions are left to stand for 1–2 minutes. Then 0.5 ml of ascorbic acid solution, 1 ml of 1% diantipyrylpropylmethane solution and 5 ml of dichloroethane are added to each funnel; the mixtures are shaken for 1 minute.

After the phases have separated, the separated extracts are filtered through dry filter paper into 1 cm cells and the optical density relative to dichloroethane is measured in an electrophotocolorimeter with a No. 3 filter or in a spectrophotometer, after which the calibration curve is constructed.

The color of the extracts is stable for a long time.

PROCEDURE

A 0.5-g sample of metallic selenium is dissolved in 15–20 ml of nitric acid, 5 ml of sulfuric acid are added and the solution is evaporated to the appearance of sulfuric acid fumes. After the residue cools, 5 ml of water are cautiously added to it and the solution is again evaporated to the appearance of sulfuric acid fumes and cooled. The residue is rinsed into a 50-ml volumetric flask with 1:1 hydrochloric acid. A 5-ml or 10-ml aliquot (depending on the tellurium content) of the solution is withdrawn and placed in a 100-ml separating funnel, 1–2 ml of 5% solution of diantipyrylpropylmethane and 1:1 hydrochloric acid to a volume of 20 ml, and 5 ml of dichloroethane are added; the mixture is shaken for 1 minute.

After phase separation the extract is decanted into a clean 50-ml separating funnel. To the residual aqueous phase are added 5 ml of dichloroethane and the mixture is again shaken for 1 minute. The extract is separated.

The dichloroethane extracts are combined (the aqueous solution is discarded) and washed twice with 10 ml of 1:1 hydrochloric acid.

The compound of tellurium is extracted from the combined extract by shaking with 10 ml of Complexone III solution.

After phase separation the dichloroethane is decanted. To the residual aqueous solution are added 5 ml of hydrobromic acid and the solution is left to stand for 1 minute, after which 0.5 ml of ascorbic acid solution, 1 ml of 1% solution of diantipyrylpropylmethane and 5 ml of dichloroethane are added; the mixture is shaken for 1 minute.

When the solution has become clear the separated extract is filtered through a dry paper into a 1-cm cell. The optical density is measured in a FEKN-57 electrophotocolorimeter with a No. 3 filter, or in a spectrophotometer, relative to dichloroethane.

The amount of tellurium is found from the calibration curve.

Extraction-photometric determination of tellurium in technical grade lead and bismuth [23, 24]

Tellurium is separated from lead and bismuth by extraction of the chloride complex of tellurium(IV) with diantipyrylpropylmethane from hydrochloric acid solution after the addition of Complexone III. Tellurium is finally determined by the extraction-photometric method as the bromide complex with pyrazolone derivatives (diantipyrylpropylmethane, diantipyrylmethane, diantipyrylmethylmethane). The yellow solutions of these bromide complexes of tellurium are stable for a long time.

The method gives accurate and reliable results. It takes relatively little time.

REAGENTS

Nitric acid, $d = 1.4\ g/cm^3$ and dilute (1:3) solution.
Sulfuric acid, dilute (1:1) solution.
Hydrochloric acid, $d = 1.19\ g/cm^3$ and dilute (1:1).
Hydrobromic acid, 40%, redistilled.
Diantipyrylpropylmethane, 5% and 1% solutions in 1:1 acetic acid. For the preparation of diantipyrylpropylmethane see p. 339.
1,2-Dichloroethane.
Solution of tellurous acid, see p. 339.
Complexone III, 0.1 *M* and 0.02 *M* solutions. If the available preparation of Complexone III is strongly contaminated, it should be purified as follows. An aqueous solution of Complexone III saturated at room temperature is prepared (about 10 g in 100 ml water). To the solution 96% ethanol is added in small portions until a permanent precipitate appears; it is then filtered off and discarded. To the filtrate an equal volume of ethanol is added and the separated precipitate of Complexone III is filtered under vacuum, washed several times with acetone and then 2–3 times with ether. The preparation is dried at 80°C to constant weight, spreading the precipitate on a watch glass in a thin layer or else in a flat-bottomed porcelain dish.

CONSTRUCTION OF CALIBRATION CURVE

Into ten 50-ml separating funnels are introduced 1, 2, 3, 4, 5, 6, 7, 8, 9 and 10 ml of the tellurous acid solution (solution B, p. 340), the solutions are diluted with water to 10 ml, and 5 ml of hydrobromic acid are added to each funnel; the solutions are left to stand for 1–2 minutes. One ml of 1% solution of diantipyrylpropylmethane and 5 ml of dichloroethane are added to each funnel, and the mixtures are shaken for 1 minute. The extracts are separated and filtered through a dry filter paper into a 1-cm cell; the optical density of the extracts relative to dichloroethane is determined in an electrophotocolorimeter or a spectrophotometer; the calibration curve is then plotted. The color of the extracts is stable for a long time.

PROCEDURE

A 0.2–0.5-g sample of lead or bismuth is dissolved in 20–25 ml of 1:3 nitric acid. The solution is evaporated almost to dryness, the walls of the flask are rinsed with water and the solution is again evaporated to dryness. The dry residue is dissolved in 15–40 ml of 0.1 *M* solution of Complexone III while heating, an equal volume of concentrated hydrochloric acid is added and the solution is transferred into a 100-ml or 200-ml separating funnel. Two ml of 5% solution of diantipyrylpropylmethane are added to the funnel, the solution is mixed and 5 ml of dichloroethane are added. The mixture is shaken for 1 minute.

After phase separation the extract is decanted into a clean 50-ml separating funnel. Another 5 ml of dichloroethane are added to the residual aqueous phase and the mixture is again shaken for 1 minute.

The dichloroethane extracts are combined. The aqueous solution is discarded. The extracts are washed twice with 10 ml of 1:1 hydrochloric acid.

The tellurium compound is reextracted by shaking the extract with 10 ml of 0.02 *M* solution of Complexone III. This operation is repeated twice.

After phase separation the dichloroethane is decanted. The resulting aqueous solutions are combined and transferred to a 50-ml or 100-ml volumetric flask. The solutions are made up to the mark with 0.02 *M* solution of Complexone III.

A 5-ml aliquot of the solution is withdrawn, 5 ml of 0.02 *M* solution of Complexone III are added, then 10 ml of hydrobromic acid, 1 ml of 1% solution of diantipyrylpropylmethane and 5 ml of dichloroethane. The mixture is shaken for 1 minute.

After separation the organic phase is filtered through a dry filter paper into a 1-cm cell. The optical density is measured in a FEKN-57 electrophotocolorimeter with a No. 3 filter or in a spectrophotometer, relative to dichloroethane, and the tellurium content is read off the calibration curve.

Note. The reagents employed must not contain iron. If iron is present in small amounts, 0.5 ml of ascorbic acid solution (1 g in 20 ml of water) must be added after the addition of hydrobromic acid.

Extraction-photometric determination of tellurium in technical grade copper [23, 24]

Small amounts of tellurium are separated from copper by dichloroethane extraction of the chloride complex of tellurium(IV) with diantipyrylpropylmethane. Tellurium is determined photometrically as the bromide complex with diantipyrylpropylmethane.

The technique of the method is simple. Other methods of determination of tellurium in copper often give unreliable results, and are lengthy and complicated.

REAGENTS

Nitric acid, dilute (1:3) solution.
Sulfuric acid, dilute (1:1) solution.
Hydrochloric acid, dilute (1:1) solution.
Diantipyrylpropylmethane, 5% and 1% solutions in acetic acid. For the preparation of the reagent see p. 339.
1,2-Dichloroethane.
Hydrobromic acid, 40%, redistilled.
Potassium bromide, saturated solution. Approximately 60 g KBr in 100 ml water.
Solution of tellurous acid, 0.00001 g Te/ml.

CONSTRUCTION OF CALIBRATION CURVE

In eight separating funnels are placed 1, 2, 3, 4, 5, 6, 7 and 8 ml of solution of tellurous acid, water is added to a volume of 10 ml and 8 ml saturated solution of potassium bromide and 0.5 ml of ascorbic acid solution are added to each funnel. The solutions are mixed, 3 ml of hydrobromic acid, 1 ml of 1% solution of diantipyrylpropylmethane and 5 ml of dichloroethane are added to each funnel; the mixtures are shaken for 1 minute and the procedure is continued as in the determination. The calibration curve is plotted from the data thus obtained.

PROCEDURE

A 10-gram sample of metallic copper is dissolved in 50 ml of nitric acid. The solution is transferred to a 200-ml volumetric flask and diluted to the mark with water. Ten ml of the solution are withdrawn, 5 ml of sulfuric acid added and the solution is evaporated to the appearance of sulfuric acid fumes, after which the walls of the beaker are rinsed with water and the solution is again evaporated to the appearance of sulfuric acid fumes. The residue is dissolved in 20 ml of hydrochloric acid, the solution is transferred to a 100-ml separating funnel, 2 ml of 5% solution of diantipyrylpropylmethane and 5 ml of dichloroethane are added; the mixture is shaken for 2 minutes.

After phase separation the extract is decanted into a clean separating funnel. Another 5 ml of dichloroethane are added to the residual aqueous solution and the mixture is shaken for 1 minute.

The dichloroethane extracts are combined (the aqueous solution is discarded) and washed with two 10-ml portions of hydrochloric acid. The tellurium compound is extracted by shaking with 10 ml of water. This operation is repeated two or three times. The aqueous solutions are collected in a 50-ml volumetric flask.

If the tellurium content is 0.02–0.01%, 5 ml of the solution are withdrawn from the volumetric flask, 8 ml of saturated solution of potassium bromide and 0.5 ml of ascorbic acid are added, the solution is stirred, and 3 ml of hydrobromic acid, 1 ml of 1% diantipyrylpropylmethane solution and 5 ml of dichloroethane are added; the mixture is shaken for 1 minute.

The extract is separated from the aqueous phase, filtered into a 1-cm cell and the optical density is measured in a FEKN-57 photocolorimeter with a No. 3 filter or on a spectrophotometer. The amount of tellurium is found from the calibration curve.

Note. If the amount of tellurium in copper is high, the determination may be completed by titrimetry.

Photometric determination of tellurium with Bismuthiol II [12]

Tellurium (IV) ions react with Bismuthiol II. In aqueous solution at approximately pH 2.3 a complex is formed with an absorption maximum at 395–400 mμ and a molar extinction coefficient of 14,700. The chloroform extract has the absorption maximum at 335–340 mμ; the molar extinction coefficient is 28,000. The ions Fe, Cu, Au (III), Pt (IV), V (V), Se and NO_2^- interfere. The remaining cations, in particular cadmium and bismuth, are masked by adding Complexone III and citrate ions.

The extraction-photometric variant of the method makes it possible to determine more than 0.01 % Te in ores directly. If tellurium is present in smaller concentrations, it is first isolated by reduction with stannous chloride in the presence of a collector.

REAGENTS

Bismuthiol II, 1 % solution in 0.1 *N* KOH.
Citric acid, 1 *M* solution.
Complexone III, 0.1 *M* solution.
Ammonia, 6 *M* solution.
Polyvinyl alcohol, 2 % solution.
Ammonium sulfate, 50 % solution, pH 2.2, in sulfuric acid.
Standard solutions of tellurous acid, 25 and 10 μg Te/ml.

CONSTRUCTION OF CALIBRATION CURVE FOR THE VARIANT NOT INVOLVING EXTRACTION

To 1, 2, 3, 4 and 5 ml of standard solution of tellurous acid with a concentration of 25 μg Te/ml are added 1 ml of citric acid solution and 1 ml of polyvinyl alcohol solution. The solutions are diluted with water to 15 ml, 2 ml of Bismuthiol II are added to each solution and the solutions are left to stand for 20–30 minutes. The pH is then adjusted to 6.5 with ammonia and the solutions are diluted to 25 ml with water. The optical density of the solutions is measured at 400 mμ. A calibration curve is constructed.

CONSTRUCTION OF CALIBRATION CURVE FOR THE EXTRACTION VARIANT OF THE DETERMINATION

Into five 125-ml separating funnels are placed 1, 2, 3, 4 and 5 ml of standard solution of tellurous acid (10 μg Te/ml); 6 ml of ammonium sulfate solution (to improve the subsequent extraction), 1 ml of citric acid solution and 3 ml of Bismuthiol II solution are added to each funnel and the solutions are left to stand for 20–30 minutes after which the pH is adjusted to 6.5 with ammonia solution. The solutions are shaken for 1 minute with 10 ml of chloroform. The optical density of the extracts is measured at 340 mμ.

PROCEDURE

The reagents as above are added to the sample solution, the optical density is determined and the content of tellurium is read off the calibration curve.

In the extraction-photometric variant, all the above reagents are added to the sample solution, the tellurium compound is extracted with chloroform, the optical density is measured at 340 mμ and the tellurium content is found from the calibration curve.

Polarographic determination of selenium and tellurium [26, 27]

In the determination of selenium and tellurium in concentrations between 0.05 and 2 millimoles/liter the following supporting electrolyte (moles/liter) is recommended:

Ammonium chloride	0.75
Ammonia	0.25
Sodium	0.1
Gelatin	0.002%

The concentration of tellurium should not exceed the concentration of selenium. The half-wave potential of tellurium is -0.83 V, that of selenium is -1.5 V (relative to saturated calomel electrode).

In the determination of selenium and tellurium in ores, both elements are first precipitated out together, and the precipitate is dissolved in hydrochloric acid containing added potassium chlorate, which does not interfere with the subsequent polarographic determination.

REAGENTS

Hydrochloric acid, $d = 1.19$ g/cm^3.
Potassium chloride, crystalline.
Sodium hydroxide, 30% solution.
Background solution, 56 g ammonium chloride, 25 g sodium sulfite, 20 ml concentrated ammonia solution and 20 mg gelatin in 1 liter.

PROCEDURE

Ten ml of hydrochloric acid are added to the precipitate of selenium and tellurium. The mixture is heated for 5–10 minutes on a water bath, with periodic additions of small amounts of potassium chlorate, until the dissolution is complete. The cooled solution is neutralized with a solution of sodium hydroxide to Phenol Red. Five drops of sodium hydroxide solution in excess are added (if the content of tellurium is more than 2 mg, 7 drops of the sodium hydroxide solution are introduced; if the selenium content is low, 1–2 drops are added), the solution is diluted to 100 ml with the background solution and a polarographic determination is made: tellurium between −0.4 and −1.0 V, selenium between −1.2 and −1.8 V.

BIBLIOGRAPHY

1. ZAIKOVSKII, F.V.—In: *Sovremennye metody analiza v metallurgii,* N.V. BUYANOV and B.A. GENEROZOV, Editors, p. 142. Metallurgizdat. 1955.
2. NIELSCH, W. and L. GIEFFER.—*Z. anal. Chem.*, **145**, 347. 1955.
3. HOSTE, J. and J. GILLIS.—*Anal. chim. acta,* **2**, 402. 1948; **12**, 158. 1955.
4. CHENG, K.L.—*Anal. Chem.*, **28**, 1738. 1956.
5. GORYUSHINA, V.G. and E.YA. BIRYUKOVA.—*Trudy GIREDMET,* **13**, 285. 1964.
6. ARIYOSHI, H., M. KINIWA, and K. TOEI.—*Talanta,* **5**, 112. 1960.
7. BUSEV, A.I. and HWANG MIN-TIAO.—*ZhAKh,* **17**, 1091. 1962.
8. BUSEV, A.I. and HWANG MIN-TIAO.—*ZhAKh,* **18**, 360. 1963.
9. BUSEV, A.I. and HWANG MIN-TIAO.—*ZhNKh,* **7**, 88. 1962.
10. GORYUSHINA, V.G., E.YA. BIRYUKOVA, and T.A. ARCHAKOVA.—*Trudy GIREDMET,* **13**, 293. 1964.
11. JANKOVSKY, J. and O. KSIR.—*Talanta,* **5**, 238. 1960.
12. CHENG, K.L.—*Talanta,* **8**, 301. 1961; **9**, 501. 1962.
13. BUSEV, A.I., N.L. BABENKO, and HWANG MIN-TIAO.—*ZhAKh,* **18**, 972. 1094. 1963.
14. SHCHERBOV, D.P. and A.I. IVANKOVA.—*Zav. Lab.*, **24**, 1346. 1958.
15. IVANKOVA, A.I. and I.A. BLYUM.—*Zav. Lab.*, **27**, 371. 1961.
16. YAKOVLEV, P.YA. and E.F. YAKOVLEVA. *Tekhnicheskii analiz v metallurgii (Technical Analysis in Metallurgy),* p. 187.—Metallurgizdat. 1963.
17. OSHMAN, V.A., Z.E. MEL'CHEKOVA, and A.YA. MASHUKOV.—In: *Metody opredleniya i analiza redkikh elementov,* pp. 596, 599. Izd. AN SSSR. 1961.
18. *Colorimetric (Photometric) Methods of Determination of Nonmetals.* [Russian translations, A.I. BUSEV, Editor, 1963]
19. JOHNSON, R.A. and B.R. ANDERSON.—*Anal. Chem.*, **27**, 120. 1955.
20. JOHNSON, R.A. and F.P. KWAN.—*Anal. Chem.*, **25**, 1017. 1953.
21. MURASHOVA, V.I.—*ZhAKh,* **17**, 80. 1962.
22. BUSEV, A.I., V.M. BYR'KO, and L.N. PROKHOROVA.—In: *Metody opredeleniya i analiza redkikh elementov,* p. 626. Izd. AN SSSR. 1961.
23. BUSEV, A.I. and N.L. BABENKO.—*ZhAKh,* **18**, 972. 1963.
24. BUSEV, A.I., N.L. BABENKO, and M.N. CHEPIK.—*ZhAKh,* **19**, 871. 1964.
25. BUSEV, A.I. and N.L. BABENKO.—*ZhAKh,* **19**, 926, 1057. 1964.
26. MASHUKOV, A.YA. and T.G. MASLAKOVA.—*Sbornik trudov VNIITsVETMET,* No. 5, 98. Metallurgizdat. 1959.
27. ZELYANSKAYA, A.I., I.E. BYKOV, and L.S. GORSHKOV.—*Trudy Instituta Metallurgii Ural'skogo Filiala AN SSSR,* No. 1, 155. 1957.

Appendixes

1. THE MOST IMPORTANT ORGANIC REAGENTS FOR RARE ELEMENTS

Name and synonyms	Formula	Elements, and methods of determination
Alizarin (1,2-dihydroxy-anthraquinone)	$C_{14}H_8O_4$ M. W. 240.22	Ga, In, Ti, Th, U
Alizarin Red S (1,2-dihydroxy-anthraquinone-3-sulfonic acid, sodium salt)	$C_{14}H_7O_7SNa \cdot H_2O$ M. W. 360.28	Photometric determination Be, Bi, Hf, Li, Sc, Tl, Ti, U, Zr Complexometric indicator for the titration of Bi and Zr
Aluminon (aurintricarboxylic acid, ammonium salt)	$C_{22}H_{23}O_9N_3$ M. W. 473.44	Detection and photometric determination of Be, Sc, Ce, La, Nd, Zr, Hf, Th, In, Ga

Appendix 1 (continued)

Name and synonyms	Formula	Elements, and methods of determination
Arsenazo I [o-(1,8-dihydroxy-3,6-disulfo-2-naphthylazo) phenylarsonic acid, sodium salt; Neotorin; Uranone]	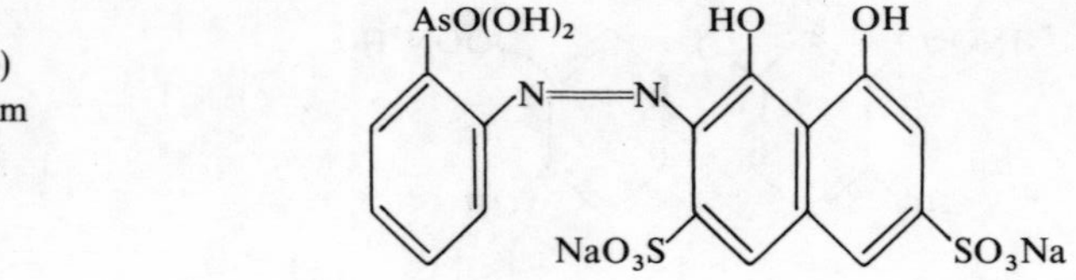 $C_{16}H_{11}O_{11}N_2S_2AsNa_2$ M. W. 592.30	Photometric determination – Th, Zr, U, Ti, Ta, REE, Be, Ga, In Complexometric, indicator for the titration of Th, Zr, U, REE
Arsenazo II [4,4′-diarsenodiphenyl-3,3′-bis-(2-azo-1,8-dihydroxynaphthalene-3,6-disulfonic acid), sodium salt; diphenyl-4,4′-diarsonic acid-bis-(3,3′-azo-2,2′)-1,8-dihydroxynaphthalene-3,6-disulfonic acid, sodium salt]	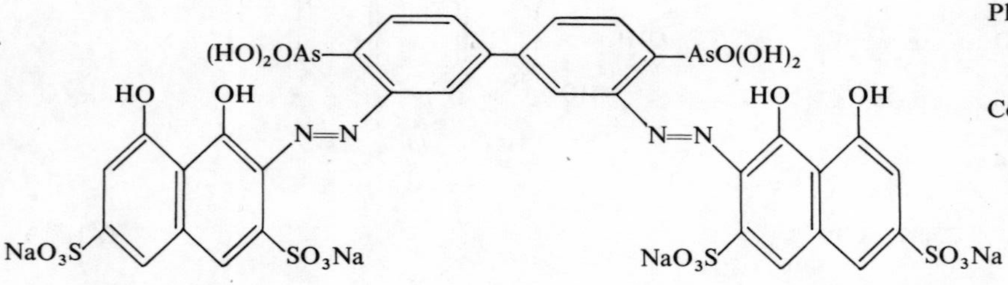 $C_{32}H_{20}O_{22}N_4S_5As_2Na_4$ M. W. 1182.59	Photometric, determination – Th, Zr, U, Ti, Ta, REE, Be, Ga, In Complexometric indicator for the titration of Th, Zr, U, REE
Arzenazo III [1,8-dihydroxynaphthalene-3,6-disulfonic acid-2,7-bis-(azo-2-phenylarsonic acid); 2,7-bis-(o-arsenophenylazo)-1,8-dihydroxynaphthalene-3,6-disulfonic acid]	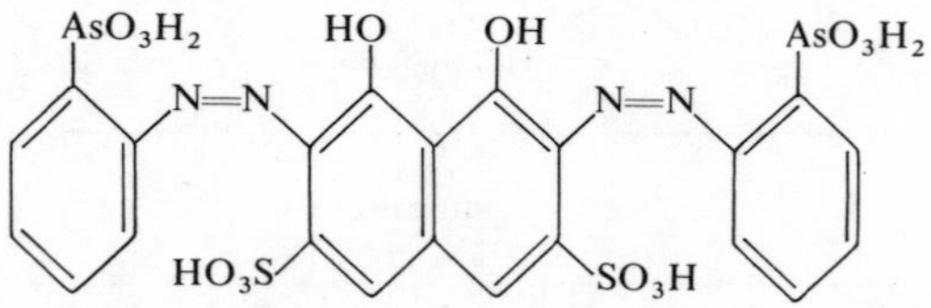 $C_{22}H_{18}O_{14}N_4S_2As_2$ M. W. 776.38	Complexometric indicator for the titration of Th, U, Zr

Appendix 1 (continued)

Name and synonyms	Formula	Elements, and methods of determination
Ascorbic acid	$C_6H_8O_6$ M. W. 176.13	Se, Ti Used as reducing agent for Fe(III) in the determination of a number of rare elements
Benzhydroxamic acid	$C_7H_7O_2N$ M. W. 137.14	Extraction-photometric determination of V
Benzidine sulfate	$C_{12}H_{12}N_2 \cdot H_2SO_4$ M. W. 282.32	V(V), W(VI), Ge(VI), Ta, Tl, Ce(IV)
N-Benzoyl-N-phenylhydroxylamine	$C_{13}H_{11}O_2N$ M. W. 213.24	Ti. Separation of Ta and Nb
α-Benzoinoxime	$C_{14}H_{13}O_2N$ M. W. 227.27	V(V), Mo(VI)

Appendix 1 (continued)

Name and synonyms	Formula	Elements, and methods of determination
Benzeneseleninic acid	$C_6H_6O_2Se$ M. W. 189.07	Gravimetric determination of Sc
Beryllone II [8-hydroxynaphthalene-3,6-disulfonic acid-(1-azo-2')-1',8'-dihydroxynaphthalene-3',6'-disulfonic acid, tetrasodium salt]	$\cdot 4H_2O$ $C_{20}H_{12}O_{15}N_2S_4Na_4 \cdot 4H_2O$ M. W. 810.60	Be
Brilliant Green (Ethyl Green, Diamond Green)	$(C_2H_5)_2N$ … $\overset{+}{N}(C_2H_5)_2Cl^-$ $\cdot ZnCl_2$ $C_{27}H_{33}N_2Cl \cdot ZnCl_2$ M. W. 557.31	Tl(III) (pH 0.1–2.6)

Appendix 1 (continued)

Name and synonyms	Formula	Elements, and methods of determination
Bromopyrogallol Red (dibromopyrogallolsulfonephthalein)	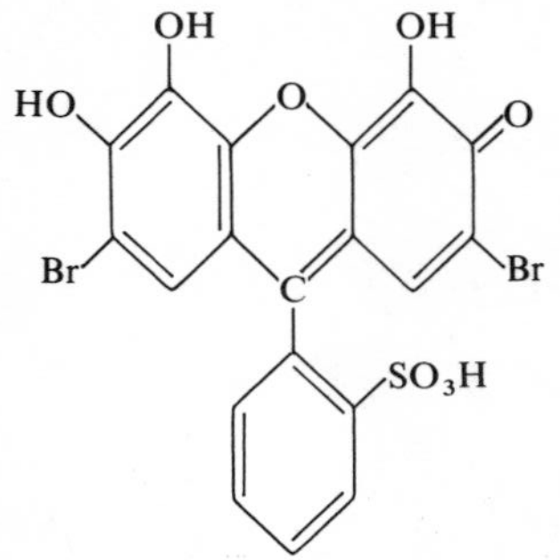 $C_{19}H_{10}O_8SBr_2$ M. W. 576.18	Complexometric indicator for the titration of Ce(III), La, Pr, Nd, Sm
Bismuthiol I (2,5-dimercapto-1,3,4-thiodiazolone)	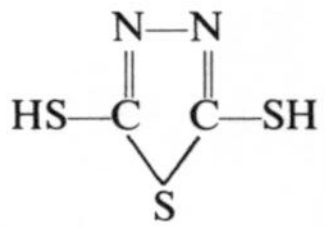 $C_2H_2N_2S_3$ M. W. 150.24	Bi, Te
Bismuthiol II (5-mercapto-3-phenyl-2-thio-1,3,4-thiodiazolone, potassium salt)	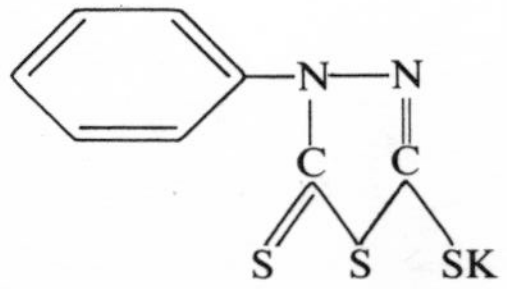 $C_8H_5N_2S_3K$ M. W. 264.44	Bi, Te

Appendix 1 (continued)

Name and synonyms	Formula	Elements, and methods of determination
Gallion (2-hydroxy-3-chloro-5-nitrobenzene-[1-azo-2′]-1′-hydroxy-8′-aminonaphthalene-3′,6-disulfonic acid)	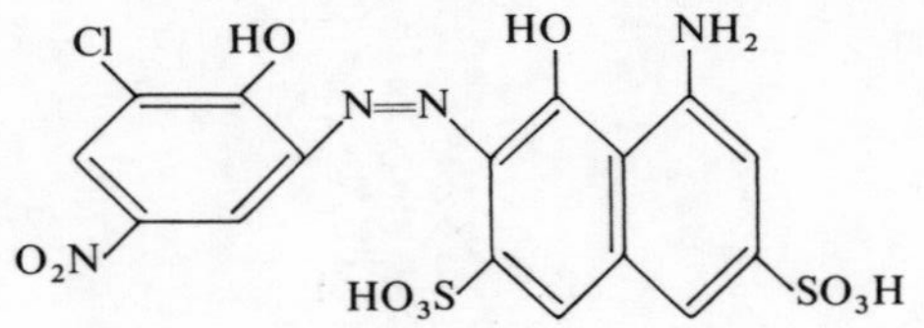 $C_{16}H_{11}O_{10}N_4S_2Cl$ M. W. 518.89	Photometric determination of Ga
Gallocyanin	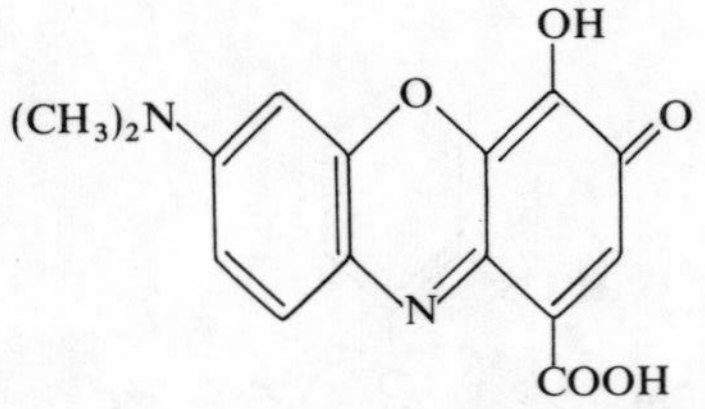 $C_{15}H_{12}O_5N_2$ M. W. 300.27	Ga
Datiscin (glycoside of 3,5,7,2′-tetrahydroxyflavone)	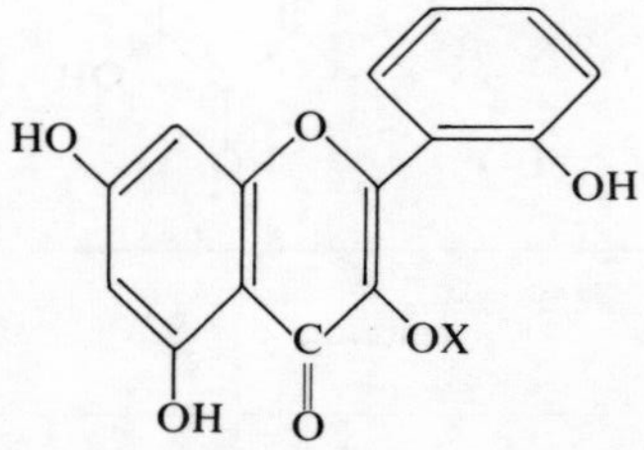 $C_{15}H_9O_6X$ (X=monosaccharide)	Fluorimetric determination of Zr

Appendix 1 (continued)

Name and synonyms	Formula	Elements, and methods of determination
3,3′-Diaminobenzidine hydrochloride (3,3′,4,4′-tetraaminodiphenyl tetrahydrochloride)	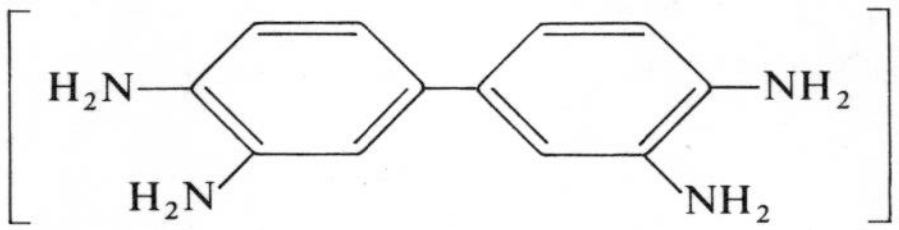 $C_{12}H_{14}N_4 \cdot 4HCl$ M. W. 360.12	Photometric determination of Se
Diantipyrylmethane	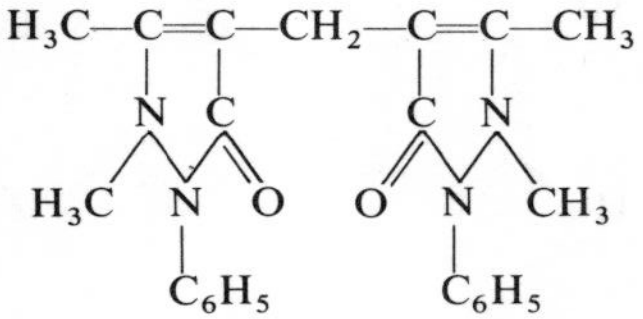 $C_{23}H_{24}O_2N_4$ M. W. 388.47	Photometric determination of Ti
Diantipyrylpropylmethane	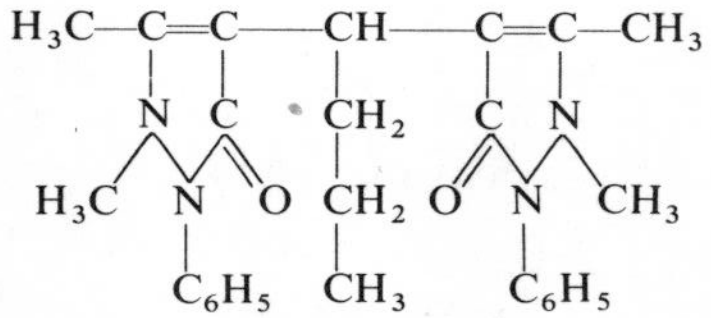 $C_{26}H_{30}O_2N_4$ M. W. 430.56	Te, Ga, Tl
5,7-Dibromo-8-hydroxyquinoline	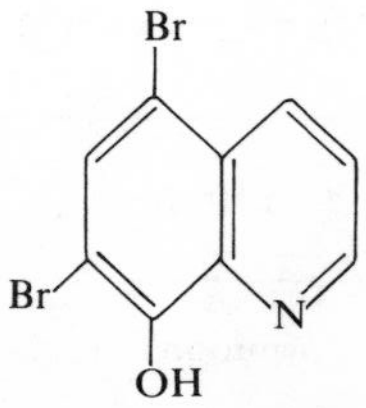 $C_9H_5ONBr_2$ M. W. 302.96	V, Ga, Ti, Zr

Appendix 1 (continued)

Name and synonyms	Formula	Elements, and methods of determination
Dimethylglyoxime	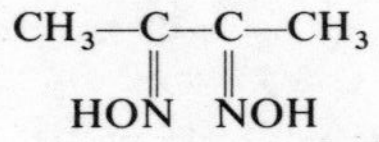 $CH_3—C(=NOH)—C(=NOH)—CH_3$ $C_4H_8O_2N_2$ M. W. 116.12	Re, V, Bi
Dimethyl-fluorone (9-p-dimethylaminophenyl 1-2,3,7-trihydroxy-6-fluorone)	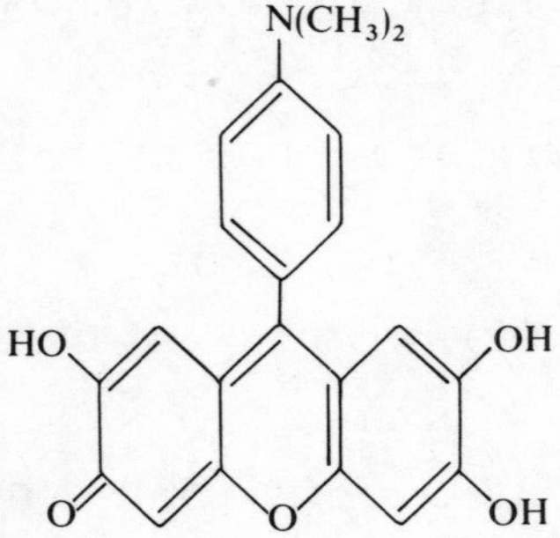 [structural formula: $N(CH_3)_2$; HO; OH; OH; O; O] $C_{21}H_{17}O_5N$ M. W. 363.37	Photometric determination of Ta
o-Dihydroxychromenols (I–6,7-dihydroxy-2,4-dimethylbenzopyrylium chloride; II–7,8-dihydroxy-2,4-dimethylbenzopyrylium chloride; III–7,8-dihydroxy-2,4-diphenylbenzopyrylium chloride; IV–7,8-dihydroxy-2,4-diphenylbenzopyrylium chloride)	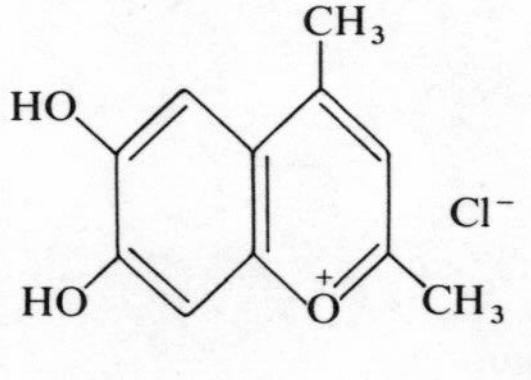 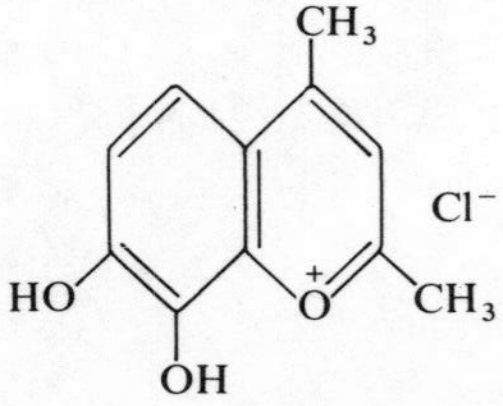 [structural formulae I and II: HO, HO, CH_3, O^+, CH_3, Cl^-; HO, OH, CH_3, O^+, CH_3, Cl^-]	Ge, Zr, Ti, Th, Mo, W, V, Ta, Nb, Sn

Appendix 1 (continued)

Name and synonyms	Formula		Elements, and methods of determination
	C_6H_5, HO, HO, $\overset{+}{O}$, Cl^-, C_6H_5 (III); C_6H_5, HO, HO, $\overset{+}{O}$, Cl^-, C_6H_5 (IV)		
	$C_{21}H_{15}O_3Cl$ III IV	M. W. 350.80	
2,2′-Dipyridyl (α,α′-dipyridyl)	N, N		V, Mo, Re
	$C_{10}H_8N_2$	M. W. 156 9	
Dipicrylamine (2,2′, 4,4′, 6,6′-hexanitrodiphenylamine)	NO_2, O_2N, O_2N—, —NH—, —NO_2, NO_2, O_2N		Rb, Cs
	$C_{12}H_5O_{12}N_7$	M. W. 439.213	
Diphenylcarbazide (1,5-diphenylcarbohydrazide)	—NH—NH—CO—NH—NH—		Photometric determination of Ti
	$C_{13}H_{14}ON_4$	M. W. 242.28	V, Mo

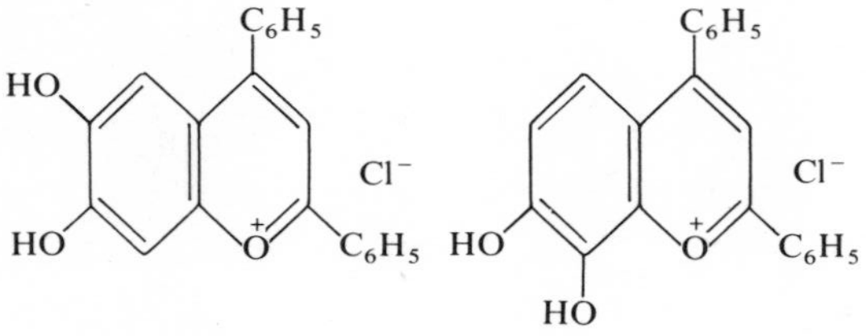

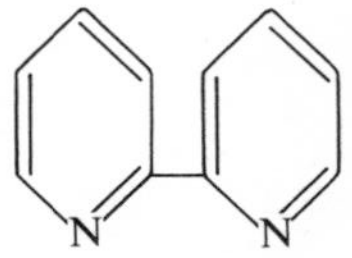

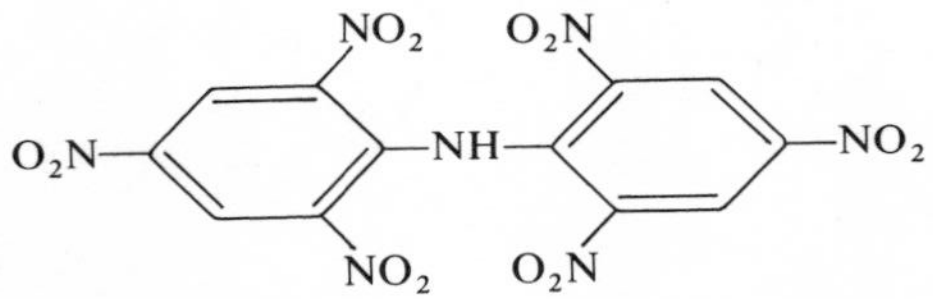

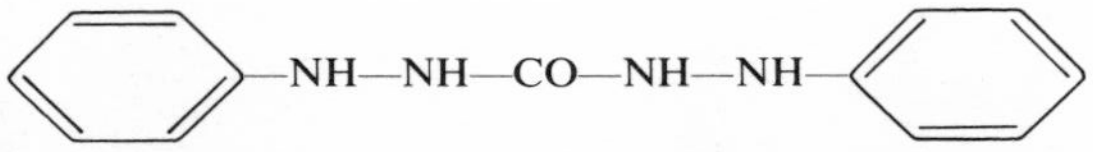

Appendix 1 (continued)

Name and synonyms	Formula	Elements, and methods of determination
2,7-Dichlorochromotropic acid (1,8-dihydroxy-2,7-dichloronaphthalene-3,6-disulfonic acid, disodium salt; dihydrolite)	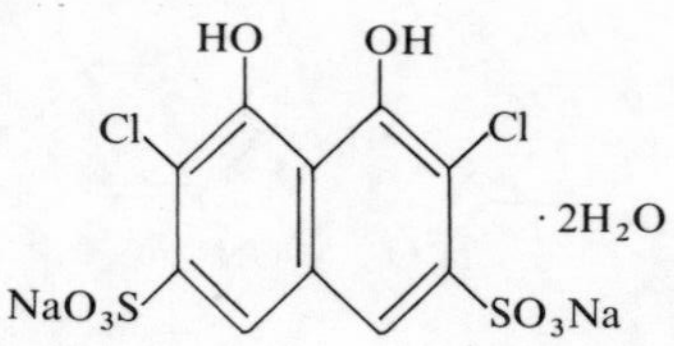 $C_{10}H_4O_8S_2Cl_2Na_2 \cdot 2H_2O$ M. W. 469.18	Photometric determination of Ti
Sodium diethyldithiocarbamate (cupral)	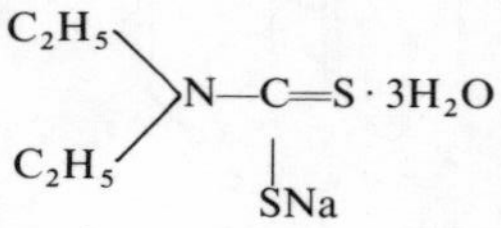 $(C_2H_5)_2N{-}C({=}S){-}SNa \cdot 3H_2O$ $C_5H_{10}NS_2Na \cdot 3H_2O$ M. W. 225.31	Bi, In, Te
Inositolhexaphosphoric acid (used as the calcium magnesium salt)	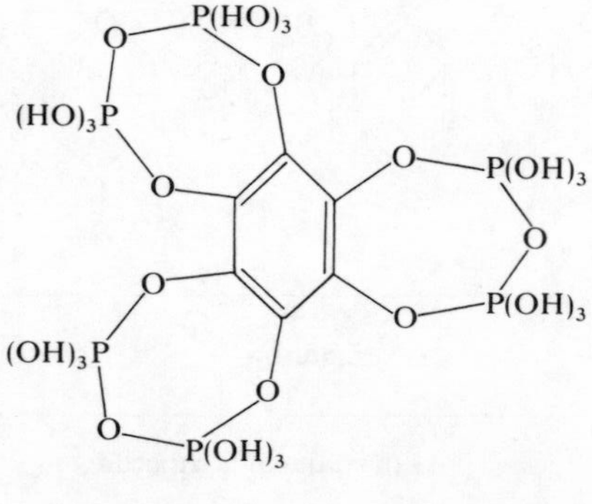 $C_6H_{24}O_{26}P_6$ M. W. 714.09	Gravimetric determination

Appendix 1 (continued)

Name and synonyms	Formula	Elements, and methods of determination
Quercetin (3,5,7,3′,4-pentahydroxyflavone)	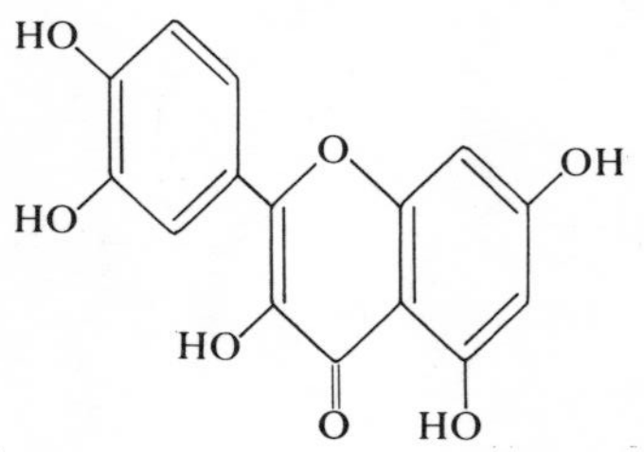 $C_{15}H_{10}O_7$ M. W. 302.24	Photometric determination of Th
Complexone III (ethylenediaminetetraacetic acid, disodium salt)	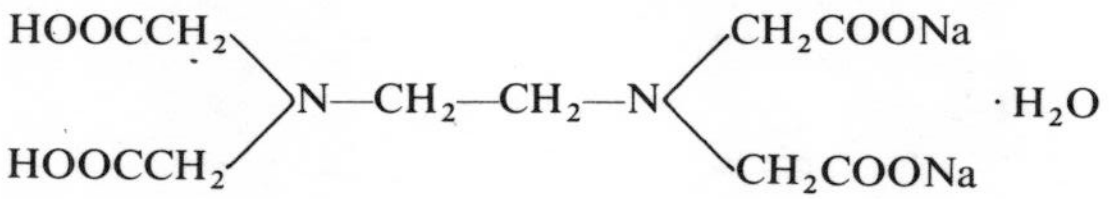 $(HOOCCH_2)_2N{-}CH_2{-}CH_2{-}N(CH_2COONa)_2 \cdot H_2O$ $C_{10}H_{14}O_8N_2Na_2 \cdot 2H_2O$ M. W. 372.24	Complexometric determination of Sc, Ga, In, Ta(III), Y, La, REE, Ti, Ge, Zr, Th, Bi, Mo
Xylenol Orange [3,3′-bis-(N,N-dicarboxymethyl)-aminomethyl-o-cresolsulfonephthalein]	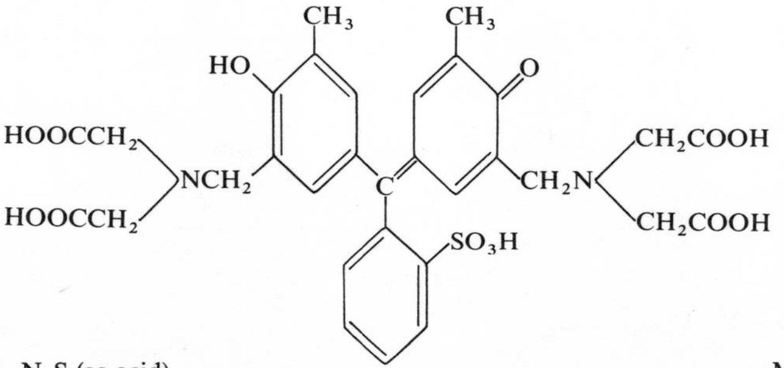 $C_{31}H_{32}O_{13}N_2S$ (as acid) M. W. 672.670	Photometric determination of V, Bi, Nb, REE, Sc, U, Zr, Hf, Ce Complexometric indicator for the determination of Ta(III)

Appendix 1 (continued)

Name and synonyms	Formula		Elements, and methods of determination
Cupferron (phenyl-α-nitroso-hydroxylamine, ammonium salt)	(phenyl)–N–ONH_4 / NO		V, Bi, Ga, Mo, Nb, Ta, Ti, Th, U, Zr
	$C_6H_9O_2N_3$	M. W. 155.16	
Lumogallion [2,2′,4′-tri-hydroxy-5-chloro-(1-azo-1′)-benzene-3-sulfonic acid]	HO_3S, OH, Cl; –N=N–; HO, –OH		Ga
	$C_{12}H_9O_6N_2SCl$	M. W. 344.73	
Malachite Green B, base (Anil Green, tetramethyldiamino-phenylcarbinol, Diamond Green B)	$(CH_3)_2N$, C, $\overset{+}{N}(CH_3)_2Cl^-$		Tl
	$C_{23}H_{25}N_2Cl$	M. W. 364.92	

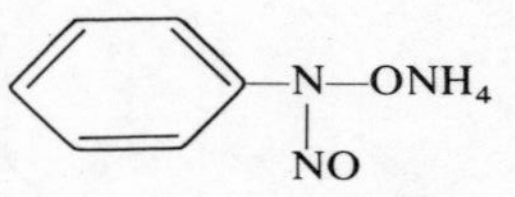

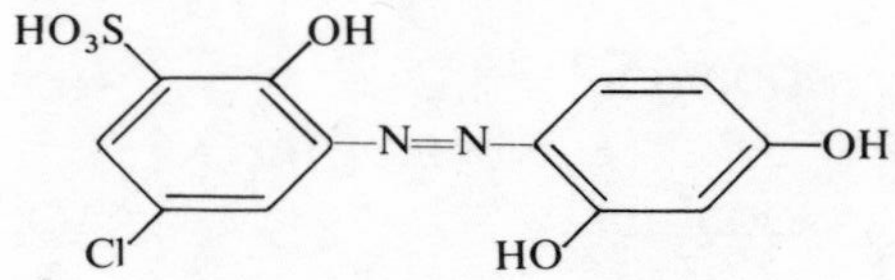

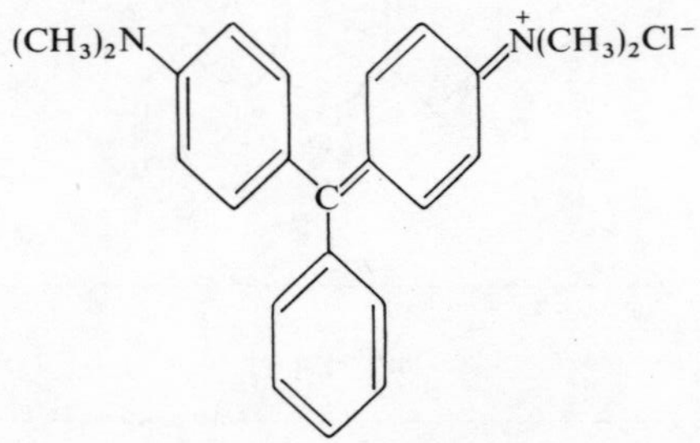

Appendix 1 (continued)

Name and synonyms	Formula	Elements, and methods of determination
2-Mercaptobenzthiazole	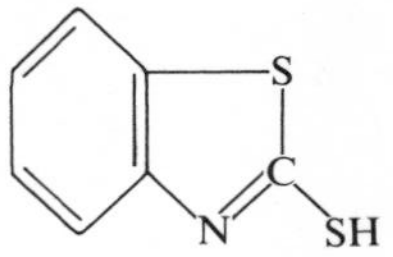 $C_7H_5NS_2$ M. W. 167.25	Bi, Tl
8-Mercaptoquinoline (thioxine), potassium salt	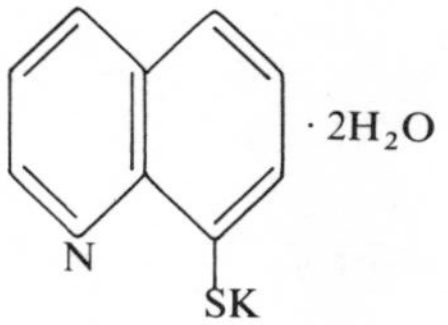 $C_9H_6NSK \cdot 2H_2O$ M. W. 235.34	Photometric determination of Re, Mo
Methyl Violet is a mixture of several substances–commercial preparations have pentamethyl para-rosaniline as the major component and varying proportions of methylated para-rosaniline as the minor component)	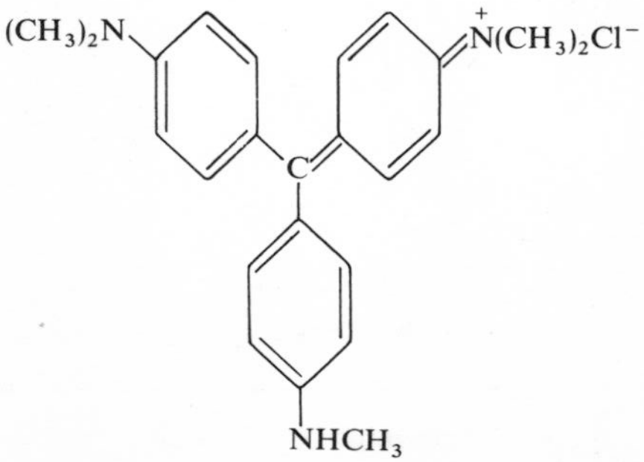 $(CH_3)_2N$, $\overset{+}{N}(CH_3)_2Cl^-$, $NHCH_3$	Photometric determination of Tl, Re, Ta

Appendix 1 (continued)

Name and synonyms	Formula	Elements, and methods of determination
Methylthymol Blue [3,3′-bis-(N-di)-carboxymethyl-(aminomethyl)-thymolsulfonephthalein]	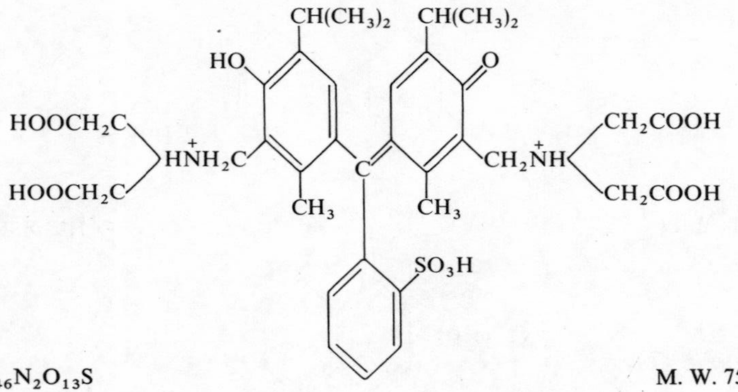 $C_{37}H_{46}N_2O_{13}S$ M. W. 758.89	Photometric determination of Hf. Complexometric indicator
9-Methyl-2,3,7-trihydroxy-6-fluorone	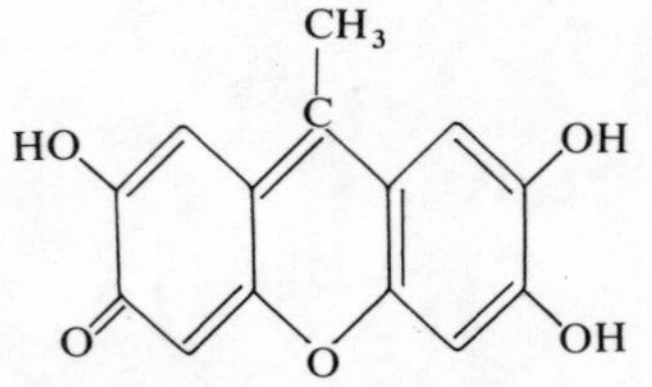 $C_{14}H_{10}O_5$ M. W. 258.23	Ge
Murexide (ammonium salt of purpuric acid)	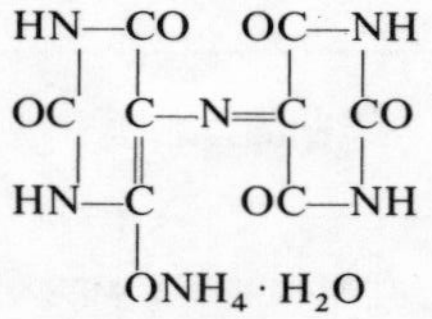 $C_8H_8O_6N_6 \cdot H_2O$ M. W. 302.21	Complexometric indicator

Appendix 1 (continued)

Name and synonyms	Formula		Elements, and methods of determination
Nitron [1,4-Diphenyl-(3,5-edoanil)-dihydro-1,2,4-triazole]	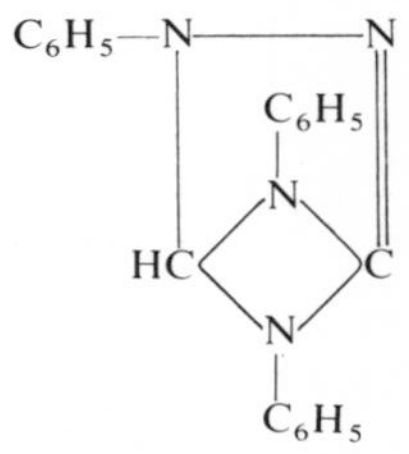 $C_{20}H_{16}N_4$	M. W. 312.38	W(VI), Re(VII)
8-Hydroxyquinoline	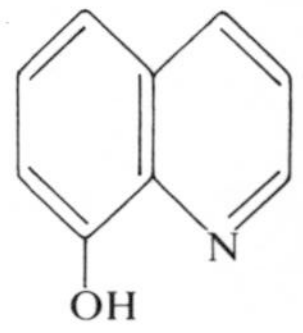 C_9H_7ON	M. W. 145.16	Be, Ga, Ge, In, La, Li, Mo, Nb, Re, Tl, Ce, Zr, U
1-(2-Pyridylazo)-resorcinol, sodium salt [4-(2-pyridylazo)-resorcinol]	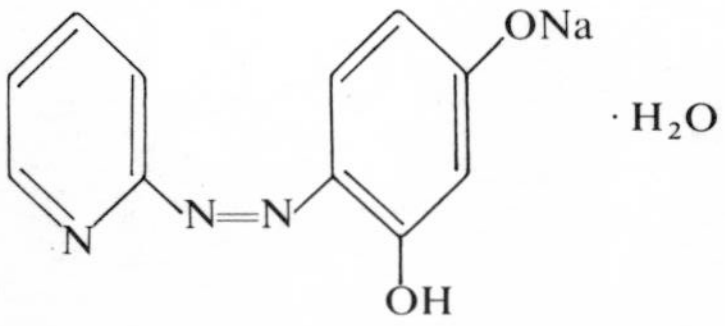 $C_{11}H_8O_2N_3Na \cdot H_2O$	M. W. 255.21	Photometric determination of Sc, In, U Complexometric titration indicator

Appendix 1 (continued)

Name and synonyms	Formula		Elements, and methods of determination
1-(2-Pyridylazo)-naphthol-2	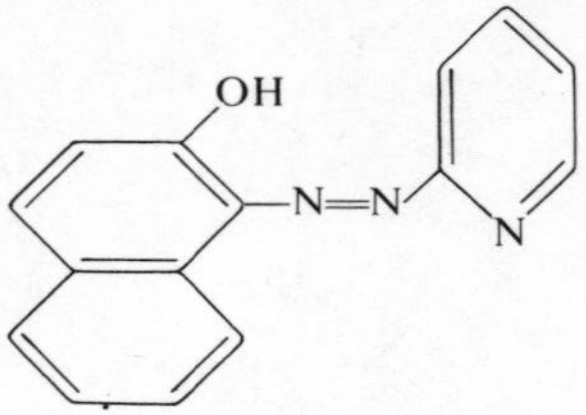 $C_{15}H_{11}ON_3$	M. W. 249.27	Photometric determination of Ga, In, Tl, Sc, Th, Nb, Ta Complexometric indicator in the titration of In
Pyrogallol (1,2,3-trihydroxybenzene)	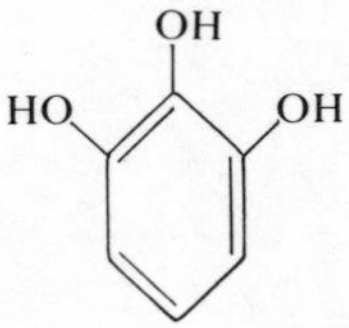 $C_6H_6O_3$	M. W. 126.12	Photometric determination of Ta
Pyrogallol Red (pyrogallolsulfonephthalein)	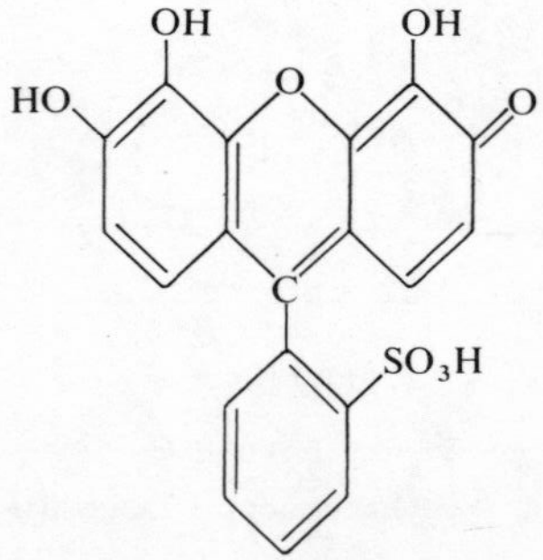 $C_{19}H_{12}O_8S$	M. W. 418.38	Complexometric indicator

Appendix 1 (continued)

Name and synonyms	Formula	Elements, and methods of determination
Pyrocatechol (1,2-dihydroxybenzene, o-dihydroxybenzene)	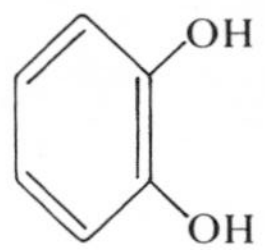 $C_6H_6O_2$ M. W. 110.11	Photometric determination of Ti, Mo, V
Pyrocatechol Violet (pyrocatecholsulfonephthalein)	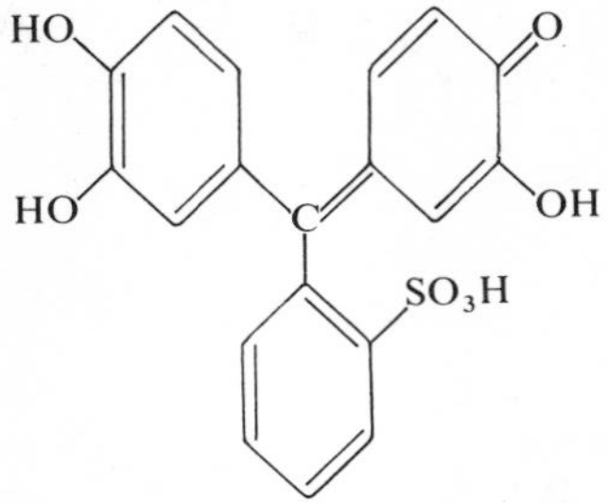 $C_{19}H_{14}O_7S$ M. W. 386.38	Photometric determination of Y; Complexometric titration indicator
Rhodamine C (Rhodamine B, tetraethyldiamino-o-carboxyphenylxanthenyl chloride)	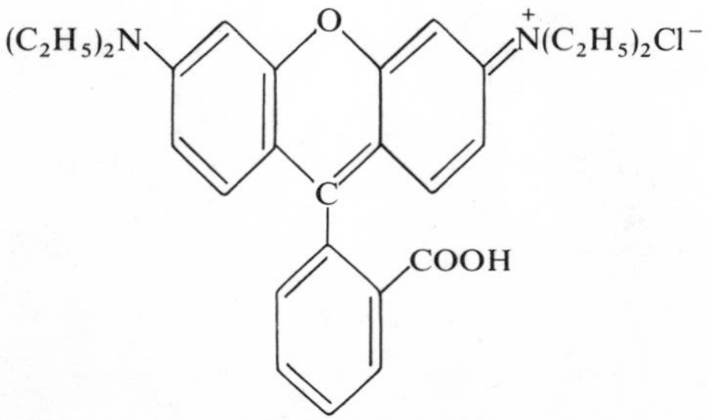 $C_{28}H_{31}O_3N_2Cl$ M. W. 479.02	Fluorimetric determination of Tl and Ga

Appendix 1 (continued)

Name and synonyms	Formula	Elements, and methods of determination
Rhodamine 6G [ethyl ester of diethylamino-o-carboxy-phenylxanthenylchloride, Rhodamine 6G Extra H]	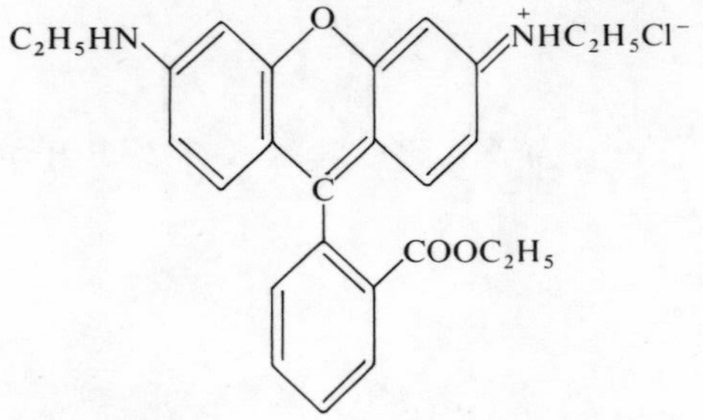 $C_{26}H_{27}O_3N_2Cl$ M. W. 450.97	Fluorimetric determination of Re, In, Tl
Sulfonazo [sulfone-bis-(4-hydroxyphenyl-(3-azo-2')-1'-hydroxy-8'-aminonaphthalene-3',6'-disulfonic acid]	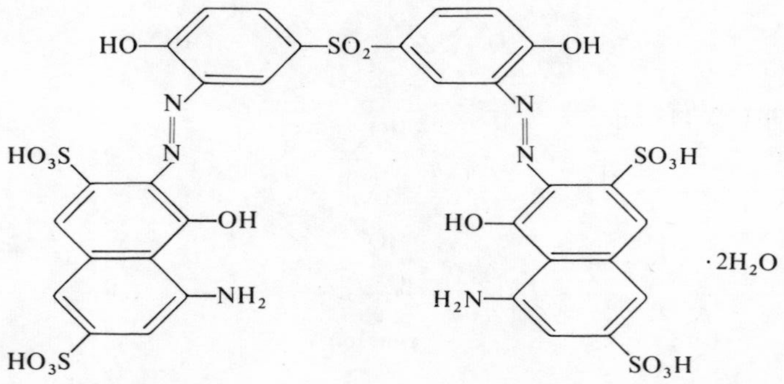 $C_{32}H_{24}O_{18}N_6S_5 \cdot 2H_2O$ M. W. 976.93	Photometric determination of In, Sc and other REE

Appendix 1 (continued)

Name and synonyms	Formula		Elements, and methods of determination
Sulfonaphtholazoresorcinol- [4-sulfo-2-naphthol-(1-azo-1′)-2′-,4′-dihydroxy-benzene]	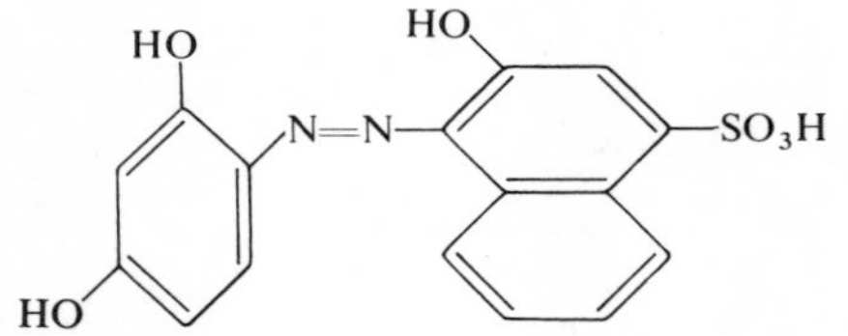 $C_{16}H_{12}O_6N_2S$	M. W. 360.35	Photometric determination of Ga
2-p-Sulfophenylazochromotropic acid, sodium salt [SPADNS; 2-(4′-sulfophenylazo)-1,8-dihydroxynaphthalene-3,6-disulfonic acid, sodium salt]	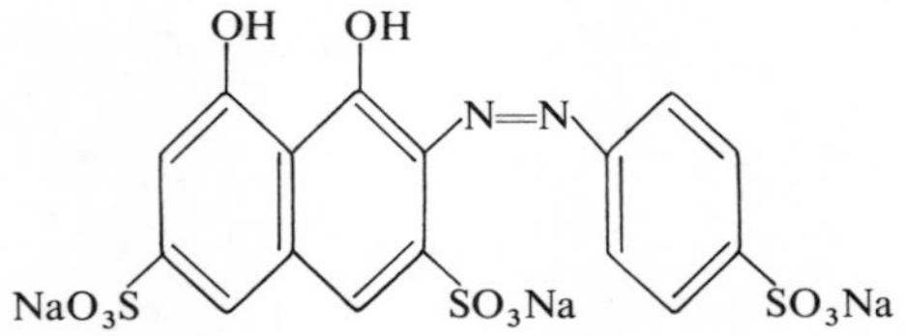 $C_{16}H_9O_{11}N_2S_3Na_3$	M. W. 570.42	Th, Zr
Tiron (2,3-dihydroxybenzene-1,5-disulfonic acid, disodium salt)	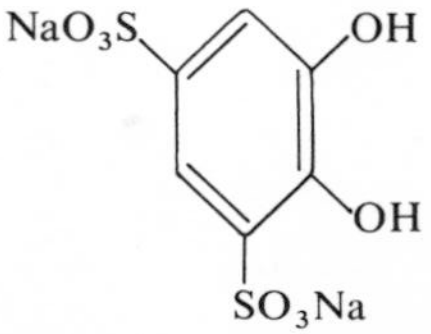 $C_6H_4O_8S_2Na$	M. W. 314.20	Mo, Ti
Thiourea (thiocarbamide)	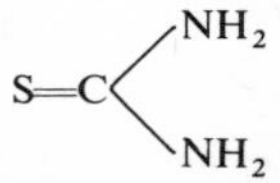 CH_4N_2S	M. W. 76.12	Bi, Re

Appendix 1 (continued)

Name and synonyms	Formula		Elements, and methods of determination
Thionalide (β-aminonaphthalmide of thioglycolic acid)	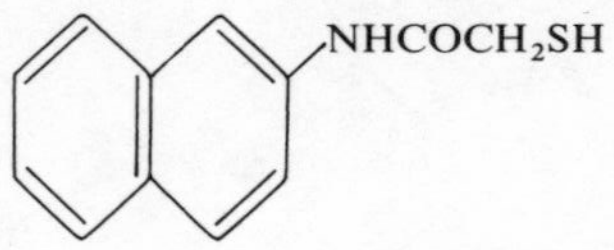 $NHCOCH_2SH$ $C_{12}H_{11}ONS$	M. W. 217.29	Bi, Ta
Toluene-3,4-dithiol	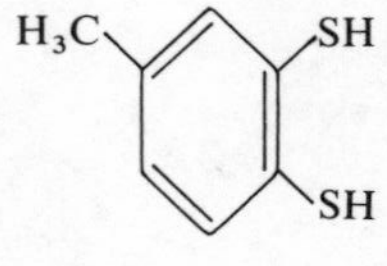 H_3C, SH, SH $C_7H_8S_2$	M. W. 156.27	W
Thoron [benzene-2-arsonic acid-(1-azo-1′)-2′-naphthol-3′-6′-disulfonic acid, disodium salt, Thoronol]	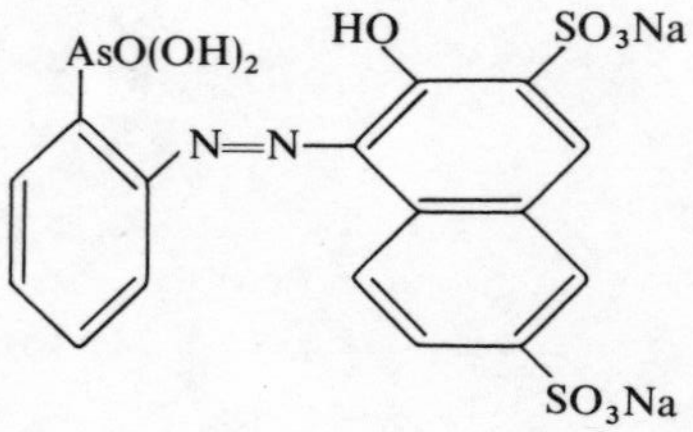 $AsO(OH)_2$, N=N, HO, SO_3Na, SO_3Na $C_{16}H_{11}O_{10}N_2S_2AsNa_2$	M. W. 576.30	Hf, Th, Zr
p-Phenetidide-1-mercaptopropionic acid	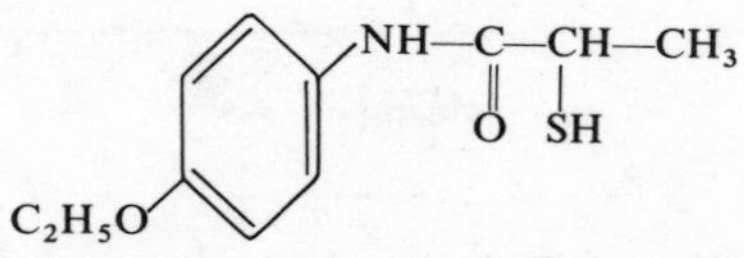 C_2H_5O, NH—C(=O)—CH(SH)—CH_3 $C_{11}H_{15}O_2S$	M. W. 225.31	Extraction-photometric determination of Mo

Appendix 1 (continued)

Name and synonyms	Formula	Elements, and methods of determination
Phenylarsonic acid	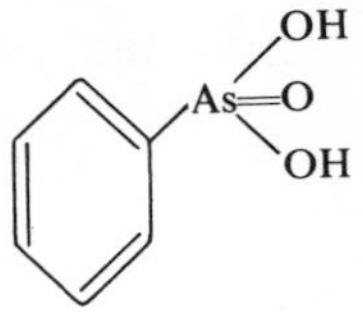 $C_6H_7O_3As$ M. W. 201.96	Gravimetric determination of Zr
Phenylfluorone (9-phenyl-2,3,7-trihydroxy-6-fluorone)	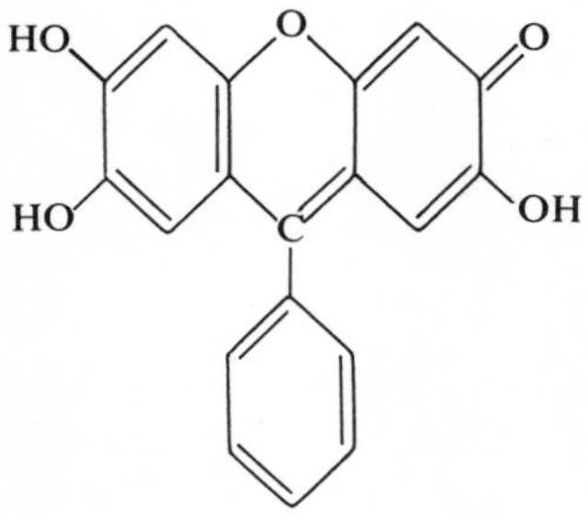 $C_{19}H_{12}O_5$ M. W. 320.30	Ge
Formamidinesulfonic acid (thiourea dioxide)	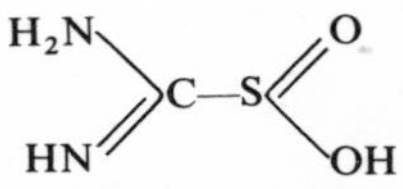 $CH_4O_2N_2S$ M. W. 108.12	Complexometric determination of U

Appendix 1 (continued)

Name and synonyms	Formula	Elements, and methods of determination
Chlorophosphonazo I [2-(4-chloro-2-phosphinic acid phenylazo)-1,8-dihydroxy-naphthalene-3,6-disulfonic acid, disodium salt]	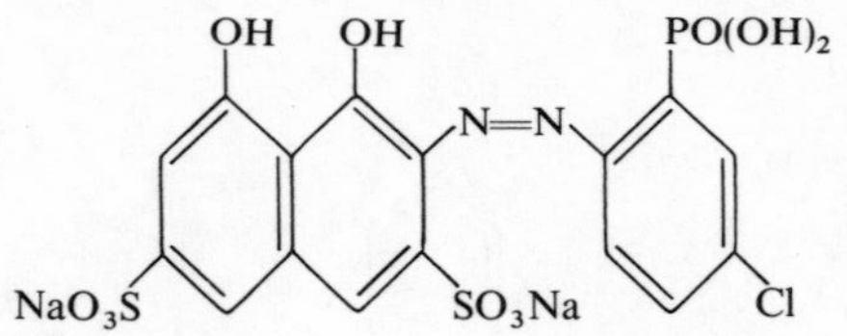 $C_{16}H_{10}O_{11}N_2S_2ClPNa_2$ M. W. 635.85	U
Chlorophosphonazo III [2,7-bis-(4-chloro-2-phosphinic acid phenylazo)-1,8-dihydroxynaphthalene-3,6-disulfonic acid	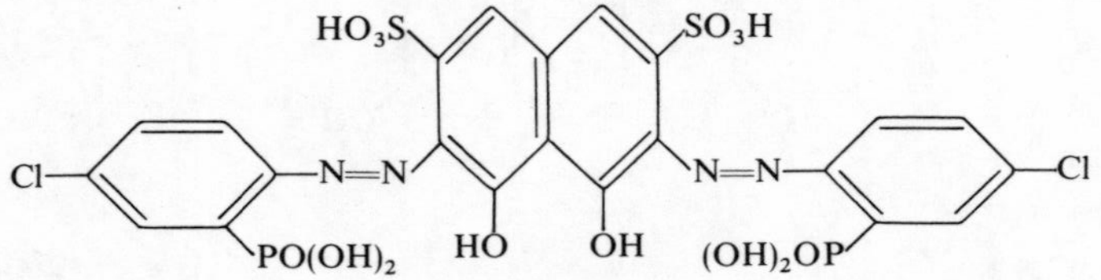 $C_{22}H_{16}O_{14}N_4S_2Cl_2P_2$ M. W. 757.38	Photometric determination of Sc
Chromotropic acid, disodium salt (1,8-dihydroxynaphthalene-3,6-disulfonic acid) disodium salt	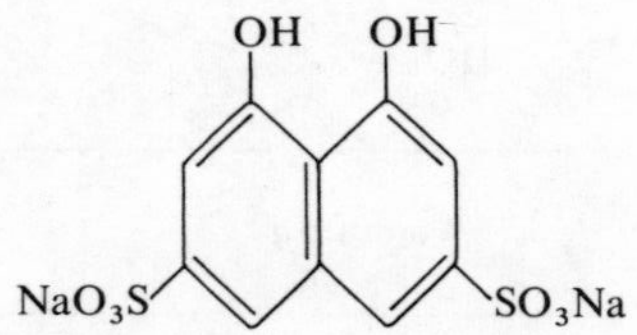 $C_{10}H_6O_8S_2Na_2$ M. W. 364.26	Ti

Appendix 1 (continued)

Name and synonyms	Formula		Elements, and methods of determination
Zincon [2-carboxy-2′-hydroxy-5′-sulfoformazylbenzene; 1-(2-hydroxy-5-sulfophenyl)-3-phenyl-5-(2-carboxyphenyl)-formazole]	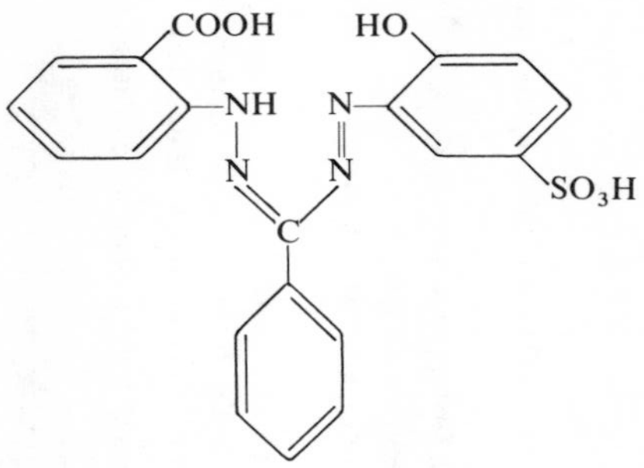		Complexometric indicator
	$C_{20}H_{16}O_6N_4S$	M. W. 440.44	
Eriochrome Black T [1,(1-hydroxy-2-naphthylazo)-6′-nitro-2′-naphthol-4′-sulfonic acid, sodium salt]	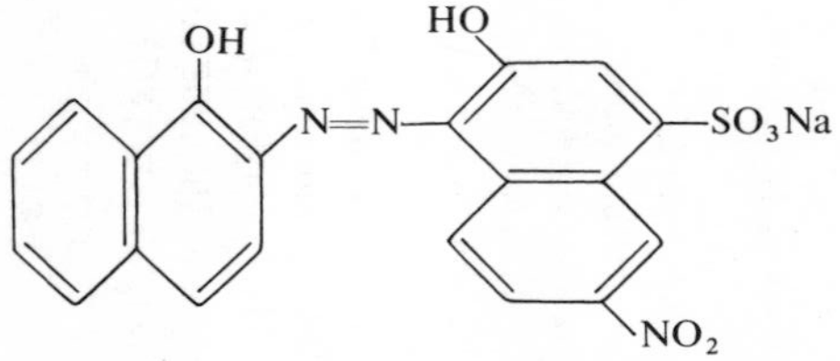		Complexometric indicator
	$C_{20}H_{12}O_7N_3SNa$	M. W. 461.39	

2. MASKING COMPOUNDS FOR CERTAIN RARE ELEMENTS

Element (ion)	Masking compound
Be	Fluoride, tartrate, sulfosalicylic acid
Bi	Aminopolycarboxylic acids, chlorides, dithizone, iodides, BAL $CH_2SH—CHSH—CH_2OH$), citrates, $Na_5P_3O_{10}$, tartrates, triethanolamine
Ce	Aminopolycarboxylic acids, citrates, fluorides, tartrates, Tiron
Ge	Oxalates, fluorides
Mo	Aminopolycarboxylic acids, ascorbic acid, citrates, oxalates, fluorides, H_2O_2, $Na_5P_3O_{10}$, hydroxylamine, thiocyanates, tartrates, Tiron
Nb	Oxalates, fluorides, H_2O_2, OH^-, tartrates
REE	Complexone III
Sc	Tartrates
Se	Sulfides, sulfites
Sr	Aminopolycarboxylic acids, citrates, sulfates, tartrates
Ta	Citrates, fluorides, H_2O_2, OH^-, tartrates
Te	Iodides
Th	Aminopolycarboxylic acids, citrates, fluorides, sulfates, 4-sulfobenzenearsonic acid, tartrates, triethanolamine
Ti	Aminopolycarboxylic acids, ascorbic acid, citrates, fluorides, H_2O_2, OH^-, $Na_5P_3O_{10}$, sulfates, sulfosalicylic acid, tartrates, triethanolamine, Tiron
Tl	Aminopolycarboxylic acids, citrates, chlorides, cyanides, hydroxylamine, tartrates, triethanolamine
U	Citrates, carbonates, oxalates, fluorides, tartrates, H_2O_2
V	Ascorbic acid, Complexone III, fluorides, H_2O_2, hydroxylamine, triethanolamine, Tiron
W	Fluorides, H_2O_2, $Na_5P_3O_{10}$, hydroxylamine, thiocyanates, tartrates, Tiron
Zr	Aminopolycarboxylic acids, tartrates, citrates, triethanolamine, oxalates, fluorides, H_2O_2, PO_4^{3-}, $P_2O_7^{4-}$, SO_4^{2-}

3. PRECIPITATION OF ELEMENTS BY THE MOST IMPORTANT REAGENTS

(Elements which can be quantitatively precipitated by the given reagents are shown in bold type)

Reagent	Conditions	Precipitated ions
Ammonia, aqueous solution	After removal of H_2S group ions, B, F	**Al,** Au, **Be,** Co*, **Cr, Cu, Fe,** Ga, **In,** Ir, **La,** Nb, Ni, Os, P*, **Pb,** lanthanides, **Sc,** Si, **Sn,** Ta, **Th, Tl, U,** V*, W*, **Y,** Zn*, **Zr**
Ammonium polysulfide, aqueous solution	After removal of H_2S and $(NH_4)_2S$ group ions, B and F	Co, Mn, Ni, Si, Te, V*, W*, Zn
Anthranilic acid, ethanolic solution	—	Ag, **Cd, Co,** Cu, Fe, **Hg, Mn, Ni, Pb, Zn**
Benzidine, ethanolic solution, 0.1 *M* HCl	—	Cd, $Fe(CN)_6^{3-}$, IO_3 **PO_4^{3-}, SO_4^{3-}, W**

Appendix 3 (continued)

Reagent	Conditions	Precipitated ions
Cinchonine, solution in 6 *M* HCl	—	Ir, Mo, Pt, **W**
Cupferron, aqueous solution	—	**Al, Bi, Cu, Fe, Ga,** lanthanides, Mo, **Nb,** Pd, Sb, **Sn,** Ta, **Th, Ti,** Tl, **U, V,** W, **Zr**
Ammonium or sodium dihydrophosphate, aqueous solution	a) Acid medium	**Bi, Co,** Hf, **In,** Ti*, **Zn, Zr**
	b) Ammoniacal medium in the presence of tartrate or citrate	Au, Ba, **Be,** Ca, Hg, In, **Mg, Mn,** Pb, lanthanides, Sr, Th, U, Zr
Dimethylglyoxime, ethanolic solution	a) Ammoniacal medium in the presence of tartrates	Ni (Fe and Co if highly concentrated)
	b) weakly acid medium	Au, **Pd,** Se
Hexamethylene tetramine, aqueous solution	—	Precipitates the same ions as NH_4OH
Hydrazine (aqueous solution)	—	Ag, Au, **Cu, Hg,** Ir, **Os,** Pd, Pt, **Rh,** Ru, **Se, Te**
Hydrogen sulfide	a) H^+-ion concentration 0.2–0.5 g-ion/liter	Ag, **As,** Au, Bi, Cd, **Cu, Ge, Hg,** In, **Ir, Mo,** Os, Pb, Pd, **Pt,** Re, **Rh,** Ru, Sb, Se, Sn, Te, Tl, V**, W*, Zn
	b) Ammoniacal medium after removal of H_2S-group ions	Co, Fe, Ga, In, Mn, Ni, Tl, U**, V**, Zn
p-Hydroxyphenylarsonic acid, aqueous solution	Dilute acid	Ce, **Sn,** Th, **Ti, Zr**
8-Hydroxyquinoline, ethanolic solution	a) Acetate buffer solution	Ag, **Al,** Bi, **Cd,** Co, Cr, **Cu, Fe, Ga,** Hg, **In,** lanthanides, **Mn, Mo,** Nb, **Ni,** Pb, Pd, Sb, Ta, Th, **Ti, U,** V, W, **Zn,** Zr
	b) Ammoniacal solution	The same ions as from acetate solutions except **Ag** and the ions Ba, **Be, Ca, Mg,** Sn, Sr
2-Mercaptobenzthiazole in acetic acid	Ammoniacal medium (except Cu, which is precipitated from acid solutions)	Ag, **Au, Bi, Cd, Cu,** Hg, **Ir, Pb, Pt, Rh,** Tl
Neocupferron, aqueous solution	—	Precipitates the same ions as cupferron
Nitron in 5% acetic acid	Solution in dilute sulfuric acid	**B,** ClO_3^-, ClO_4^-, NO_3^-, ReO_4^-, W
1-Nitroso-2-naphthol in very dilute alkali	Acid medium	Ag, Au, B, **Co,** Cr, **Cu, Fe,** Mo, Pd, Ti, V, W, Zr
Oxalic acid, aqueous solution	Dilute acid	**Ag, Au,** Cu, **Hg,** lanthanides, Ni, **Pb, Sc, Th,** U(IV), **Zn**
Phenylarsonic acid, aqueous solution	Acid medium	**Bi,** Ce(IV), Fe, **Hf, Nb, Sn, Ta, Th,** Ti, U(IV), W, Zr
Phenylthiohydantoic acid, aqueous or ethanolic solution	Acid medium	Bi, Cd, **Co,** Cu, Fe, Hg, Ni, Pb, Sb
Picrolonic acid, aqueous solution	Neutral medium	**Ca,** Mg, **Pb,** Th

Appendix 3 (continued)

Reagent	Conditions	Precipitated ions
Pyridine and thiocyanate ions	Dilute acid	**Ag, Cd, Cu, Mn, Ni**
Quinaldic acid, aqueous solution	Dilute acid	Ag, **Cd,** Co, **Cu,** Fe, Hg, Mo, Ni, Pb, Pd, Pt(II), **U,** W, **Zn**
Salicylaldoxime, ethanolic solution	Dilute acid	**Ag, Bi,** Cd, Co, **Cu,** Fe, Hg, Mg, Mn, Ni, **Pb, Pd,** V, Zn
Silver nitrate	Dilute nitric acid solution (HNO_3)	**As(V), Br^-, CN^-, OCN^-, SCN^-, Cl^-, I^-, IO_3^-, Mo(VI), N_3^-, S^{2-}, V(V)**
Tannin, aqueous solution	Ammoniacal medium, in the presence of tartrate and electrolytes	**Al, Be,** Cr, **Ga, Ge, Nb,** Sb, **Sn, Ta, Th, Ti,** U, V, **W, Zr**
Tartaric acid	—	**Ca, K, Nb, Sc, Sr, Ta**
Tetraphenylarsonium chloride, aqueous solution	—	**Re, Te**
β-Aminonaphthalide of thioglycolic acid, ethanolic solution	—	Ag, As, Au, Bi, **Cu, Hg, Os, Pb,** Pd, **Rh, Ru,** Sb, Sn, Tl

* Usually precipitated only in the presence of other ions of this group.

** In the presence of tartrates the precipitation does not take place.

4. SOLUBILITY PRODUCTS OF CERTAIN SPARINGLY SOLUBLE COMPOUNDS OF RARE ELEMENTS AT 18–25°

Compound	Index of solubility product, p*L*	Solubility product, *L*
Ag_2MoO_4	11.55	$2.8 \cdot 10^{-12}$
AgSeCN	15.40	$4.0 \cdot 10^{-16}$
Ag_2SeO_3	15.01	$9.7 \cdot 10^{-16}$
Ag_2SeO_4	7.25	$5.6 \cdot 10^{-8}$
$AgVO_3$	6.3	$5 \cdot 10^{-7}$
Ag_2HVO_4	13.7	$2 \cdot 10^{-14}$
Ag_2WO_4	11.26	$5.5 \cdot 10^{-12}$
$Be(NbO_3)_2$	15.92	$1.2 \cdot 10^{-16}$
$Be(OH)_2$	21.2	$7 \cdot 10^{-22}$
$BiAsO_4$	9.36	$4.4 \cdot 10^{-10}$
BiI_3	18.09	$8.1 \cdot 10^{-19}$
BiOOH	9.4	$4 \cdot 10^{-10}$
$BiPO_4$	22.89	$1.3 \cdot 10^{-23}$
Bi_2S_3	97	$1 \cdot 10^{-97}$
$Ca(NbO_3)_2$	17.06	$8.8 \cdot 10^{-18}$
$CaSeO_3$	5.53	$3.0 \cdot 10^{-6}$
$CaWO_4$	8.06	$8.7 \cdot 10^{-9}$

Appendix 4 (continued)

Compound	Index of solubility product, pL	Solubility product, L
$CdSeO_3$	8.89	$1.3 \cdot 10^{-9}$
$Ce_2(C_2O_4)_3 \cdot 9H_2O$	28.5	$3 \cdot 10^{-29}$
$Ce_2(C_4H_4O_6)_3 \cdot 9H_2O$	19.01	$9.7 \cdot 10^{-20}$
$Ce(IO_3)_3$	9.50	$3.2 \cdot 10^{-10}$
$Ce(OH)_3$	19.8	$2 \cdot 10^{-20}$
Ce_2S_3	10.22	$6.0 \cdot 10^{-11}$
$Ce_2(SeO_3)_3$	24.43	$3.7 \cdot 10^{-25}$
$CoSeO_3$	6.8	$2 \cdot 10^{-7}$
$CsClO_4$	2.40	$4.0 \cdot 10^{-3}$
$CuSeO_3$	7.68	$2.1 \cdot 10^{-8}$
$Er(OH)_3$	23.39	$4.1 \cdot 10^{-24}$
$Eu(OH)_3$	23.05	$8.9 \cdot 10^{-24}$
$Fe_2(SeO_3)_3$	30.7	$2 \cdot 10^{-31}$
$Ga_4[Fe(CN)_6]_3$	33.82	$1.5 \cdot 10^{-34}$
$Ga(OH)_3$	35.15	$7.1 \cdot 10^{-36}$
$Gd(HCO_3)_3$	1.7	$2 \cdot 10^{-2}$
$Gd(OH)_3$	22.74	$1.8 \cdot 10^{-23}$
$Hf(OH)_4$	25.4	$4 \cdot 10^{-26}$
$HgSe$	59	$1 \cdot 10^{-59}$
Hg_2SeO_3	14.2	$6 \cdot 10^{-15}$
$HgSeO_3$	13.82	$1.5 \cdot 10^{-14}$
Hg_2WO_4	16.96	$1.1 \cdot 10^{-17}$
$In_4[Fe(CN)_6]_3$	43.72	$1.9 \cdot 10^{-44}$
$In(OH)_3$	33.2	$6 \cdot 10^{-34}$
KUO_2AsO_4	22.60	$2.5 \cdot 10^{-23}$
$La_2(C_2O_4)_3 \cdot 9H_2O$	26.60	$2.5 \cdot 10^{-27}$
$La(IO_3)_3$	11.21	$6.2 \cdot 10^{-12}$
$La(OH)_3$	18.7	$2 \cdot 10^{-19}$
La_2S_3	12.70	$2.0 \cdot 10^{-13}$
$LiUO_2AsO_4$	18.82	$1.5 \cdot 10^{-19}$
$Lu(OH)_3$	23.72	$1.9 \cdot 10^{-24}$
$Mg(NbO_3)_2$	16.64	$2.3 \cdot 10^{-17}$
$MgSeO_3$	4.89	$1.3 \cdot 10^{-5}$
$MnSeO_3$	6.9	$1 \cdot 10^{-7}$
$NH_4UO_2AsO_4$	23.77	$1.7 \cdot 10^{-24}$
$NaUO_2AsO_4$	21.87	$1.3 \cdot 10^{-22}$
$Nd(OH)_3$	21.49	$3.2 \cdot 10^{-22}$
$NiSeO_3$	5	$1 \cdot 10^{-5}$
$Pb(NbO_3)_2$	16.62	$2.4 \cdot 10^{-17}$
$PbSeO_3$	11.5	$3 \cdot 10^{-12}$
$PbSeO_4$	6.84	$1.4 \cdot 10^{-7}$
$Pr(OH)_3$	21.17	$6.8 \cdot 10^{-22}$
$Pu(OH)_3$	19.7	$2 \cdot 10^{-20}$
$RaSO_4$	10.37	$4.3 \cdot 10^{-11}$
$RbClO_4$	2.60	$2.5 \cdot 10^{-3}$

Appendix 4 (continued)

Compound	Index of solubility product, pL	Solubility product, L
$Sc(OH)_3$	30.1	$8 \cdot 10^{-31}$
$Sm(OH)_3$	22.08	$8.3 \cdot 10^{-23}$
$Sr_3(AsO_4)_2$	18.09	$8.1 \cdot 10^{-19}$
$SrCO_3$	9.96	$1.1 \cdot 10^{-10}$
$SrC_2O_4 \cdot H_2O$	6.80	$1.6 \cdot 10^{-7}$
$SrCrO_4$	4.44	$3.6 \cdot 10^{-5}$
SrF_2	8.61	$2.5 \cdot 10^{-9}$
$Sr(IO_3)_2$	6.48	$3.3 \cdot 10^{-7}$
$Sr(NbO_3)_2$	17.38	$4.2 \cdot 10^{-18}$
$SrSO_4$	6.49	$3.2 \cdot 10^{-7}$
$SrSeO_3$	5.74	$1.8 \cdot 10^{-6}$
$Th(OH)_4$	44.4	$4 \cdot 10^{-45}$
$Th(HPO_4)_2$	20	$1 \cdot 10^{-20}$
$Ti(OH)_3$	40	$1 \cdot 10^{-40}$
$TiO(OH)_2$	29	$1 \cdot 10^{-29}$
$TlBr$	5.47	$3.4 \cdot 10^{-6}$
$TlBrO_3$	4.07	$8.5 \cdot 10^{-5}$
$Tl_2C_2O_4$	3.7	$2 \cdot 10^{-4}$
$TlCl$	3.76	$1.7 \cdot 10^{-4}$
Tl_2CrO_4	12.01	$9.8 \cdot 10^{-13}$
TlI	7.19	$6.5 \cdot 10^{-8}$
$TlIO_3$	5.51	$3.1 \cdot 10^{-6}$
TlN_3	3.66	$2.2 \cdot 10^{-4}$
Tl_2S	20.3	$5 \cdot 10^{-21}$
$TlSCN$	3.77	$1.7 \cdot 10^{-4}$
UO_2HAsO_4	10.50	$3.0 \cdot 10^{-11}$
UO_2KAsO_4	22.60	$2.5 \cdot 10^{-23}$
UO_2LiAsO_4	18.82	$1.5 \cdot 10^{-19}$
$UO_2NH_4AsO_4$	23.77	$1.7 \cdot 10^{-24}$
UO_2NaAsO_4	21.87	$1.3 \cdot 10^{-22}$
$UO_2C_2O_4 \cdot 3H_2O$	3.7	$2 \cdot 10^{-4}$
$(UO_2)_2[Fe(CN)_6]$	13.15	$7.1 \cdot 10^{-14}$
$UO_2(OH)_2$	21.95	$1.1 \cdot 10^{-22}$
UO_2HPO_4	10.67	$2.1 \cdot 10^{-11}$
$VO(OH)_2$	22.13	$7.4 \cdot 10^{-23}$
$(VO)_3(PO_4)_2$	24.1	$8 \cdot 10^{-25}$
$Y(OH)_3$	22.1	$8 \cdot 10^{-23}$
$Yb(OH)_3$	23.6	$3 \cdot 10^{-24}$
$ZnSeO_3$	6.59	$2.6 \cdot 10^{-7}$
$Zr(OH)_4$	52	$1 \cdot 10^{-52}$

5. ISOLATION OF RARE ELEMENTS BY EXTRACTION*

Element extracted	Elements from which separation is effected; Sample materials	Initial aqueous solution	Extractant
Ac	Ra	pH 5.5	TTA (0.25 *M*)–benzene
Be	Al	2-Methyl-8-hydroxyquinoline (pH 8.0)	Chloroform
	Al, Cu, Fe	Butyric acid, Complexone III, KCl (pH 9.3–9.5)	Chloroform
	Al, Sr, Y	pH 6–7	TTA (0.02 *M*)–benzene
	Al, Zn	pH 2	Acetylacetone
	Bivalent ions; bronze	Acetylacetone, Complexone III	Carbon tetrachloride
	Large number of elements	NH_4SCN (7 *M*); hydrochloric acid (0.5 *M*)	Ether
	Large number of elements	Complexone III (pH 5–7)	Acetylacetone–benzene
		Sodium benzoate (pH 7)	Ethyl acetate, butanol or amyl alcohol
		EDTA (weakly acid medium)	Acetylacetone
		2-Methyl-8-hydroxyquinoline, pH 8.1	Chloroform
Bk	—	pH $\approx$ 2.5	TTA (0.2 *M*)–benzene
Ce	Ferrous alloys	KCN (pH 9.9–10.5)	Hydroxyquinoline (0.3 %)–chloroform
	Fission products	Nitric acid (9 *M*)	Methyl isobutyl ketone
		Sulfuric acid (0.5 *M*)	TTA (0.5 *M*)–xylene
	Numerous metals	Complexone III (pH > 4)	Acetylacetone
		Cupferron (pH 2)	Butyl acetate
		Nitric acid (8–10 *M*) + $NaNO_3$ (3 *M*)	TBP
		Nitric acid (8 *M*)	Ether
		Sodium tartrate (pH 8.5)	5,7-Dibromo-8-hydroxyquinoline (0.1 %)
Cm	—	pH 3.5	TTA (0.2 *M*)–benzene
Cs	—	Solution of iodine in potassium iodide	Nitromethane–benzene
		0.4 *M* KPF_6	Nitromethane
		$NaB(C_6H_5)_4$(0.001 *M*), pH 6.6	Nitrobenzene
Er	—	pH $\geqq$ 8.3	0.02 *M* 5,7-Dichloro-8-hydroxyquinoline–chloroform
Ga	Al, In	pH 1.2	Acetylacetone
		pH 3.0	1 % Hydroxyquinoline–chloroform

Appendix 5 (continued)

Element extracted	Elements from which separation is effected; Sample materials	Initial aqueous solution	Extractant
Ga	Al, In, Sb, Tl, W, etc.	Rhodamine C, 6 *M* HCl	Benzene
	Fe	Hydrochloric acid (6.5 *M*), $TiCl_3$	Diisopropyl ether
	Many elements	HBr	Diethyl ether
		HCl	Diethyl ether, diisopropyl ether or TBP
		NH_4SCN (3–7 *M*), 0.5 *M* HCl	Diethyl ether
	Silicate rocks	2-Methyl-8-hydroxyquinoline, hydroxylamine hydrochloride, sodium acetate (pH 3.9)	Chloroform
		Sodium benzoate	Ethyl acetate, butanol or amyl alcohol
		HCl	Amyl acetate
		Erio OS	Chloroform
		5,7-Dibromo-8-hydroxyquinoline, potassium biphthalate, hydroxylamine hydrochloride	Chloroform
		7 *M* HCl	0.1 *M* TOPO–cyclohexane
Ge	As, Hg, Sb, etc.	9 *M* HCl	Benzene, carbon tetrachloride, etc.
		Cupferron in in weakly acid medium	Methyl isobutyl ketone
		Ammonium molybdate, sulfuric acid	Isoamyl alcohol
Hf	Zr	SCN^- ions	Diethyl ether or methyl isobutyl ketone
		2 *M* $HClO_4$	TTA 0.1 *M*–benzene
		1 *M* HCl	0.1 *M* TOPO–cyclohexane
In	Al	pH 3	1% hydroxyquinoline–chloroform
	Al, Be, Bi, Fe, Ga, Mo, W	1.5 *M* KI, 0.75 *M* H_2SO_4	Diethyl ether
	Al, Ga	pH 3	Acetylacetone–chloroform 1 : 1
	Al, Ga, Tl, Zn, etc.	0.5–6 *M* HBr	Diethyl ether or diisopropyl ether
	Be	2-Methyl-8-hydroxyquinoline (pH 5.5)	Chloroform
	Cd	8-Hydroxyquinoline	Chloroform
	Cu	CN^- ions + NH_4OH	Dithizone–chloroform
	Cu, Fe	DDTC, NaCN (pH 9.0)	Carbon tetrachloride
	Fe	HBr, $TiCl_3$	Diethyl ether

Element extracted	Elements from which separation is effected; Sample materials	Initial aqueous solution	Extractant
In	Ga	5 *M* HBr	Butyl acetate
		KI (0.25 *M*), sulfuric acid (0.05 *M*)	Cyclohexanone
	Th	0.5 *M* NaI, 1 *M* $HClO_4$	Methyl isobutyl ketone
	Fission products	Oxalic acid, hydrogen peroxide, sulfuric acid (1 *M*), $(NH_4)_2SO_4$ (2.5 *M*)	Dibutylphosphoric acid (0.6 *M*)–di-n-butyl ether
	Numerous elements	2–3 *M* NH_4SCN, 0.5 *M* HCl	Diethyl ether
		Sodium benzoate (pH 7)	Ethyl acetate, butanol and amyl alcohol
		Rhodamine C, 2.5 *M* HBr	Benzene
La	—	Phenylcarboxylic acids	Chloroform or methyl isobutyl ketone
		pH 7	BPHA (0.1 *M*)–chloroform
		pH 8	5,7-Dichloro-8-hydroxyquinoline–chloroform
	—	Solution in acetoacetic ester (pH 5)	0.1 *M* TTA–methyl isobutyl ketone
Li	K, Na	Cl^- ions (concentrated solution)	Acetone, ethanol–diethyl ether, n-butanol, amyl alcohol or 2-ethyl-1-hexanol
		KOH (1 *M*)	Dipivaloylmethane (0.1 *M*)–diethyl ether
		Solution of iodine in KI	Nitromethane–benzene
Mo	Ag, Al, As, Cr, Cu, Fe, Hg, Pb, Pu, Ti, Tl, U, Zn, Zr	6 *M* HCl, 0.4 *M* HF	Methyl isobutyl ketone
	Re	KSCN, $Hg_2(NO_3)_2$	Diethyl ether
	Ti, W, other steel components	KSCN, Complexone III, NaF, HCl	Amyl alcohol–carbon tetrachloride
	U	KSCN, $NaNO_2$, $SnCl_2$	Diethyl ether–kerosene (2:1)
		Toluene-3,4-dithiol, hydrazine sulfate	Carbon tetrachloride
	W	Phosphoric acid, hydrochloric acid	Diethyl ether
		Thioglycolic acid, KSCN, H_2SO_4	Butyl acetate
		KSCN, NaF, $SnCl_2$	Butyl acetate
		Toluene-3,4-dithiol, phosphoric acid, citric acid	Petroleum ether
	W, other steel components	Strongly acid medium	Toluene-3,4-dithiol–amyl acetate
	Zr	KSCN (acid medium)	Butyl acetate

Appendix 5 (continued)

Element extracted	Elements from which separation is effected; Sample materials	Initial aqueous solution	Extractant
Mo	All elements except W	Acid medium	α-Benzoinoxime–chloroform
	Iron alloys	Complexone III (pH 1.5)	1% 8-Hydroxyquinoline–chloroform
		3 *M* sulfuric acid	Acetylacetone–chloroform (1:1)
	Many elements	NH_4SCN (1–7 *M*), HCl (0.5 *M*)	Diethyl ether
	Most metals	NaF or EDTA, HCl (0.1–0.5 *M*)	Morin–n-butanol
		SO_4^{2-} (pH 0.85)	TOPO (0.1 *M*)–kerosene + caprylic alcohol (2%)
	Silicate rocks	α-Benzoinoxime, up to 1.8 *M* HCl	Chloroform
		6 *M* HCl	Diethyl ether
		DDTC, HCl	Chloroform
		1 *M* HCl	TOPO (0.1 *M*)–cyclohexane
		Potassium ethylxanthate (pH 1.11–1.56)	Chloroform, chlorobenzene or toluene
Nb	Al, Fe, Ga, Mn, Sn, Ti, U, Zr	10 *M* HF, 2.2 *M* NH_4F, 6 *M* H_2SO_4	Methyl isobutyl ketone
	Pa	6 *M* HF, 6 *M* H_2SO_4	Diisobutylcarbinol
	Ta	Concentrated HCl	Methyldioctylamine–xylene or tribenzylamine–chloroform (dichlororethane)
	Fission products	Hydrofluoric acid, sulfuric acid	Tributyl phosphate
		Oxalic acid, sulfuric acid (1 *M*), $(NH_4)_2SO_4$ (2.5 *M*)	0.6 *M* dibutylphosphoric acid–di-n-butyl ether
Nb	Many elements	20 *M* HF	Diethyl ether
		F^- (pH 1)	BPHA–chloroform
		Cupferron, acid medium	Chloroform
		9 *M* HF, 6 *M* H_2SO_4	Diisopropyl ketone
		1 *M* solution of ammonia	8-Hydroxyquinoline–chloroform
		KSCN, $SnCl_2$, HCl	Diethyl ether
Nd	—	pH $\geqq$ 8.3	5,7-Dichloro-8-hydroxyquinoline–chloroform

Appendix 5 (continued)

Element extracted	Elements from which separation is effected; Sample materials	Initial aqueous solution	Extractant
Np	Am, Bk, Cf, Cm, Pu, U, fission products	HCl (1 *M*)	TTA–xylene
		Nitric acid (6–9 *M*)	Ether, methyl isobutyl ketone or dibutylcarbinol
		Nitric acid saturated with ammonium nitrate	Diethyl ether
Pa	Al, Ba, Cr, Mg, Mn, Th, Ti, V	0.6 *M* HF, 8 *M* HCl, saturated with $AlCl_3$	Diisopropyl ketone
	Mn, Ti, U, Zr	Cupferron, HCl (0.1–4 *M*)	Chloroform, benzene or amyl acetate
	Nb, Th	HCl (6 *M*)	Diisobutyl carbinol
	Ti, Zn and many other elements	6 *M* HCl, saturated with $MgCl_2$	β,β'-Dichloroethyl ether
Pa^{233}	Nb^{95}	Oxalic acid, 6 *M* HCl	Diisobutylcarbinol
Pa(IV)	Pa(V)	HCl (6 *M*)	Hexone or TBP–benzene
	All elements except Nb and Zr	Nitric acid (4 *M*)	0.4 *M* TTA–benzene
Po	Bi, Pb	HCl (6 *M*)	20% TBP–dibutyl ether
	Pa	HCl (7 *M*)	Diisopropyl ketone
		Hydrogen perixode, 8 *M* HNO_3	Diethyl ether, hexone or diisopropyl ether, n-amyl and isoamyl alcohols
		pH 0.2–5	Dithizone–chloroform
		pH 1.5–2.0	0.25 *M* TTA–benzene
Pu	Al, Nb, Ru, Th, Zr, REE, alkali metals and alkaline earth metals	$Al(NO_3)_3$ 5 *M* nitric acid	TBP
		HCl (4.8 *M*)	Triisooctylamine–5% xylene
	Fission products	Nitric acid, which may or may not contain $Ca(NO_3)_2$	Methyl isobutyl ketone
		pH 4.5	Acetylacetone–chloroform
		pH 2.5–4.5	Cinnamic acid–amyl acetate
		DDTC (pH 3)	Amyl acetate

Appendix 5 (continued)

Element extracted	Elements from which separation is effected; Sample materials	Initial aqueous solution	Extractant
Pu	Fission products	Ammonium nitrate saturated nitric acid	Diethyl ether
		pH 4–8	8-Hydroxyquinoline–amyl acetate
		Nitric acid (0.5 *M*)	TTA–benzene
		Nitric acid (1 *M*), $NaNO_2$, NH_2OH	0.5 *M* TTA–xylene
Rb	—	Solution of iodine in KI	Nitromethane–benzene
		$NaB(C_6H_5)_4$	Nitrobenzene
Re	Mo	α-Benzoinoxime 6 *M* sulfuric acid	Benzyl alcohol
		4-Hydroxy-3-mercaptotoluene, 6 *M* HCl	Chloroform–isobutanol
	Mo, W	$(C_6H_5)_4AsCl$ (pH 9)	Chloroform
	Many elements	20 *M* HF	Diethyl ether
		DDTC, conc. HCl	Ethyl acetate
		2,4-Diphenylthiosemicarbazide, 6 *M* HCl	Chloroform
		NaOH (4 *M*)	Pyridine
		KSCN, HCl, $SnCl_2$	Butanol
		Toluene-3,4-dithiol in acid medium	Amyl acetate or chloroform
Sc	Al, Be, Cr, REE, Ti	Hydrogen peroxide, HCl	TBP
	Al, Ca, Mg, Na, REE, Y	6 *M* HCl	TBP
	REE, Y	pH 1.5	0.5 *M* TTA–benzene
	Many elements	1 *M* nitric acid saturated with $LiNO_3$ at 35°C	Diethyl ether
		7 *M* NH_4SCN, 0.5 *M* HCl	Diethyl ether
	Ores	SCN^- ions	Diethyl ether
		Sodium benzoate (pH 7)	Ethyl acetate, butanol or amyl alcohol
		8-Hydroxyquinoline (pH 9.7–10.5)	Benzene
		pH 8.0–8.5	8-Hydroxyquinoline–chloroform
		Quinalizarin in alkaline medium	Isoamyl alcohol
Se	Cu, Fe, Te	3,3′-Diaminobenzidine, EDTA (pH 6–7)	Toluene

Appendix 5 (continued)

Element extracted	Elements from which separation is effected; Sample materials	Initial aqueous solution	Extractant
Sr	Many metals	DDTC, EDTA (pH 5–6)	Carbon tetrachloride
		pH 11.3	1 *M* 8-Hydroxyquinoline–chloroform
		pH > 10	TTA–benzene
Ta	Al, Fe, Ga, Mn, Sn, Ti, U, Zr	10 *M* HF, 2.2 *M* NH_4F, Se, 6 *M* sulfuric acid	Methyl isobutyl ketone
	Cr, Ge, Nb, Sb, Ti	HF, HCl	Hexone
	Hf, Mn, Nb, Se, Si, Sn, Ti, Zr	0.4 *M* HF, 6 *M* HCl	Diisopropyl ketone
	Nb, Ti	20% pyrocatechol, ammonium oxalate (pH 3)	n-Butanol
	Nb, Zr	HF, H_2SO_4	Cyclohexanone
	All metals except Pb	HF, HNO_3, ammonium sulfate	Acetone–isobutanol
	All elements except free	0.4 *M* HF, 6 *M* H_2SO_4	Methyl isobutyl ketone
	halogens, Nb, Se, Te	0.4 *M* HF, 6 *M* H_2SO_4	Diisopropyl ketone
	Many elements		
		20 *M* HF	Ether
		F^- (pH 1)	BPHA–chloroform
		1 *M* HNO_3	0.6 *M* dibutylphosphoric acid–n-dibutyl ketone
Tc	Mo, U	Methyl violet, HF (pH 2.3)	Benzene
		$(C_6H_5)_4AsCl$ (pH 10–11)	Chloroform
Te	Bi, Cd, Cu	NaOH (4 *M*)	Pyridine
	Se, other metals	HCl, $SnCl_2$	Ethyl acetate
	Te(IV), Te(VI)	DDTC, NaCN, EDTA (pH 8.5–8.8)	Carbon tetrachloride
	Many elements	2–10 *M* HCl	TBP
		0.6 *M* NaI, 1 *M* HCl	Ether–n-amyl alcohol
		3–6 *M* HCl	DDTC–TBP
Th	Al	pH 1	Dithizone–carbon tetrachloride
	Am, Cm, Np, Pu, Ra, U	$HNO_3 + LiNO_3$	Mesityl oxide
		Nitric acid (pH 1.4–1.5)	0.5 *M* TTA–xylene

Appendix 5 (continued)

Element extracted	Elements from which separation is effected; Sample materials	Initial aqueous solution	Extractant
Th	Ce, V, Y	$Al(NO_3)_3$	Mesityl oxide
	La, U(VI)	pH 2	0.1 *M* BPHA–chloroform
	REE	NH_4SCN	n-Amyl alcohol
	REE	6 *M* NH_4NO_3 + 0.3 *M* nitric acid	Ether–dibutoxytetraethyl glycol (1:2)
	REE, Zr	Saturated solution of $Th(NO_3)_4$	Methyl n-hexyl ketone, n-hexanol, methyl isobutyl ketone or ethyl butyrate
	All elements except Po	pH 1	0.25 *M* TTA–benzene
	Bivalent metals	Cupferron (pH 0.3–0.8)	Benzene–isoamyl alcohol
	Many metals	$Ca(NO_3)_2$ + nitric acid	TBP
	Ores	$Al(NO_3)_3$ + nitric acid	Mesityl oxide or hexone–TBP
		pH > 5.8	Acetylacetone–benzene
		Cupferron, 0.25 *M* sulfuric acid	Butyl acetate
		1 *M* nitric acid saturated with Al, Ca, Fe, Li, Mg or Zn nitrates	Ether
		Phenylcarboxylic acids	Chloroform or methyl isobutyl ketone
		Quercetin (pH 6.5)	Isoamyl alcohol
		pH 4	5,7-Dichloro-8-hydroxyquinoline–chloroform
		pH > 4.9	8-Hydroxyquinoline–chloroform
		pH 2.0	0.5 *M* TTA–CCl_4
Ti	Al	8-Hydroxyquinoline, H_2O_2 (pH 2.2)	Chloroform
	Al, Cr, Ga, V	Cupferron, 1:9 HCl	Chloroform
	Al, Fe	pH 5.3	2-Methyl-8-hydroxyquinoline–chloroform
	Co, Ni, Zn	pH 1.6	Acetylacetone–chloroform
	Cu	Salicylaldoxime, thiourea (pH 5.3)	Isobutanol
	Nb		
	Nb, Ta	Pyrocatechol (20%), $C_2O_4^{2-}$ (pH 3)	n-Butanol

Appendix 5 (continued)

Element extracted	Elements from which separation is effected; Sample materials	Initial aqueous solution	Extractant
Ti	Many metals	Cupferron, ammonium tartrate (pH 5)	Isoamyl alcohol
		Complexone III (pH 8–9)	8-Hydroxyquinoline–chloroform
		NH_4SCN (3 *M*), hydrochloric acid (0.5 *M*)	Ether
		Cupferron, Complexone III (pH 5.5)	Hexone
		Morin in acid medium	Amyl alcohol
		Hydrochloric acid (7 *M*)	0.1 *M* TOPO–cyclohexane
		pH 2.4–4	SSA, tributylammonium acetate or chloroform
Tl	Ag, Au, Cu, Fe, Hg, Pd, Sb, W, Zn, Bi, Mo, W	NaCN (pH 9–12)	Dithizone–CCl_4
		bis-(Dimethylaminophenyl)-antipyridylcarbinol	Benzene–carbon tetrachloride (2:3)
	All elements, except Bi	DDTC, NaCN + Complexone III (pH 11)	CCl_4
	Many elements	HBr, 1–6 *M*	Ether
	Many elements	SSA, Methyl Violet, hydrochloric acid	Benzene
	Many elements	Rhodamine B, bromine water	Benzene
		Hydrochloric acid (6 *M*)	Ether
		0.5 *M* HI	Ether
		pH 6.5–7.0	8-Hydroxyquinoline–chloroform
		pH 3.8	0.25 *M* TTA–benzene
U	Bi, Th	Complexone III, pH 7	8-Hydroxyquinoline–hexone
	Bi and ores	Nitric acid (4.7 *M*)	TBP–ether
	Co, Cu, Cr, Fe, Mn, Mo, Ni, Pb, Th	$Al(NO_3)_3$ (pH 0–3)	Methyl isobutyl ketone
	Fe, V	1-Nitroso-2-naphthol	Isoamyl alcohol
	La	pH 3.5	BPHA (0.1 *M*)–chloroform
	Nd, Pr	Acetate ions, pH 5.5–6	Diethylammonium diethyldithiocarbamate–chloroform
	Th	Nitrate ions, pH 1.0	TOPO (0.1 *M*)–CCl_4 or kerosene
	Th, etc.	1-(2-Pyridylazo)-2-naphthol, Complexone III	o-Dichlorobenzene

Appendix 5 (continued)

Element extracted	Elements from which separation is effected; Sample materials	Initial aqueous solution	Extractant
U	Th, monazites		
	All elements, except Be	Hydrochloric acid (7 *M*)	TBP–methyl isobutyl ketone
	All elements, except Mo and V	$Ca(NO_3)_2$, Complexone III, pH 7	Dibenzoylmethane–ethyl acetate
		Sulfuric acid, pH 0.85	0.1 *M* TOPO–kerosene, containing 2% caprylic alcohol
	Fission products	Nitric acid + $Ca(NO_3)_2$	Methyl isobutyl ketone
	Many elements	DDTC, Complexone III, pyridine	Benzene, chloroform, ethyl acetate or isoamyl alcohol–benzene
		Nitric acid + salting-out agents ($Al(NO_3)_3$, NH_4NO_3, $NaNO_3$, etc.)	Ether, ethyl acetate, nitromethane–methyl isobutyl ketone; TBP–chloroform or benzene (isooctane); hexone, 2-methylcyclohexanone or dibutylcarbinol
	Many elements	Tetrapropylammonium nitrate, $Al(NO_3)_3$	Hexone
	Many elements	Hydrochloric acid (1 *M*)	0.1 *M* TOPO–cyclohexane
	Most metals	Hydrochloric acid (7 *M*)	Triisooctylamine (5%)–xylene
		Complexone III (pH 4.6)	Acetylacetone
		Complexone III (pH 8.8)	8-Hydroxyquinoline–chloroform
	Monovalent and some bivalent cations	Perfluorobutyric acid	Ether
		2-Acetoacetylpyridine	Butyl acetate
		Ascorbic acid, NH_4SCN, EDTA	TBP–CCl_4
		Phenylcarboxylic acids	Chloroform or methyl isobutyl ketone
		Dilute sulfuric acid	Cupferron–ether
		DDTC, pH 6	Chloroform
		Weakly acid medium	5,7-Dichloro-8-hydroxyquinoline–chloroform
		pH 3	0.2 *M* TTA–benzene

Appendix 5 (continued)

Element extracted	Elements from which separation is effected; Sample materials	Initial aqueous solution	Extractant
V	Al, Co, Cr, Fe, Mn, Ni	NaF (pH 3.8–4.5)	0.3% 8-hydroxyquinoline–isobutanol
	Mo, U	Salicylhydroxamic acid (pH 3.0–3.5)	Ethyl acetate
	Ti	DDTC (pH 4.5–5.0)	Chloroform
	U	DDTC, tartrate ions, pH 0.4–0.5	Amyl acetate
	Ferrous alloys	pH 2	Acetylacetone–chloroform
	Many elements	Cupferron, 1:9 hydrochloric acid or sulfuric acid	Ethyl acetate, ether or chloroform
	Many metals	Ca–EDTA, pH 5	8-Hydroxyquinoline–chloroform
		Benzhydrosamic acid in acid solution	n-Hexanol
		BPHA in acid solution	Benzene
		8-Hydroxyquinoline, acetate ions (pH 3.5–4.5)	Chloroform or isoamyl alcohol
		KSCN, $SnCl_2$	Ethyl acetate or hexone
		Hydrochloric acid (7 *M*)	0.1 *M* TOPO–cyclohexane
		Nitric acid (pH 1.5–2.0)	0.6 *M* TOPO–kerosene
W	Al, Cr, Mn, Nb, Ni, S, Ta, Ti, V	Toluene-3,4-dithiol H_2SO_4, dilute HCl	Amyl acetate
	Mo	Hydrochloric acid, $SnCl_2$	Toluene-3,4-dithiol–amyl acetate
	All elements except Mo	α-Benzoinoxime in acid medium	Chloroform
	Many elements	Phosphoric acid, 3 *M* sulfuric acid	n-Amyl alcohol
	Most metals	Complexone III, pH 2.4	8-Hydroxyquinoline–chloroform
		Cupferron, 1:9 hydrochloric acid	Ethyl acetate
		DDTC, pH 1–1.5	Ethyl acetate
		Toluene-3,4-dithiol, hydrochloric acid	Butyl acetate
		KSCN (0.15 *M*), hydrochloric acid (6 *M*)	Ether
		SCN^-, $SnCl_2$, hydrochloric acid	Ether, diisopropyl ketone or isoamyl alcohol–chloroform
Y	Al, Mg, REE	Nitric acid (15.6 *M*)	TBP
	Cerium group of REE	Nitric acid (1 *M*), hydrogen peroxide	0.6 *M* dibutylphosphoric acid–di-n-butyl ketone
	REE	pH > 6	TTA–benzene

Appendix 5 (continued)

Element extracted	Elements from which separation is effected; Sample materials	Initial aqueous solution	Extractant
Y	Sr	Nitric acid (0.1 *M*)	Dibutylphosphoric acid–chloroform
		pH 8.5	8-Hydroxyquinoline–chloroform
Zr	Al, Fe, REE	Hydrochloric acid (6 *M*)	0.5 *M* TTA–xylene
	Ce, V, Y	$Al(NO_3)_3$	Mesityl oxide
	Hf	Nitric acid	TBP–dibutyl ether
	Nb, fission products	$H_2C_2O_4$, H_2O_2, sulfuric acid, (1 *M*), $(NH_4)_2SO_4$ (2.5 *M*)	0.06 *M* dibutylphosphoric acid–di-n-butyl ether
	Bivalent metals	Cupferron (pH 0.3–1.0)	Benzene–isoamyl alcohol
	Many elements	Hydrochloric acid (1 *M*)	0.1 *M* TOPO–cyclohexane
		Hydrochloric acid (7 *M*)	TOPO–cyclohexane
		Ammonium thiocyanate, 7 *M* HCl	TOPO–cyclohexane
		Perchloric acid (> 5 *M*)	TBP–CCl_4
		HCl	TBP–benzene
		Nitric acid	TBP–kerosene
		Cupferron, 1:9 sulfuric acid	Ethyl acetate
		Acetate buffer solution	8-Hydroxyquinoline–chloroform

6. CHOICE OF EXPERIMENTAL CONDITIONS FOR SELECTIVE COMPLEXOMETRIC TITRATION OF METAL IONS

In choosing the conditions for the complexometric titration of any metal ion (pH, indicator, selectivity of titration, the need for additives to prevent precipitation of basic salts) it is convenient to use the following diagram, which has been compiled from experimental data. In this way the sequence of titration of several ions, or the possibility of titrating the sum of the ions, can be readily established.

For legend, see p. 390.

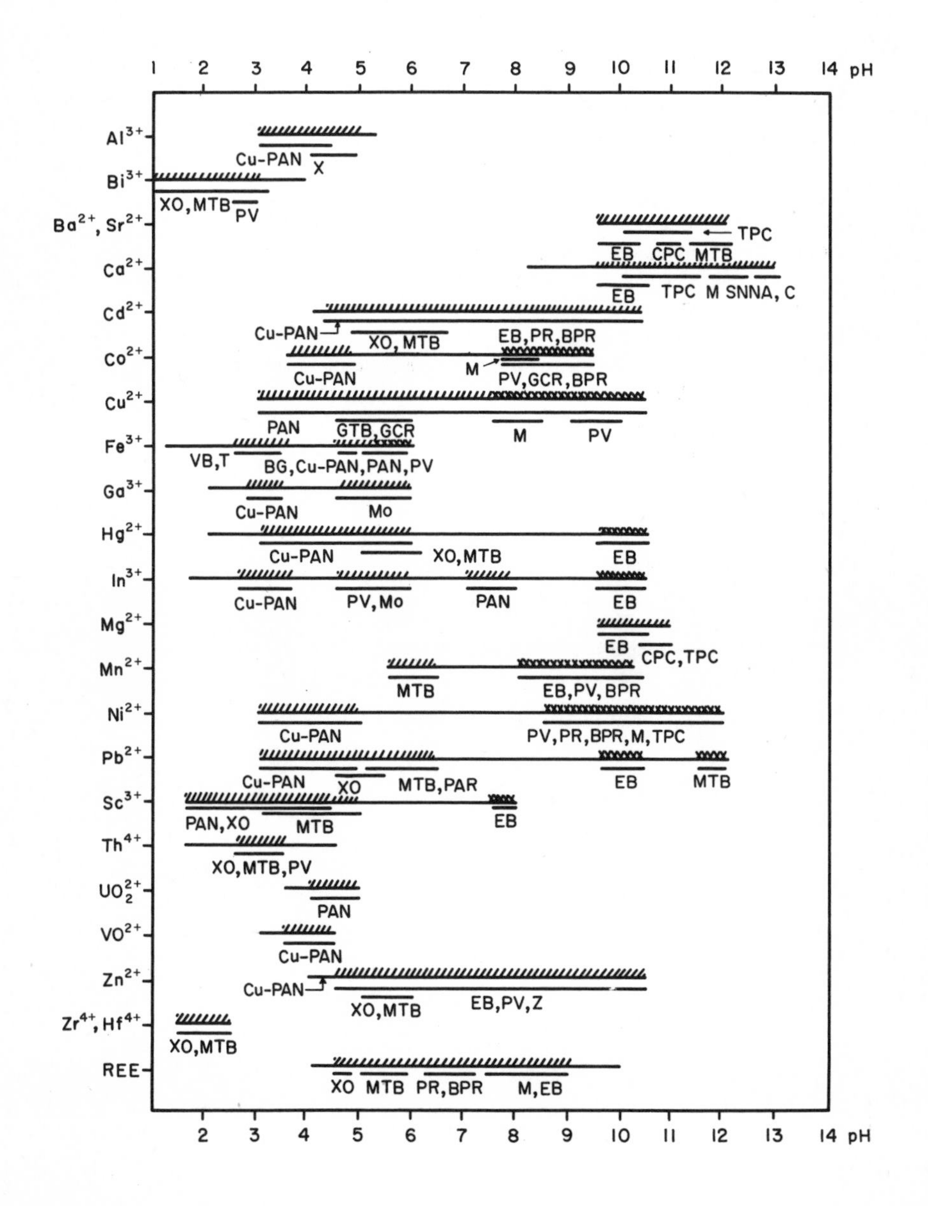

LEGEND

——— pH range in which the metal ions quantitatively react with Complexone III.

/ / / / / / pH range in which the metal may in principle be titrated against complexone, in the presence of indicators.

\ \ \ \ \ \ pH range in which suitable compounds must be added during direct complexometric titration of the given ion in order to prevent precipitation of the basic salt or of the hydroxide of the metal.

$\overline{\text{Ind}}$—pH range in which the metal ions may be titrated against Complexone III with the use of the indicator shown under the bar.

BPR — Bromopyrogallol Red
VB — Variamine Blue B, base
GCR — Glycinecresol Red
GTB — Glycinethymol Blue
BG — Bindschedler's Green, leuco base
C — calcein
XO — Xylenol Orange
CPC — o-cresolphthalein complexone
M — murexide
Mo — morin
MB — Methylthymol Blue
SNNA — 2-hydroxy-1-(2-hydroxy-4-sulfo-1-naphthylazo)-3-naphthoic acid
PAN — pyridylazonaphthol
PAR — pyridylazoresorcinol
PR — Pyrogallol Red
PV — Pyrocatechol Violet
T — Tiron
TPC — thymolphthalein complexone
CC — Chromazurol C
Z — Zincon
EB — Eriochrome Black T
Cu-PAN — PAN + copper complexonate

Subject Index

Printed in Israel
Manufactured at the Israel Program for Scientific Translations, Jerusalem